Introductory Mathematics for Engineering Applications

Kuldip S. Rattan

Wright State University

Nathan W. Klingbeil

Wright State University

WILEY

VP & Publisher: Don Fowley
Editor: Dan Sayre
Editorial Assistant: Jessica Knecht
Marketing Manager: Chris Ruel
Marketing Assistant: Marissa Carroll
Designer: Kenji Ngieng
Associate Production Manager: Joyce Poh

This book was set by Aptara Corporation.

Founded in 1807, John Wiley & Sons, Inc. has been a valued source of knowledge and understanding for more than 200 years, helping people around the world meet their needs and fulfill their aspirations. Our company is built on a foundation of principles that include responsibility to the communities we serve and where we live and work. In 2008, we launched a Corporate Citizenship Initiative, a global effort to address the environmental, social, economic, and ethical challenges we face in our business. Among the issues we are addressing are carbon impact, paper specifications and procurement, ethical conduct within our business and among our vendors, and community and charitable support. For more information, please visit our website: www.wiley.com/go/citizenship.

ISBN 978-1-118-14180-9 (Paperback)

Printed in the United States of America

10 9 8 7 6 5 4 3 2 1

Contents

Preface

This book is intended to provide first-year engineering students with a comprehensive introduction to the application of mathematics in engineering. This includes math topics ranging from precalculus and trigonometry through calculus and differential equations, with all topics set in the context of an engineering application. Specific math topics include linear and quadratic equations, trigonometry, 2-D vectors, complex numbers, sinusoids and harmonic signals, systems of equations and matrices, derivatives, integrals, and differential equations. However, these topics are covered only to the extent that they are *actually used* in core first- and second-year engineering courses, including physics, statics, dynamics, strength of materials, and electric circuits, with occasional applications from upper-division courses. Additional motivation is provided by a wide range of worked examples and homework problems representing a variety of popular engineering disciplines.

While this book provides a *comprehensive introduction* to both the math topics and their engineering applications, it provides *comprehensive coverage of neither*. As such, it is not intended to be a replacement for any traditional math or engineering textbook. It is more like an advertisement or movie trailer. Indeed, everything covered in this book will be covered again in either an engineering or mathematics classroom. This gives the instructor an enormous amount of freedom – the freedom to integrate math and physics by *immersion*. The freedom to leverage student intuition, and to introduce new physical contexts for math without the constraint of prerequisite knowledge. The freedom to let the physics help explain the math and the math to help explain the physics. The freedom to teach math to engineers the way it really ought to be taught – within a context, and for a *reason*.

Ideally, this book would serve as the primary text for a first-year engineering mathematics course, which would replace traditional math prerequisite requirements for core sophomore-level engineering courses. This would allow students to advance through the first two years of their chosen degree programs without first completing the required calculus sequence. Such is the approach adopted by Wright State University and a growing number of institutions across the country, which are now enjoying significant increases not only in engineering student retention but also in engineering student performance in their first required calculus course.

Alternatively, this book would make an ideal reference for any freshman engineering program. Its organization is highly compartmentalized, which allows instructors to pick and choose which math topics and engineering applications to cover. Thus, any institution wishing to increase engineering student preparation and motivation for the required calculus sequence could easily integrate selected topics into an existing freshman engineering course without having to find room in the curriculum for additional credit hours. Finally, this book would provide an outstanding resource for nontraditional students returning to school from the workplace, for students who are undecided or are considering a switch to engineering from another major, for math and science teachers or education majors seeking physical contexts for their students, or for upper-level high school students who are thinking about studying engineering in college. For all of these students, this book represents a one-stop shop for how math is really used in engineering.

Acknowledgement

The authors would like to thank all those who have contributed to the development of this text. This includes their outstanding staff of TA's, who have not only provided numerous suggestions and revisions, but also played a critical role in the success of the first-year engineering math program at Wright State University. The authors would also like to thank their many colleagues and collaborators who have joined in their nationwide quest to change the way math is taught to engineers. Special thanks goes to Jennifer Serres, Werner Klingbeil and Scott Molitor, who have contributed a variety of worked examples and homework problems from their own engineering disciplines. Thanks also to Josh Deaton, who has provided detailed solutions to all end-of-chapter problems. Finally, the authors would like to thank their wives and families, whose unending patience and support have made this effort possible.

This material is based upon work supported by the National Science Foundation under Grant Numbers EEC-0343214, DUE-0618571, DUE-0622466 and DUE-0817332. Any opinions, findings, and conclusions or recommendations expressed in this material are those of the authors and do not necessarily reflect the views of the National Science Foundation.

Straight Lines in Engineering

In this chapter, the applications of straight lines in engineering are introduced. It is assumed that the students are already familiar with this topic from their high school algebra course. This chapter will show, with examples, why this topic is so important for engineers. For example, the velocity of a vehicle while braking, the voltage-current relationship in a resistive circuit, and the relationship between force and displacement in a preloaded spring can all be represented by straight lines. In this chapter, the equations of these lines will be obtained using both the slope-intercept and the point-slope forms.

1.1 VEHICLE DURING BRAKING

The velocity of a vehicle during braking is measured at two distinct points in time, as indicated in Fig. 1.1.

t, s	$v(t)$, m/s
1.5	9.75
2.5	5.85

Figure 1.1 A vehicle while braking.

The velocity satisfies the equation

$$v(t) = at + v_o \tag{1.1}$$

where v_o is the initial velocity in m/s and a is the acceleration in m/s^2.

(a) Find the equation of the line $v(t)$ and determine both the initial velocity v_o and the acceleration a.

(b) Sketch the graph of the line $v(t)$ and clearly label the initial velocity, the acceleration, and the total stopping time on the graph.

The equation of the velocity given by equation (1.1) is in the slope-intercept form $y = mx + b$, where $y = v(t)$, $m = a$, $x = t$, and $b = v_o$. The slope m is given by

$$m = \frac{\Delta y}{\Delta x} = \frac{y_2 - y_1}{x_2 - x_1}.$$

Therefore, the slope $m = a$ can be calculated using the data in Fig. 1.1 as

$$a = \frac{v_2 - v_1}{t_2 - t_1} = \frac{5.85 - 9.75}{2.5 - 1.5} = -3.9 \text{ m/s}^2.$$

The velocity of the vehicle can now be written in the slope-intercept form as

$$v(t) = -3.9\, t + v_o.$$

The y-intercept $b = v_o$ can be determined using either one of the data points. Using the data point $(t, v) = (1.5, 9.75)$ gives

$$9.75 = -3.9\,(1.5) + v_o.$$

Solving for v_o gives

$$v_o = 15.6 \text{ m/s}.$$

The y-intercept $b = v_o$ can also be determined using the other data point $(t, v) = (2.5, 5.85)$, yielding

$$5.85 = -3.9\,(2.5) + v_o.$$

Solving for v_o gives

$$v_o = 15.6 \text{ m/s}.$$

The velocity of the vehicle can now be written as

$$v(t) = -3.9\, t + 15.6 \text{ m/s}.$$

The total stopping time (time required to reach $v(t) = 0$) can be found by equating $v(t) = 0$, which gives

$$0 = -3.9\, t + 15.6.$$

Solving for t, the stopping time is found to be $t = 4.0$ s. Figure 1.2 shows the velocity of the vehicle after braking. Note that the stopping time $t = 4.0$ s and the initial velocity

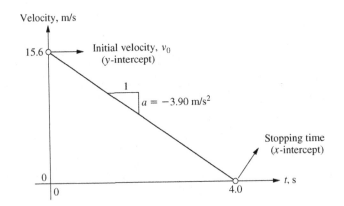

Figure 1.2 Velocity of the vehicle after braking.

$v_o = 15.6$ m/s are the x- and y-intercepts of the line, respectively. Also, note that the slope of the line $m = -3.90$ m/s^2 is the acceleration of the vehicle during braking.

1.2   VOLTAGE-CURRENT RELATIONSHIP IN A RESISTIVE CIRCUIT

For the resistive circuit shown in Fig. 1.3, the relationship between the applied voltage V_s and the current I flowing through the circuit can be obtained using **Kirchhoff's voltage law (KVL)** and **Ohm's law**. For a closed-loop in an electric circuit, KVL states that the sum of the voltage rises is equal to the sum of the voltage drops, i.e.,

$$\textbf{Kirchhoff's voltage law:} \Rightarrow \quad \sum \textit{Voltage rise} = \sum \textit{Voltage drop}.$$

V_s, V	I, A
10.0	0.1
20.0	1.1

Figure 1.3   Voltage and current in a resistive circuit.

Applying KVL to the circuit of Fig. 1.3 gives

$$V_s = V_R + V. \tag{1.2}$$

Ohm's law states that the voltage drop across a resistor V_R in volts (V) is equal to the current I in amperes (A) flowing through the resistor multiplied by the resistance R in ohms (Ω), i.e.,

$$V_R = I\,R. \tag{1.3}$$

Substituting equation (1.3) into equation (1.2) gives a linear relationship between the applied voltage V_s and the current I as

$$V_s = I\,R + V. \tag{1.4}$$

The objective is to find the value of R and V when the current flowing through the circuit is known for two different voltage values given in Fig. 1.3.

The voltage-current relationship given by equation (1.4) is the equation of a straight line in the slope-intercept form $y = mx + b$, where $y = V_s, x = I, m = R$, and $b = V$. The slope m is given by

$$m = R = \frac{\Delta y}{\Delta x} = \frac{\Delta V_s}{\Delta I}.$$

Using the data in Fig. 1.3, the slope R can be found as

$$R = \frac{20 - 10}{1.1 - 0.1} = 10\ \Omega.$$

Therefore, the source voltage can be written in slope-intercept form as

$$V_s = 10\,I + b.$$

The y-intercept $b = V$ can be determined using either one of the data points. Using the data point $(V_s, I) = (10, 0.1)$ gives

$$10 = 10\,(0.1) + V.$$

Solving for V gives

$$V = 9 \text{ V}.$$

The y-intercept V can also be found by finding the equation of the straight line using the point-slope form of the straight line $(y - y_1) = m(x - x_1)$ as

$$V_s - 10 = 10(I - 0.1) \;\Rightarrow\; V_s = 10\,I - 1.0 + 10.$$

Therefore, the voltage-current relationship is given by

$$V_s = 10\,I + 9. \tag{1.5}$$

Comparing equations (1.4) and (1.5), the values of R and V are given by

$$R = 10\,\Omega, \quad V = 9 \text{ V}.$$

Figure 1.4 shows the graph of the source voltage V_s versus the current I. Note that the slope of the line $m = 10$ is the resistance R in Ω and the y-intercept $b = 9$ is the voltage V in volts.

    The values of R and V can also be determined by switching the interpretation of x and y (the independent and dependent variables). From the voltage-current relationship $V_s = I\,R + V$, the current I can be written as a function of V_s as

$$I = \frac{1}{R} V_s - \frac{V}{R}. \tag{1.6}$$

This is an equation of a straight line $y = m\,x + b$, where x is the applied voltage V_s, y is the current I, $m = \frac{1}{R}$ is the slope, and $b = -\frac{V}{R}$ is the y-intercept. The slope and

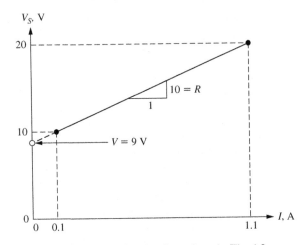

Figure 1.4   Voltage-current relationship for the data given in Fig. 1.3.

y-intercept can be found from the data given in Fig. 1.3 using the slope-intercept method as

$$m = \frac{\Delta y}{\Delta x} = \frac{\Delta I}{\Delta V_s}.$$

Using the data in Fig. 1.3, the slope m can be found as

$$m = \frac{1.1 - 0.1}{20 - 10} = 0.1.$$

Therefore, the current I can be written in slope-intercept form as

$$I = 0.1 \, V_s + b.$$

The y-intercept b can be determined using either one of the data points. Using the data point $(V_s, I) = (10, 0.1)$ gives

$$0.1 = 0.1 \, (10) + b.$$

Solving for b gives

$$b = -0.9.$$

Therefore, the equation of the straight line can be written in the slope-intercept form as

$$I = 0.1 \, V_s - 0.9. \tag{1.7}$$

Comparing equations (1.6) and (1.7) gives

$$\frac{1}{R} = 0.1 \quad \Rightarrow \quad R = 10 \, \Omega$$

and

$$-\frac{V}{R} = -0.9 \quad \Rightarrow \quad V = 0.9 \, (10) = 9 \, \text{V}.$$

Figure 1.5 is the graph of the straight line $I = 0.1 V_s - 0.9$. Note that the y-intercept is $-\dfrac{V}{R} = -0.9$ A and the slope is $\dfrac{1}{R} = 0.1$.

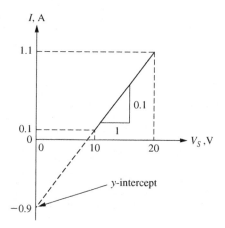

Figure 1.5   Straight line with I as independent variable for the data given in Fig. 1.3.

1.3 FORCE-DISPLACEMENT IN A PRELOADED TENSION SPRING

The force-displacement relationship for a spring with a preload f_o is given by

$$f = ky + f_o, \tag{1.8}$$

where f is the force in *Newtons* (N), y is the displacement in *meters* (m), and k is the spring constant in N/m.

f, N	y, m
1	0.1
5	0.9

Figure 1.6 Force-displacement in a preloaded spring.

The objective is to find the spring constant k and the preload f_o, if the values of the force and displacement are as given in Fig. 1.6.

Method 1 Treating the displacement y as an independent variable, the force-displacement relationship $f = ky + f_o$ is the equation of a straight line $y = mx + b$, where the independent variable x is the displacement y, the dependent variable y is the force f, the slope m is the spring constant k, and the y-intercept is the preload f_o. The slope m can be calculated using the data given in Fig. 1.6 as

$$m = \frac{5-1}{0.9-0.1} = \frac{4}{0.8} = 5.$$

The equation of the force-displacement equation in the slope-intercept form can therefore be written as

$$f = 5y + b.$$

The y-intercept b can be found using one of the data points. Using the data point $(f, y) = (5, 0.9)$ gives

$$5 = 5(0.9) + b.$$

Solving for b gives

$$b = 0.5 \text{ N}.$$

Therefore, the equation of the straight line can be written in slope-intercept form as

$$f = 5y + 0.5. \tag{1.9}$$

Comparing equations (1.8) and (1.9) gives

$$k = 5N/m, \quad f_o = 0.5N.$$

Method 2 Now treating the force f as an independent variable, the force-displacement relationship $f = ky + f_o$ can be written as $y = \dfrac{1}{k}f - \dfrac{f_o}{k}$. This relationship is the equation of a straight line $y = mx + b$, where the independent variable x is the force f, the dependent variable y is the displacement y, the slope m is the reciprocal of the spring constant $\dfrac{1}{k}$, and the y-intercept is the negated preload divided by the spring constant $-\dfrac{f_o}{k}$. The slope m can be calculated using the data given in Fig. 1.6 as

$$m = \frac{0.9 - 0.1}{5 - 1} = \frac{0.8}{4} = 0.2.$$

The equation of the displacement y as a function of force f can therefore be written in slope-intercept form as

$$y = 0.2f + b.$$

The y-intercept b can be found using one of the data points. Using the data point $(y, f) = (0.9, 5)$ gives

$$0.9 = 0.2\,(5) + b.$$

Solving for b gives

$$b = -0.1.$$

Therefore, the equation of the straight line can be written in the slope-intercept form as

$$y = 0.2f - 0.1. \tag{1.10}$$

Comparing equation (1.10) with the expression $y = \dfrac{1}{k}f - \dfrac{f_o}{k}$ gives

$$\frac{1}{k} = 0.2 \quad \Rightarrow \quad k = 5 \text{ N/m}$$

and

$$-\frac{f_o}{k} = -0.1 \quad \Rightarrow \quad f_o = 0.1\,(5) = 0.5 \text{ N}.$$

Therefore, the force-displacement relationship for a preloaded spring given in Fig. 1.6 is given by

$$f = 5y + 0.5.$$

1.4 FURTHER EXAMPLES OF LINES IN ENGINEERING

Example 1-1

The velocity of a vehicle follows the trajectory shown in Fig. 1.7. The vehicle starts at rest (zero velocity) and reaches a maximum velocity of 10 m/s in 2 s. It then cruises at a constant velocity of 10 m/s for 2 s before coming to rest at 6 s. Write the equation of the function $v(t)$; in other words, write the expression of $v(t)$ for times between 0 and 2 s, between 2 and 4 s, between 4 and 6 s, and greater than 6 s.

Figure 1.7 Velocity profile of a vehicle.

Solution

The velocity profile of the vehicle shown in Fig. 1.7 is a piecewise linear function with three different equations. The first linear function is a straight line passing through the origin starting at time 0 sec and ending at time equal to 2 s. The second linear function is a straight line with zero slope (cruise velocity of 10 m/s) starting at 2 s and ending at 4 s. Finally, the third piece of the trajectory is a straight line starting at 4 s and ending at 6 s. The equation of the piecewise linear function can be written as

(a) $0 \leq t \leq 2$:

$$v(t) = mt + b$$

where $b = 0$ and $m = \dfrac{10 - 0}{2 - 0} = 5$. Therefore,

$$v(t) = 5t \text{ m/s}.$$

(b) $2 \leq t \leq 4$:

$$v = 10 \text{ m/s}.$$

(c) $4 \leq t \leq 6$:

$$v(t) = mt + b,$$

where $m = \dfrac{0 - 10}{6 - 4} = -5$ and the value of b can be calculated using the data point $(t, v(t)) = (6, 0)$ as

$$0 = -5(6) + b \quad \Rightarrow \quad b = 0 + 30 = 30.$$

The value of b can also be calculated using the point-slope formula for the straight line

$$v - v_1 = m(t - t_1),$$

where $v_1 = 0$ and $t_1 = 6$. Thus,

$$v - 0 = -5(t - 6).$$

Therefore,

$$v(t) = -5(t - 6).$$

or

$$v(t) = -5t + 30 \text{ m/s}.$$

(d) $t > 6$:

$$v(t) = 0 \text{ m/s}.$$

Example 1-2

The velocity of a vehicle is given in Fig. 1.8.

(a) Determine the equation of $v(t)$ for

    (i) $0 \le t \le 3$ s

    (ii) $3 \le t \le 6$ s

    (iii) $6 \le t \le 9$ s

    (iv) $t \ge 9$

(b) Knowing that the acceleration of the vehicle is the slope of velocity, plot the acceleration of the vehicle.

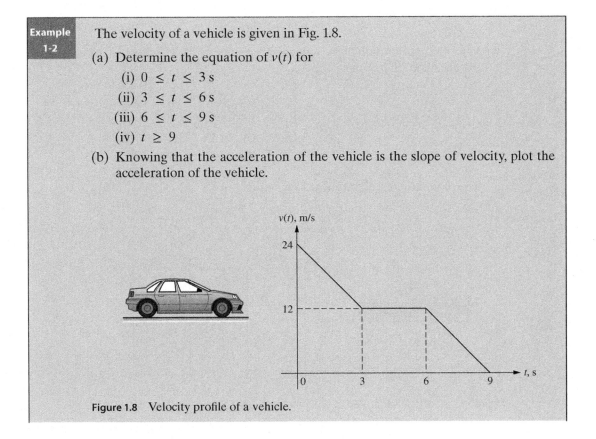

Figure 1.8   Velocity profile of a vehicle.

Solution (a) The velocity of the vehicle for different intervals can be calculated as

(i) $0 \leq t \leq 3$ s:

$$v(t) = mt + b,$$

where $m = \dfrac{12 - 24}{3 - 0} = -4$ m/s^2 and $b = 24$ m/s. Therefore,

$$v(t) = -4t + 24 \text{ m/s}.$$

(ii) $3 \leq t \leq 6$ s:

$$v(t) = 12 \text{ m/s}.$$

(iii) $6 \leq t \leq 9$ s:

$$v(t) = mt + b,$$

where $m = \dfrac{0 - 12}{9 - 6} = -4$ m/s^2 and b can be calculated in slope-intercept form using point $(t,\ v(t)) = (9,\ 0)$ as

$$0 = -4(9) + b.$$

Therefore, $b = 36$ m/s and

$$v(t) = -4t + 36 \text{ m/s}.$$

(iv) $t > 9$ s:

$$v(t) = 0 \text{ m/s}.$$

(b) Since the acceleration of the vehicle is the slope of the velocity in each interval, the acceleration a in m/s^2 is given by

$$a = \begin{cases} -4; & 0 \leq t \leq 3 \text{ s} \\ 0; & 3 \leq t \leq 6 \text{ s} \\ -4; & 6 \leq t \leq 9 \text{ s} \\ 0; & t > 9 \text{ s} \end{cases}$$

The plot of the acceleration is shown in Fig. 1.9.

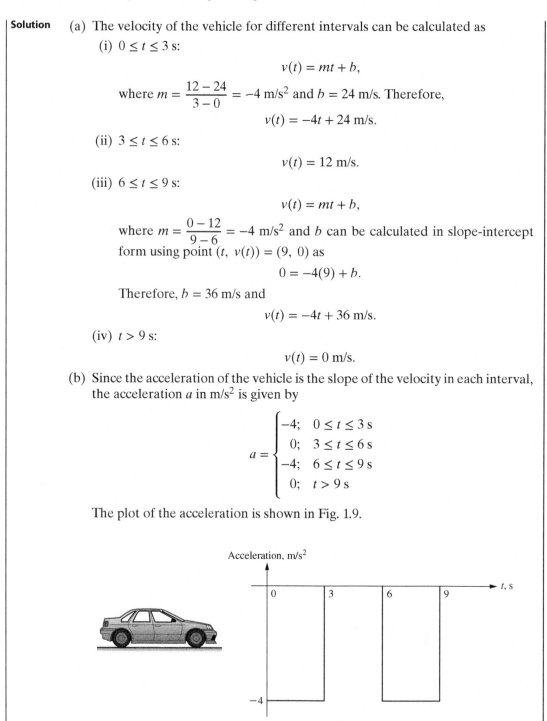

Figure 1.9 Acceleration profile of the vehicle in Fig. 1.8.

Example 1-3

In a bolted connector shown in Fig. 1.10, the force in the bolt F_b is related to the external load P as

$$F_b = CP + F_i,$$

where C is the joint constant and F_i is the preload in the bolt.

(a) Determine the joint constant C and the preload F_i given the data in Fig. 1.10.

(b) Plot the bolt force F_b as a function of the external load P, and label C and F_i on the graph.

P (l(b)	F_b (l(b)
400	500
800	600

Figure 1.10   External force applied to a bolted connection.

Solution

(a) The force-load relationship $F_b = CP + F_i$ is the equation of a straight line, $y = mx + b$. The slope m is the joint constant C, which can be calculated as

$$C = \frac{\Delta F_b}{\Delta P} = \frac{600 - 500}{800 - 400} = \frac{100 \text{ lb}}{400 \text{ lb}} = 0.25.$$

Therefore,

$$F_b(P) = 0.25\,P + F_i. \tag{1.11}$$

Now, the y-intercept F_i can be calculated by substituting one of the data points into equation (1.11). Substituting the second data point $(F_b, P) = (600, 800)$ gives

$$600 = 0.25 \times 800 + F_i.$$

Solving for F_i yields

$$F_i = 600 - 200 = 400 \text{ lb}.$$

Therefore, $F_b = 0.25\,P + 400$ is the equation of the straight line, where $C = 0.25$ and $F_i = 400$ lb. Note that the joint constant C is dimensionless!

(b) The plot of the force F_b in the bolt as a function of the external load P is shown in Fig. 1.11.

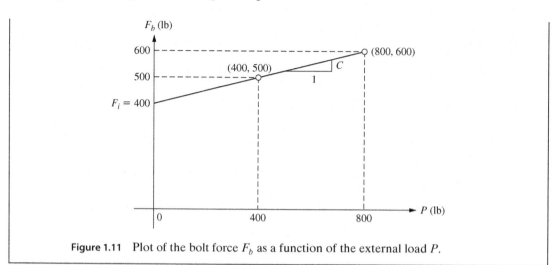

Figure 1.11   Plot of the bolt force F_b as a function of the external load P.

| Example 1-4 | For the electric circuit shown in Fig. 1.12, the relationship between the voltage V and the applied current I is given by $V = (I + I_o)R$. Find the values of R and I_0 if the voltage across the resistor V is known for the two different values of the current I as shown in Fig. 1.12. |

I, amp	V, volt
0.1	1.2
0.2	2.2

Figure 1.12   Circuit for Example 1-4.

Solution   The voltage-current relationship $V = R\,I + R\,I_o$ is the equation of a straight line $y = mx + b$, where the slope $m = R$ can be found from the data given in Fig. 1.12 as

$$R = \frac{\Delta V}{\Delta I} = \frac{2.2 - 1.2}{0.2 - 0.1} = \frac{1}{0.1}\frac{\text{volt}}{\text{amp}} = 10\,\Omega.$$

Therefore,

$$V = 10(I) + 10\,I_0. \tag{1.12}$$

The y-intercept $b = 10\,I_0$ can be found by substituting the second data point $(2.2, 0.2)$ in equation (1.12) as

$$2.2 = 100 \times 0.2 + 10\,I_0.$$

Solving for I_0 gives

$$10\,I_0 = 2.2 - 2 = 0.2,$$

which gives

$$I_0 = 0.02 \text{ A}.$$

Therefore, $V = 10\,I + 0.2$; and $R = 10\ \Omega$ and $I_0 = 0.02$ A.

Example 1-5

The output voltage v_o of the operational amplifier (Op–Amp) circuit shown in Fig. 1.13 satisfies the relationship $v_o = \left(-\dfrac{100}{R}\right) v_{in} + \left(1 + \dfrac{100}{R}\right) v_b$, where R in $k\Omega$ is the unknown resistance and v_b is the unknown voltage. Fig. 1.13 gives the values of the output voltage for two different values of the input voltage.

(a) Determine the value of R and v_b.

(b) Plot the output voltage v_o as a function of the input voltage v_{in}. On the plot, clearly indicate the value of the output voltage when the input voltage is zero (y-intercept) and the value of the input voltage when the output voltage is zero (x-intercept).

v_{in}, V	v_o, V
5	5
10	-5

Figure 1.13   An Op–Amp circuit as a summing amplifier.

Solution   (a) The input-output relationship $v_o = \left(-\dfrac{100}{R}\right) v_{in} + \left(1 + \dfrac{100}{R}\right) v_b$ is the equation of a straight line, $y = mx + b$, where the slope $m = -\dfrac{100}{R}$ can be found from the data given in Fig. 1.13 as

$$-\frac{100}{R} = \frac{\Delta v_o}{\Delta v_{in}} = \frac{-5 - 5}{10 - 5} = \frac{-10}{5} = -2.$$

Solving for R gives $R = 50\ \Omega$. Therefore,

$$v_0 = \left(-\frac{100}{50}\right) v_{in} + \left(1 + \frac{100}{50}\right) v_b$$

$$= -2\,v_{in} + 3\,v_b. \tag{1.13}$$

The y-intercept $b = 3\,v_b$ can be found by substituting the first data point $(v_0, v_{in}) = (5, 5)$ in equation (1.13) as

$$5 = -2 \times 5 + 3\,v_b.$$

Solving for v_b yields

$$3\, v_b = 5 + 10 = 15,$$

which gives $v_b = 5$ V. Therefore, $v_o = -2\, v_{in} + 15$, $R = 50\ \Omega$, and $v_b = 5$ V. The x-intercept can be found by substituting $v_o = 0$ in the equation $v_o = -2\, v_{in} + 15$ and finding the value of v_{in} as

$$0 = -2\, v_{in} + 15,$$

which gives $v_{in} = 7.5$ V. Therefore, the x-intercept occurs at $V_{in} = 7.5$ V.

(b) The plot of the output voltage of the Op–Amp as a function of the input voltage if $v_b = 5$ V is shown in Fig. 1.14.

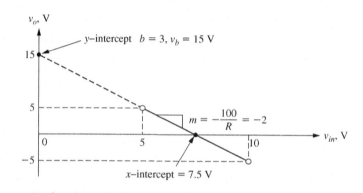

Figure 1.14   An Op–Amp circuit as a summing amplifier.

Example 1-6

An actuator used in a prosthetic arm (Fig. 1.15) can produce a different amount of force by changing the voltage of the power supply. The force and voltage satisfy the linear relation $F = kV$, where V is the voltage applied and F is the force produced by the prosthetic arm. The maximum force the arm can produce is $F = 44.5$ N when supplied with $V = 12$ volts.

(a) Find the force produced by the actuator when supplied with $V = 7.3$ volts.

(b) What voltage is needed to achieve a force of $F = 6.0$ N?

(c) Using the results of parts (a) and (b), sketch the graph of F as a function of voltage V. Use the appropriate scales and clearly label the slope and the results of parts (a) and (b) on your graph.

Figure 1.15   Prosthetic arm.

Solution   (a) The input-output relationship $F = kV$ is the equation of a straight line $y = mx$, where the slope $m = k$ can be found from the given data as

$$k = \frac{44.5}{12} = 3.71 \text{ N/V}.$$

Therefore, the equation of the straight line representing the actuator force F as a function of applied voltage V is given by

$$F = 3.71 \, V. \tag{1.14}$$

Thus, the force produced by the actuator when supplied with 7.3 volts is found by substituting $V = 7.3$ in equation (1.14) as

$$F = 3.71 \times 7.3$$
$$= 27.08 \text{ N}.$$

(b) The voltage needed to achieve a force of 6.0 N can be found by substituting $F = 6.0$ N in equation (1.14) as

$$6.0 = 3.71 \, V$$
$$V = \frac{6.0}{3.71}$$
$$= 1.62 \text{ volts}. \tag{1.15}$$

(c) The plot of force F as a function of voltage V can now be drawn as shown in Fig. 1.16.

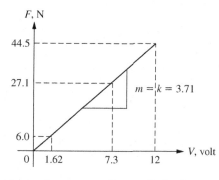

Figure 1.16   Plot of the actuator force verses the applied voltage.

Example 1-7

The electrical activity of muscles can be monitored with an electromyogram (EMG). The following root mean square (RMS) value of the amplitude measurements of the EMG signal were taken when a woman was using her hand grip muscles to ensure a lid was tight on a jar.

A, V	F, N
0.0005	110
0.00125	275

Figure 1.17   Amplitude measurements of the EMG signal.

The RMS amplitude of the EMG signal satisfies the linear equation

$$A = mF + b \tag{1.16}$$

where A is the RMS value of the EMG amplitude in V, F is the applied muscle force in N, and m is the slope.

(a) Determine the value of m and b.

(b) Plot the RMS amplitude A as a function of the applied muscle force F.

(c) Using the equation of the line from part (a), find the RMS value of the amplitude for a muscle force of 200 N.

Solution

(a) The input-output relationship $A = mF + b$ is the equation of a straight line $y = mx + b$, where the slope m can be found from the EMG data given in the table (Fig. 1.17) as

$$m = \frac{\Delta A}{\Delta F} = \frac{0.00125 - 0.0005}{275 - 110} = \frac{0.00075}{165} = 4.55 \times 10^{-6} \ \frac{V}{N}.$$

The y-intercept b can be found by substituting the first data point $(A, F) = (0.0005, 110)$ in equation (1.16) as

$$0.0005 = 4.55 \times 10^{-6}(110) + b.$$

Solving for b yields

$$b = 5 \times 10^{-7} \approx 0.$$

Therefore, the equation of the straight line representing the RMS amplitude as a function of applied force is given by

$$A = 4.55 \times 10^{-6} \ F. \tag{1.17}$$

(b) The plot of the RMS amplitude as a function of the applied muscle force is shown in Fig. 1.18.

Figure 1.18 Plot of the RMS amplitude verses the applied muscle force.

(c) The RMS value of the amplitude for a muscle force of 200 N can be found by substituting $F = 200$ N in equation (1.17) as

$$A = 4.55 \times 10^{-6} \times (200) = 0.91 \times 10^{-3} \text{V}.$$

Example 1-8

A civil engineer needs to establish the elevation of the cornerstone for a building located between two benchmarks, B1 and B2, of known elevations as shown in Fig. 1.19.

	l_1 (m)	E (m)
B1	0	428.4
B2	1001.2	476.8

Figure 1.19 Elevations along a uniform grade.

The elevation E along the grade satisfies the linear relationship

$$E = ml + E_1 \tag{1.18}$$

where E_1 is the elevation of B1, l is the distance from B1 along the grade, and m is the average slope of the grade.

(a) Find the equation of the line E and determine the slope m of the grade.

(b) Using the equation of the line from part (a), find the elevation of the cornerstone E^* if it is located at a distance $l = 565$ m from B1.

(c) Sketch the graph of E as a function of l and clearly indicate both the slope m and elevation E_1 of B1.

Solution   (a) The equation of elevation given by equation (1.18) is a straight line in the slope-intercept form $y = mx + b$, where the slope m can be found from the elevation data given in Fig. 1.19 as

$$m = \frac{\Delta E}{\Delta l} = \frac{476.8 - 428.4}{1001.2 - 0} = \frac{48.4}{1001.2} = 0.0483.$$

The y-intercept E_1 can be found by substituting the first data point $(E, l) = (428.4, 0)$ in equation (1.18) as

$$428.4 = 0.0483 \times (0) + E_1$$

Solving for E_1 yields

$$E_1 = 428.4 \text{ m}.$$

Therefore, the equation of the straight line representing the elevation as a function of distance l is given by

$$E = 0.0483\, l + 428.4 \text{ m}. \qquad (1.19)$$

(b) The elevation E^* of the cornerstone can be found by substituting $l = 565$ m in equation (1.19) as

$$E^* = 0.0483 \times (565) + 428.4 = 455.7 \text{ m}.$$

(c) The plot of the elevation as a function of the length is shown in Fig. 1.20.

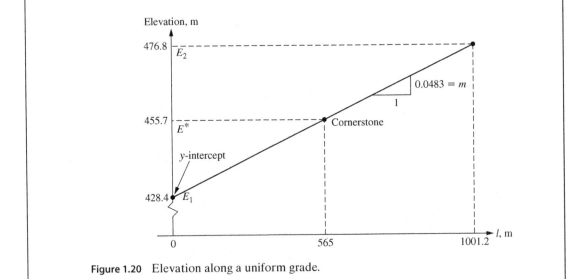

Figure 1.20   Elevation along a uniform grade.

PROBLEMS

1-1. A constant force $F = 2$ N is applied to a spring and the displacement x is measured as 0.2 m. If the spring force and displacement satisfy the linear relation $F = kx$, find the stiffness k of the spring.

F (N)	x (m)
2	0.2

Figure P1.1 Displacement of a spring in problem P1-1.

1-2. The spring force F and displacement x for a close-wound tension spring are measured as shown in Fig. P1.2. The spring force F and displacement x satisfy the linear equation $F = kx + F_i$, where k is the spring constant and F_i is the preload induced during manufacturing of the spring.

(a) Using the given data in Fig. P1.2, find the equation of the line for the spring force F as a function of the displacement x, and determine the values of the spring constant k and preload F_i.

(b) Sketch the graph of F as a function of x. Use appropriate axis scales and clearly label the preload F_i, the spring constant k, and both given data points on your graph.

F (N)	x (cm)
34.5	1.5
57.0	3.0

Figure P1.2 Close-wound tension spring for problem P1-2.

1-3. The spring force F and displacement x for a close-wound tension spring are measured as shown in Fig. P1.3.

The spring force F and displacement x satisfy the linear equation $F = kx + F_i$, where k is the spring constant and F_i is the preload induced during manufacturing of the spring.

(a) Using the given data, find the equation of the line for the spring force F as a function of the displacement x, and determine the values of the spring constant k and preload F_i.

(b) Sketch the graph of F as a function of x and clearly indicate both the spring constant k and preload F_i.

F (N)	x (cm)
135	25
222	50

Figure P1.3 Close-wound tension spring for problem P1-3.

1-4. In a bolted connection shown in Fig. P1.4, the force in the bolt F_b is given in terms of the external load P as $F_b = CP + F_i$.

(a) Given the data in Fig. P1.4, determine the joint constant C and the preload F_i.

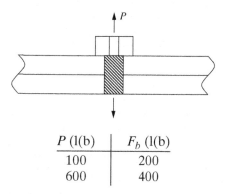

P (lb)	F_b (lb)
100	200
600	400

Figure P1.4 Bolted connection for problem P1-4.

(b) Plot the bolt force F_b as a function of the load P and label C and F_i on the graph.

1-5. Repeat problem P1-4 for the data given in Fig. P1.5.

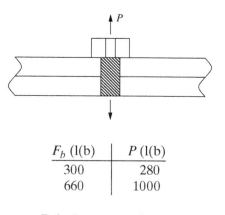

F_b (l(b))	P (l(b))
300	280
660	1000

Figure P1.5   Bolted connection for problem P1-5.

1-6. The velocity $v(t)$ of a ball thrown upward satisfies the equation $v(t) = v_o + at$, where v_o is the initial velocity of the ball in ft/s and a is the acceleration in ft/s^2.
(a) Given the data in Fig. P1.6, find the equation of the line representing the velocity $v(t)$ of the ball, and determine both the initial velocity v_o and the acceleration a.
(b) Sketch the graph of the line $v(t)$, and clearly indicate both the initial velocity and the acceleration on your graph. Also determine the time at which the velocity is zero.

$v(t)$ (ft/s)	t (s)
67.8	1.0
3.4	3.0

Figure P1.6   A ball thrown upward with an initial velocity $v_o(t)$ in problem P1-6.

1-7. The velocity $v(t)$ of a ball thrown upward satisfies the equation $v(t) = v_o + at$, where v_o is the initial velocity of the ball in m/s and a is the acceleration in m/s^2.
(a) Given the data in Fig. P1.7, find the equation of the line representing the velocity $v(t)$ of the ball, and determine both the initial velocity v_o and the acceleration a.
(b) Sketch the graph of the line $v(t)$, and clearly indicate both the initial velocity and the acceleration on your graph. Also determine the time at which the velocity is zero.

$v(t)$ (m/s)	t (s)
3.02	0.1
1.06	0.3

Figure P1.7   A ball thrown upward with an initial velocity $v_o(t)$ in problem P1-7.

1-8. A model rocket is fired in the vertical plane. The velocity $v(t)$ is measured as shown in Fig. P1.8. The velocity satisfies the equation $v(t) = v_o + at$, where v_o is the initial velocity of the rocket in m/s and a is the acceleration in m/s^2.
(a) Given the data in Fig. P1.8, find the equation of the line representing the velocity $v(t)$ of the rocket, and determine both the initial velocity v_o and the acceleration a.

$v(t)$ (m/s)	t (s)
34.3	0.5
19.6	2.0

Figure P1.8   A model rocket fired in the vertical plane in problem P1-8.

(b) Sketch the graph of the line $v(t)$ for $0 \leq t \leq 8$ seconds, and clearly indicate both the initial velocity and the acceleration on your graph. Also determine the time at which the velocity is zero (i.e., the time required to reach the maximum height).

1-9. A model rocket is fired in the vertical plane. The velocity $v(t)$ is measured as shown in Fig. P1.9. The velocity satisfies the equation $v(t) = v_o + at$, where v_o is the initial velocity of the rocket in ft/s and a is the acceleration in ft/s^2.

(a) Given the data in Fig. P1.9, find the equation of the line representing the velocity $v(t)$ of the rocket, and determine both the initial velocity v_o and the acceleration a.

(b) Sketch the graph of the line $v(t)$ for $0 \leq t \leq 10$ seconds, and clearly indicate both the initial velocity and the acceleration on your graph. Also determine the time at which the velocity is zero (i.e., the time required to reach the maximum height).

$v(t)$ (ft/s)	t (s)
128.8	1.0
32.2	4.0

Figure P1.9 A model rocket fired in the vertical plane in problem P1-9.

1-10. The velocity of a vehicle is measured at two distinct points in time as shown in Fig. P1.10. The velocity satisfies the relationship $v(t) = v_o + at$, where v_o is the initial velocity in m/s and a is the acceleration in m/s^2.

(a) Find the equation of the line $v(t)$, and determine both the initial velocity v_o and the acceleration a.

(b) Sketch the graph of the line $v(t)$, and clearly label the initial velocity, the acceleration, and the total stopping time on the graph.

$v(t)$ (m/s)	t (s)
30	1.0
10	2.0

Figure P1.10 Velocity of a vehicle during braking in problem P1-10.

1-11. The velocity of a vehicle is measured at two distinct points in time as shown in Fig. P1.11. The velocity satisfies the relationship $v(t) = v_o + at$, where v_o is the initial velocity in ft/s and a is the acceleration in ft/s^2.

(a) Find the equation of the line $v(t)$, and determine both the initial velocity v_o and the acceleration a.

(b) Sketch the graph of the line $v(t)$, and clearly label the initial velocity, the acceleration, and the total stopping time on the graph.

$v(t)$ (ft/s)	t (s)
112.5	0.5
37.5	1.5

Figure P1.11 Velocity of a vehicle during braking in problem P1-11.

1-12. The velocity $v(t)$ of a vehicle during braking is given in Fig. P1.12. Determine the equation for $v(t)$ for

(a) $0 \leq t \leq 2$ s

(b) $2 \leq t \leq 4$ s

(c) $4 \leq t \leq 6$ s

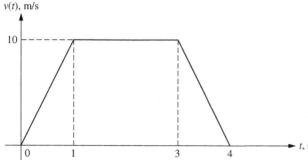

Figure P1.12 Velocity of a vehicle during braking in problem P1-12.

1-13. A linear trajectory is planned for a robot to pick up a part in a manufacturing process. The velocity of the trajectory of one of the joints is shown in Fig. P1.13. Determine the equation of $v(t)$ for
   (a) $0 \le t \le 1$ s
   (b) $1 \le t \le 3$ s
   (c) $3 \le t \le 4$ s

Figure P1.13 Velocity of a robot trajectory.

1-14. The acceleration of the linear trajectory of problem P1-13 is shown in Fig. P1.14. Determine the equation of $a(t)$ for

   (a) $0 \le t \le 1$ s
   (b) $1 \le t \le 3$ s
   (c) $3 \le t \le 4$ s

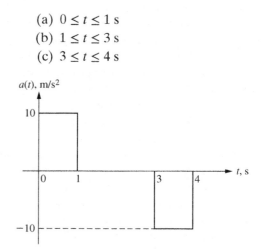

Figure P1.14 Acceleration of the robot trajectory.

1-15. The temperature distribution in a well-insulated axial rod varies linearly with respect to distance when the temperature at both ends is held constant as shown in Fig. P1.15. The temperature satisfies the equation of a line $T(x) = C_1 x + C_2$, where C_1 and C_2 are constants of integration with units of °F/ft and °F, respectively
   (a) Find the equation of the line $T(x)$, and determine both constants C_1 and C_2.
   (b) Sketch the graph of the line $T(x)$ for $0 \le x \le 1.5$ ft, and clearly label C_1 and C_2 on your graph. Also, clearly indicate the temperature at the center of the rod ($x = 0.75$ ft).

$T(x)$ (°F)	x (ft)
30	0.0
70	1.5

Figure P1.15 Temperature distribution in a well-insulated axial rod in problem P1-15.

1-16. The temperature distribution in a well-insulated axial rod varies linearly with respect to distance when the temperature at both ends is held constant as shown in Fig. P1.16. The temperature satisfies the equation of a line $T(x) = C_1 x + C_2$, where C_1 and C_2 are constants of integration with units of °C/m and °C, respectively.

(a) Find the equation of the line $T(x)$, and determine both constants C_1 and C_2.

(b) Sketch the graph of the line $T(x)$ for $0 \leq x \leq 0.5$ m, and clearly label C_1 and C_2 on your graph. Also, clearly indicate the temperature at the center of the rod ($x = 0.25$ m).

$T(x)$ (°C)	x (m)
0	0.0
20	0.5

Figure P1.16 Temperature distribution in a well-insulated axial rod in problem P1-16.

1-17. The voltage-current relationship for the circuit shown in Fig. P1.17 is given by Ohm's law as $V = IR$, where V is the applied voltage in volts, I is the current in amps, and R is the resistance of the resistor in ohms.

(a) Sketch the graph of I as a function of V if the resistance is 5 Ω.

Figure P1.17 Resistive circuit for problem P1-17.

(b) Find the current I if the applied voltage is 10 V.

1-18. A voltage source V_s is used to apply two different voltages (12V and 18V) to the single-loop circuit shown in Fig. P1.18. The values of the measured current are shown in Fig. P1.18. The voltage and current satisfy the linear relation $V_s = IR + V$, where R is the resistance in ohms, I is the current in amps, and V_s is the voltage in volts.

(a) Using the data given in Fig. P1.18, find the equation of the line for V_s as a function of I, and determine the values of R and V.

(b) Sketch the graph of V_s as a function of I and clearly indicate the resistance R and voltage V on the graph.

V_s (volt)	I (amp)
12.0	0.75
18.0	1.5

Figure P1.18 Single-loop circuit for problem P1-18.

1-19. Repeat problem P1-18 for the data shown in Fig. P1.19.

V_s (volt)	I (amp)
9.0	2.0
18.0	5.0

Figure P1.19 Single-loop circuit for problem P1-19.

1-20. Repeat problem P1-18 for the data shown in Fig. P1.20.

V_s (volt)	I (amp)
1.5	24
2.5	32

Figure P1.20   Single-loop circuit for problem P1-20.

1-21. A linear model of a diode is shown in Fig. P1.21, where R_d is the forward resistance of the diode and V_{ON} is the voltage that turns the diode ON. To determine the resistance R_d and voltage V_{ON}, two voltage values are applied to the diode and the corresponding currents are measured. The applied voltage V_S and the measured current I are given in Fig. P1.21. The applied voltage and the measured current satisfy the linear equation $V_s = I R_d + V_{ON}$.

(a) Find the equation of the line for V_s as a function of I and determine the resistance R_d and the voltage V_{ON}.

V_s (volt)	I (amp)
5.0	0.086
10.0	0.186

Figure P1.21   Linear model of a diode for problem P1-21.

(b) Sketch the graph of V_S as a function of I, and clearly indicate the resistance R_d and the voltage V_{ON} on the graph.

1-22. Repeat problem P1-21 for the data given in Fig. P1.22.

V_s (volt)	I (amp)
2.0	0.035
6	0.135

Figure P1.22   Linear model of a diode for problem P1-22.

1-23. The output voltage, v_o, of the Op–Amp circuit shown in Fig. P1.23 satisfies the relationship $v_o = \left(1 + \frac{100}{R}\right)\left(\frac{v_{in}}{2}\right) - \left(\frac{100}{R}\right) v_b$, where R is the unknown resistance in kΩ and v_b is the unknown voltage in volts. Fig. P1.23 gives the values of the output voltage for two different values of the input voltage.

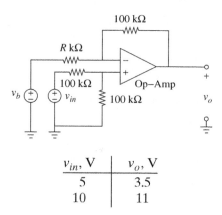

v_{in}, V	v_o, V
5	3.5
10	11

Figure P1.23   An Op–Amp circuit as a summing amplifier for problem P1.23.

(a) Determine the equation of the line for v_o as a function of v_{in} and find the values of R and v_b.

(b) Plot the output voltage v_o as a function of the input voltage v_{in}. On the plot, clearly indicate the value of the output voltage when the input voltage is zero (y-intercept) and the value of the input voltage when the output voltage is zero (x-intercept).

1-24. The output voltage, v_o, of the Op–Amp circuit shown in Fig. P1.24 satisfies the relationship $v_o = -\left(v_2 + \dfrac{100}{R} v_{in}\right)$, where R is the unknown resistance in kΩ, v_{in} is the input voltage, and v_2 is the unknown voltage. Fig. P1.24 gives the values of the output voltage for two different values of the input voltage v_{in}.

(a) Find the equation of the line for v_o as a function of v_{in} and determine the values of R and v_2.

(b) Plot the output voltage v_o as a function of the input voltage v_{in}. Clearly indicate the value of the output voltage when the input voltage is zero (y-intercept) and the value of the input voltage when the output voltage is zero (x-intercept).

1-25. A DC motor is driving an inertial load J_L shown in Fig. P1.25. To maintain a constant speed, two different values of the voltage e_a are applied to the motor. The voltage e_a and the current i_a flowing through the armature winding of the motor satisfy the relationship $e_a = i_a R_a + e_b$, where R_a is the resistance of the armature winding in ohms and e_b is the back-emf in volts. Figure P1.25 gives the values of the current for two different values of the input voltage applied to the armature of the DC motor.

(a) Find the equation of the line for e_a as a function of i_a and determine the values of R_a and e_b.

(b) Plot the applied voltage e_a as a function of the current i_a. Clearly indicate the value of the back-emf e_b and the winding resistance R_a.

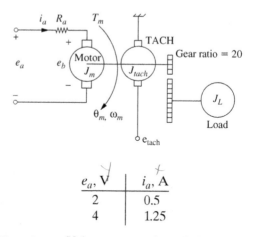

e_a, V	i_a, A
2	0.5
4	1.25

Figure P1.25 Voltage-current data of a DC motor for problem P1-25.

v_{in}, V	v_o, V
5	−4.5
10	−7.0

Figure P1.24 An Op–Amp circuit for problem P1-24.

1-26. Repeat problem P1-25 for the data shown in Fig. P1.26.

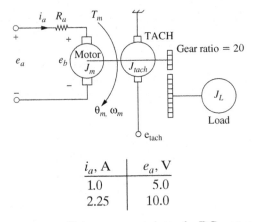

i_a, A	e_a, V
1.0	5.0
2.25	10.0

Figure P1.26 Voltage-current data of a DC motor in problem P1-26.

1-27. In the active region, the output voltage v_o of the n-channel enhancement-type MOSFET (NMOS) circuit shown in Fig. P1.27 satisfies the relationship $v_o = V_{DD} - R_D i_D$, where R_D is the unknown drain resistance and V_D is the unknown drain voltage. Fig. P1.27 gives

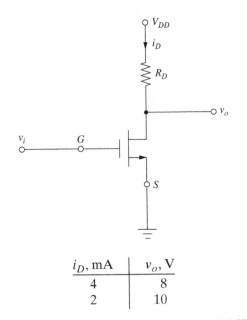

i_D, mA	v_o, V
4	8
2	10

Figure P1.27 n-channel enhancement-type MOSFET.

the values of the output voltage for two different values of the drain current. Plot the output voltage v_o as a function of the input drain current i_D. On the plot, clearly indicate the values of R_D and V_{DD}.

1-28. Repeat problem P1-27 for the data given in Fig. P1.28.

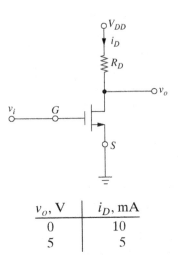

v_o, V	i_D, mA
0	10
5	5

Figure P1.28 NMOS for P1-28.

1-29. An actuator used in a prosthetic arm can produce different amounts of force by changing the voltage of the power supply. The force and voltage satisfy the linear relation $F = kV$, where V is the voltage applied and F is the force produced by the prosthetic arm. The maximum force the arm can produce is 30.0 N when supplied with 10 V.

(a) Find the force produced by the actuator when supplied with 6.0 V.

(b) What voltage is needed to achieve a force of 5.0 N?

(c) Using the results of parts (a) and (b), sketch the graph of F as a function of voltage V. Use the appropriate scales and clearly label the slope and the results of parts (a) and (b).

1-30. The following two measurements of maximum heart rate R (in beats per minute, bpm) were recorded in an exercise physiology laboratory.

R, bpm	A, years
183	30
169.5	45

The maximum heart rate R and age A satisfy the linear equation

$$R = mA + B$$

where R is the heart rate in beats per minute and A is the age in years.
(a) Using the data provided, find the equation of the line for R.
(b) Sketch R as a function of A.
(c) Using the relationship developed in part (a), find the maximum heart rate of a 60-year-old person.

1-31. The electrical activity of muscles can be monitored with an electromyogram (EMG). The RMS amplitude measurements of the EMG signal when a person is using the hand grip muscle to tighten the lid on a jar is given in the table below:

A, V	F, N
0.5 E-3	110
1.25 E-3	275

The RMS amplitude of the EMG signal satisfies the linear equation $R = mA + B$, where A is the RMS amplitude in volts, F is the applied muscle force in N, and m is the slope of the line.
(a) Using the data provided in the table, find the equation of the line for $A(F)$.
(b) Sketch A as a function of F.
(c) Using the relationship developed in part (a), find the RMS amplitude for a muscle force of 200 N.

1-32. A civil engineer needs to establish the elevation of the cornerstone for a building located between two benchmarks, B1 and B2, of known elevations, as shown in Fig. P1.32.

	l_1 (m)	E (m)
B1	0	500
B2	500	600

Figure P1.32   Elevations along a uniform grade for exercise P1-32.

The elevation E along the grade satisfies the linear relationship

$$E = ml + E_1 \qquad (1.20)$$

where E_1 is the elevation of B1, l is the taped distance from B1 along the grade, and m is the rate of change of E with respect to l.
(a) Find the equation of the line E and determine the slope m of the linear relationship.
(b) Using the equation of the line from part (a), find the elevation of the cornerstone E^* if it is located at a distance $l = 300$ m from B1.
(c) Sketch the graph of E as a function of l and clearly indicate both the slope m and elevation E_1 of B1.

1-33. A thermocouple is a temperature measurement device, which produces a voltage V proportional to the temperature at the junction of two dissimilar metals. The voltage across a thermocouple is calibrated using the boiling point of water (100°C) and the freezing point of Zinc (420°C), as shown in Fig. P1.33.

T (°C)	V (mV)
100	5.0
420	17.8

Figure P1.33   Thermocouple to measure temperature in Celsius.

The junction temperature T and the voltage across the thermocouple V satisfy the linear equation $T = \frac{1}{\alpha} V + T_R$, where α is the thermocouple sensitivity in mV/°C and T_R is the reference temperature in °C.

(a) Using the calibration data given in Fig. P1.33, find the equation of the line for the measured temperature T as a function of the voltage V and determine the value of the sensitivity α and the reference temperature T_R.

(b) Sketch the graph of T as a function of V and clearly indicate both the reference temperature T_R and the sensitivity α on the graph.

1-34. The voltage across a thermocouple is calibrated using the boiling point of water (373°K) and the freezing point of Silver (1235°K), as shown in Fig. P1.34.

T (°K)	V (mV)
373	10.0
1235	70.0

Figure P1.34   Thermocouple to measure temperature in Kelvin.

The junction temperature T and the voltage across the thermocouple V satisfy the linear equation $T = \frac{1}{\alpha} V + T_R$, where α is the thermocouple sensitivity in mV/°K and T_R is the reference temperature in °K.

(a) Using the calibration data given in Fig. P1.34, find the equation of the line for the measured temperature T as a function of the voltage V and determine the value of the sensitivity α and the reference temperature T_R.

(b) Sketch the graph of T as a function of V and clearly indicate both the reference temperature T_R and the sensitivity α on the graph.

1-35. An iron-costantan thermocouple is calibrated by inserting its junction in boiling water (100°C) and measuring a voltage $V = 5.27$ mV, and then inserting the juction in silver chloride at its melting point (455°C) and measuring $V = 24.88$ mV.

(a) Find the equation of the line for the measured temperature T as a function of the voltage V and determine the value of the sensitivity α and the reference temperature T_R.

(b) Sketch the graph of T as a function of V and clearly indicate both the reference temperature T_R and the sensitivity α on the graph.

(c) If the thermocouple is mounted in a chemical reactor and the voltage is observed to go from 10.0 mV to 13.6 mV, what is the change in temperature of the reactor?

1-36. Strain is a measure of the deformation of an object. It can be measured using a foil strain gauge shown in Fig. P1.36.

The strain being measured (ϵ) and resistance of the sensor satisfy the linear equation $R = R_o + R_o S_\epsilon \epsilon$, where R_o is the initial resistance (measured in ohms, Ω) of the sensor with no strain, and S_ϵ is the gauge factor (a multiplier with NO units).

(a) Using the given data, find the equation of the line for the sensor's resistance R as a function of the strain ϵ, and determine the values of the gauge factor S_ϵ and initial resistance R_0.

(b) Sketch the graph of R as a function of ϵ, and clearly indicate R_o on your graph.

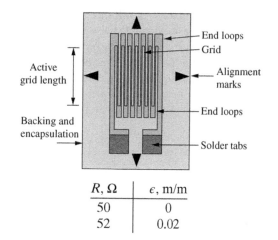

R, Ω	ϵ, m/m
50	0
52	0.02

Figure P1.38 Foil strain gauge to measure strain in problem P1-38.

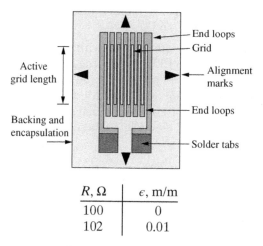

R, Ω	ϵ, m/m
100	0
102	0.01

Figure P1.36 Foil strain gauge to measure strain in problem P1-36.

1-37. Repeat problem P1-36 for the data shown in Fig. P1.37.

1-38. To determine the concentration of a purified protein sample, a graduate student used spectrophotometry to measure the absorbance given in the table below:

c (μg/ml)	a
3.50	0.342
8.00	0.578

The concentration-absorbance relationship for this protein satisfies a linear equation $a = mc + a_i$, where c is the concentration of a purified protein, a is the absorbance of the sample, m is the the rate of change of absorbance a with respect to concentration c, and a_i is the y-intercept.

(a) Find the equation of the line that describes the concentration-absorbance relationship for this protein and determine the slope m of the linear relationship.

(b) Using the equation of the line from part (a), find the concentration of the sample if this sample had an absorption of 0.486.

(c) If the sample is diluted to a concentration of 0.00419 μg/ml, what would you expect the absorption to be? Would this value be accurate?

(d) Sketch the graph of absorbance a as a function of concentration c and clearly indicate both the slope m and the y-intercept.

1-39. Repeat problem P1-38 if the absorption-concentration data for another sample is obtained as

c (μg/ml)	a
4.00	0.35
8.50	0.65

1-40. A chemistry student is performing an experiment to determine the temperature-volume behavior of a gas mixture at constant pressure and quantity. Due to technical difficulties, he could only obtain values at two temperatures as shown in the table below:

T (°C)	V (L)
50	1.08
98	1.24

The student knows that the gas volume directly depends on temperature, that is, $V(T) = m T + K$, where V is the volume in L, T is the temperature in °C, K is the y-intercept in L, and m is the slope of the line in L/°C.

(a) Find the equation of the line that describes the temperature-volume relationship of the gas mixture and determine the slope m of the linear relationship.

(b) Using the equation of the line from part (a), find the temperature of the gas mixture if the volume is 1.15 L.

(c) Using the equation of the line from part (a), find the volume of the gas if the temperature is 70°C.

(d) Sketch the graph of the volume-temperature relationship for the gas mixture from -300°C to 100°C and clearly indicate both the slope m and the y-intercept. What is the significance of the temperature when $V = 0$ L?

1-41. To obtain the linear relationship between the Fahrenheit and Celsius temperature scales, the freezing and boiling point of water is used as given in the table below:

T (°F)	T (°C)
32	0
212	100

The relationship between the temperatures in Fahrenheit and Celsius scales satisfies the linear equation T (°F) $= a T$ (°C) $+ b$.

(a) Using the given data, find the equation of the line relating the Fahrenheit and Celsius scales.

(b) Sketch the graph of T (°F) as a function of T (°C), and clearly indicate a and b on your graph.

(c) Using the graph obtained in part (b), find the temperature interval in °C if the temperature is between 20°F and 80°F.

1-42. A thermostat control with dial marking from 0 to 100 is used to regulate the temperature of an oil bath. To calibrate the thermostat, the data for the temperature T (°F) versus the dial setting R was obtained as shown in the table below:

T (°F)	R
110.0	20.0
40.0	40.0

The relationship between the temperature T in Fahrenheit and the dial setting R satisfies the linear equation T (°F) $= a R + b$.

(a) Using the given data, find the equation of the line relating the temperature to the dial setting.

(b) Sketch the graph of T (°F) as a function of R, and clearly indicate a and b on your graph.

(c) Calculate the thermostat setting needed to obtain a temperature of 320°F.

1-43. In a pressure-fed journal bearing, forced cooling is provided by a pressurized lubricant flowing along the axial direction of the shaft (the x-direction) as shown in Fig. P1.43. The lubricant

$p(x)$, psi	x, in.
24	0.5
12	1.5

Figure P1.43 Pressure-fed journal bearing.

pressure satisfies the linear equation

$$p(x) = \frac{p_s}{l} x + p_s,$$

where p_s is the supply pressure and l is the length of the bearing.

(a) Using the data given in the table (Fig. P1.43), find the equation of the line for the lubricant pressure $p(x)$ and determine the values of the supply pressure p_s and the bearing length l.

(b) Calculate the lubricant pressure $p(x)$ if x is 1.0 in.

(c) Sketch the graph of the lubricant pressure $p(x)$, and clearly indicate both the supply pressure p_s and the bearing length l on the graph.

CHAPTER 2

Quadratic Equations in Engineering

In this chapter, the applications of quadratic equations in engineering are introduced. It is assumed that students are familiar with this topic from their high school algebra course. A quadratic equation is a second-order polynomial equation in one variable that occurs in many areas of engineering. For example, the height of a ball thrown in the air can be represented by a quadratic equation. In this chapter, the solution of quadratic equations will be obtained by three methods: factoring, the quadratic formula, and completing the square.

2.1 A PROJECTILE IN A VERTICAL PLANE

Suppose a ball thrown upward from the ground with an initial velocity of 96 ft/s reaches a height $h(t)$ after time t s as shown in Fig. 2.1. The height is expressed by the quadratic equation $h(t) = 96\,t - 16\,t^2$ ft. Find the time t in seconds when $h(t) = 80$ ft.

$$h(t) = 96\,t - 16\,t^2$$

Figure 2.1 A ball thrown upward to a height of $h(t)$.

Solution:

$$h(t) = 96\,t - 16\,t^2 = 80$$

or

$$16\,t^2 - 96\,t + 80 = 0. \qquad (2.1)$$

Equation (2.1) is a quadratic equation of the form $ax^2 + bx + c = 0$ and will be solved using three different methods.

32

Method 1: Factoring   Dividing equation (2.1) by 16 yields

$$t^2 - 6t + 5 = 0. \tag{2.2}$$

Equation (2.2) can be factored as

$$(t-1)(t-5) = 0.$$

Therefore, $t - 1 = 0$ or $t = 1$ s and $t - 5 = 0$ or $t = 5$ s. Hence, the ball reaches the height of 80 ft at 1 s and 5 s.

Method 2: Quadratic Formula   If $ax^2 + bx + c = 0$, then the quadratic formula to solve for x is given by

$$x = \frac{-b \pm \sqrt{b^2 - 4ac}}{2a}. \tag{2.3}$$

Using the quadratic formula in equation (2.3), the quadratic equation (2.2) can be solved as

$$t = \frac{6 \pm \sqrt{36 - 20}}{2}$$
$$= \frac{6 \pm 4}{2}.$$

Therefore, $t = \dfrac{6-4}{2} = 1$ s and $t = \dfrac{6+4}{2} = 5$ s. Hence, the ball reaches the height of 80 ft at 1 s and 5 s.

Method 3: Completing the Square   First, rewrite the quadratic equation (2.2) as

$$t^2 - 6t = -5. \tag{2.4}$$

Adding the square of $\left(\dfrac{-6}{2}\right)$ (one-half the coefficient of the first-order term) to both sides of equation (2.4) gives

$$t^2 - 6t + \left(\frac{-6}{2}\right)^2 = -5 + \left(\frac{-6}{2}\right)^2,$$

or

$$t^2 - 6t + 9 = -5 + 9. \tag{2.5}$$

Equation (2.5) can now be written as

$$(t-3)^2 = (\pm\sqrt{4})^2$$

or

$$t - 3 = \pm 2.$$

Therefore, $t = 3 \pm 2$ or $t = 1, 5$ s. To check if the answer is correct, substitute $t = 1$ and $t = 5$ into equation (2.1). Substituting $t = 1$ s gives

$$16 \times 1 - 96 \times 1 + 80 = 0,$$

which gives $0 = 0$. Therefore, $t = 1$ s is the correct time when the ball reaches a height of 80 ft. Now, substitute $t = 5$ s,

$$16 \times 5^2 - 96 \times 5 + 80 = 0,$$

which again gives $0 = 0$. Therefore, $t = 5$ s is also the correct time when the ball reaches a height of 80 ft.

It can be seen from Fig. 2.2 that the height of the ball is 80 ft at both 1 s and 5 s. The ball is at 80 ft and going up at 1 s, and it is at 80 ft and going down at 5 s. Hence, the maximum height of ball must be halfway between 1 and 5 s, which is $1 + ((5 - 1)/2) = 3$ s. Therefore, the maximum height can be found by substituting $t = 3$ s in $h(t)$, which is $h(3) = 96(3) - 16(3)^2 = 144$ ft. These three points (height at $t = 1, 3$, and 5 s) can be used to plot the trajectory of the ball. However, to plot the trajectory accurately, additional data points can be added. The height of the ball at $t = 0$ is zero since the ball is thrown upward from the ground. To check this, substitute $t = 0$ in $h(t)$. This gives $h(0) = 96(0) - 16(0)^2 = 0$ ft. The time when the ball hits the ground again can be calculated by equating $h(t) = 0$. Therefore,

$$96t - 16t^2 = 0$$
$$6t - t^2 = 0$$
$$t(6 - t) = 0.$$

Therefore, $t = 0$ and $6 - t = 0$ or $t = 6$ s. Since the ball is thrown in the air from the ground ($h(t) = 0$) at $t = 0$, it will hit the ground again at $t = 6$ s. Using these data points, the trajectory of the ball thrown upward with an initial velocity of 96 ft/s is shown in Fig. 2.2.

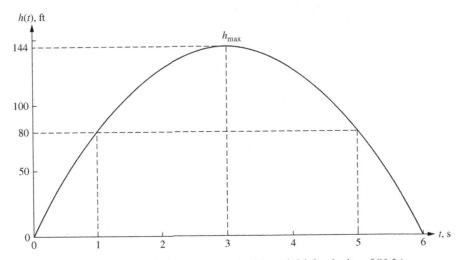

Figure 2.2   The height of the ball thrown upward with an initial velocity of 96 ft/s.

Suppose now you wish to find the time t in seconds when the height of the ball reaches 144 ft. Setting $h(t) = 144$ gives

$$h(t) = 96t - 16t^2 = 144.$$

Therefore,

$$16t^2 - 96t + 144 = 0$$

or

$$t^2 - 6t + 9 = 0. \tag{2.6}$$

The quadratic equation given in equation (2.6) can also be solved using the three methods as

Factoring	Quad Formula	Completing the Square
$t^2 - 6t + 9 = 0$ $(t-3)(t-3) = 0$ $t - 3 = 0$ $t = 3$ s	$t^2 - 6t + 9 = 0$ $t = \dfrac{6 \pm \sqrt{36 - 36}}{2}$ $t = 3 \pm 0$ $t = 3, 3$ $t = 3$ s	$t^2 - 6t + 9 = 0$ $t^2 - 6t = -9$ $t^2 - 6t + \left(\dfrac{-6}{2}\right)^2 = -9 + \left(\dfrac{-6}{2}\right)^2$ $t^2 - 6t + 9 = 9 - 9$ $= 0$ $(t-3)^2 = 0$ $t - 3 = \pm 0$ $t = 3, 3$ $t = 3$ s

Now suppose you wish to find the time t when the height of the ball reaches $h(t) = 160$ ft. Setting $h(t) = 160$ gives

$$h(t) = 96t - 16t^2 = 160.$$

Therefore,

$$16t^2 - 96t + 160 = 0$$

or

$$t^2 - 6t + 10 = 0. \tag{2.7}$$

The quadratic equation given in equation (2.7) can be solved using the three methods as

Factoring	Quad Formula	Completing the Square
$t^2 - 6t + 10 = 0$ cannot be factored using real integers	$t^2 - 6t + 10 = 0$ $t = \dfrac{6 \pm \sqrt{36 - 40}}{2}$ $t = \dfrac{6 \pm \sqrt{-4}}{2}$ $t = 3 \pm \sqrt{-1}$ $t = 3 \pm j$	$t^2 - 6t + 10 = 0$ $t^2 - 6t = -10$ $t^2 - 6t + \left(\dfrac{-6}{2}\right)^2 = -10 + \left(\dfrac{-6}{2}\right)^2$ $t^2 - 6t + 9 = -1$ $(t-3)^2 = -1$ $t - 3 = \pm\sqrt{-1}$ $t = 3 \pm j$

In the above solution $i = j = \sqrt{-1}$ is the imaginary number, therefore the roots of the quadratic equation are complex. Hence, the ball never reaches the height of 160 ft. The maximum height achieved is 144 ft at 3 s.

2.2 CURRENT IN A LAMP

A 100 W lamp and a 20 Ω resistor are connected in series to a 120 V power supply as shown in Fig. 2.3. The current I in amperes satisfies a quadratic equation as follows. Using KVL,

$$120 = V_L + V_R.$$

Figure 2.3 A lamp and a resistor connected to a 120 V supply.

From Ohm's law, $V_R = 20\,I$. Also, since the power is the product of voltage and current, $P_L = V_L\,I = 100$ W, which gives $V_L = \dfrac{100}{I}$. Therefore,

$$120 = \frac{100}{I} + 20\,I. \tag{2.8}$$

Multiplying both sides of equation (2.8) by I yields

$$120\,I = 100 + 20\,I^2. \tag{2.9}$$

Dividing both sides of equation (2.9) by 20 and rearranging gives

$$I^2 - 6\,I + 5 = 0. \tag{2.10}$$

The quadratic equation given in equation (2.10) can be solved using the three methods as

Factoring	Quad Formula	Completing the Square
$I^2 - 6I + 5 = 0$ $(I-1)(I-5) = 0$ $I = 1, 5$ A	$I^2 - 6I + 5 = 0$ $I = \dfrac{6 \pm \sqrt{36 - 20}}{2}$ $I = 3 \pm 2$ $I = 1, 5$ A	$I^2 - 6I + 5 = 0$ $I^2 - 6I + \left(\dfrac{-6}{2}\right)^2 = -5 + \left(\dfrac{-6}{2}\right)^2$ $I^2 - 6I + 9 = -5 + 9$ $(I-3)^2 = 4$ $I - 3 = \pm 2$ $I = 3 \pm 2$ $I = 1, 5$ A

Note that the two solutions correspond to two lamp choices.

Case I: For $I = 1$ A,

$$V_L = \frac{100}{I} = \frac{100}{1} = 100 \text{ V}.$$

Case II: For $I = 5$ A,

$$V_L = \frac{100}{5} = 20 \text{ V}.$$

Case I corresponds to a lamp rated at 100 V, and Case II corresponds to a lamp rated at 20 V.

2.3 EQUIVALENT RESISTANCE

Suppose two resistors are connected in **parallel**, as shown in Fig. 2.4. If the equivalent resistance $\boldsymbol{R} = \dfrac{\boldsymbol{R_1 R_2}}{\boldsymbol{R_1 + R_2}} = 100 \ \Omega$ and $R_1 = 4R_2 + 100 \ \Omega$, find R_1 and R_2.

Figure 2.4 Equivalent resistance of two resistors connected in parallel.

The equivalent resistance of two resistors connected in parallel as shown in Fig. 2.4 is given by

$$\frac{R_1 R_2}{R_1 + R_2} = 100 \ \Omega. \tag{2.11}$$

Substituting $R_1 = 4R_2 + 100 \ \Omega$ in equation (2.11) gives

$$100 = \frac{(4 R_2 + 100)(R_2)}{(4 R_2 + 100) + R_2} = \frac{4 R_2^2 + 100 R_2}{5 R_2 + 100}. \tag{2.12}$$

Multiplying both sides of equation (2.12) by $5R_2 + 100$ yields

$$100 (5 R_2 + 100) = 4 R_2^2 + 100 R_2. \tag{2.13}$$

Simplifying equation (2.13) gives

$$4 R_2^2 - 400 R_2 - 10{,}000 = 0. \tag{2.14}$$

Dividing both sides of equation by (2.14) by 4 gives

$$R_2^2 - 100 R_2 - 2500 = 0. \tag{2.15}$$

Equation (2.15) is a quadratic equation in R_2 and cannot be factored with whole numbers. Therefore, R_2 is solved using the quadratic formula as

$$R_2 = \frac{100 \pm \sqrt{10{,}000 - 4(-2500)}}{2} = \frac{100 \pm \sqrt{2(10{,}000)}}{2}.$$

Therefore,

$$R_2 = \frac{100 \pm 100\sqrt{2}}{2} = 50 \pm 50\sqrt{2}.$$

Since R_2 canot be negative,

$$R_2 = 50 + 50\sqrt{2} = 120.7 \ \Omega.$$

Substituting the value of R_2 in $R_1 = 4R_2 + 100 \ \Omega$ yields

$$R_1 = 4(120.7) + 100 = 582.8 \ \Omega.$$

Therefore, $R_1 = 582.8 \ \Omega$ and $R_2 = 120.7 \ \Omega$.

2.4   FURTHER EXAMPLES OF QUADRATIC EQUATIONS IN ENGINEERING

Example 2-1

A model rocket is fired into the air from the ground with an initial velocity of 98 m/s as shown in Fig. 2.5. The height $h(t)$ satisfies the quadratic equation

$$h(t) = 98\,t - 4.9\,t^2 \text{ m.} \tag{2.16}$$

(a) Find the time when $h(t) = 245$ m.

(b) Find the time it takes the rocket to hit the ground.

(c) Use the results of parts (a) and (b) to sketch $h(t)$ and determine the maximum height.

$$h(t) = 98\,t - 4.9\,t^2$$

Figure 2.5   A rocket fired vertically in the air.

Solution   (a) Substituting $h(t) = 245$ in equation (2.16), the quadratic equation is given by

$$-4.9\,t^2 + 98\,t - 245 = 0. \tag{2.17}$$

Dividing both sides of equation (2.17) by -4.9 gives

$$t^2 - 20\,t + 50 = 0. \tag{2.18}$$

The quadratic equation given in equation (2.18) can be solved using the three methods used in Section 2.1 as

Factoring	Quad Formula	Completing the Square
$t^2 - 20t + 50 = 0$ can't be factored with whole numbers	$t^2 - 20t + 50 = 0$ $t = \dfrac{20 \pm \sqrt{400 - 200}}{2}$ $t = 10 \pm \sqrt{50}$ $t = 10 \pm 7.07$ $t = 2.93, 17.07 \ \text{s}$	$t^2 - 20t + 50 = 0$ $t^2 - 20t = -50$ $t^2 - 20t + 100 = -50 + 100$ $(t - 10)^2 = 50$ $t - 10 = \pm\sqrt{50}$ $t = 10 \pm 7.07$ $t = 2.93, 17.07 \ \text{s}$

(b) Since the rocket hit the ground at $h(t) = 0$,

$$h(t) = 98\,t - 4.9\,t^2 = 0$$
$$4.9\,t\,(20 - t) = 0.$$

Therefore, $t = 0$ s and $t = 20$ s. Since the rocket is fired from the ground at $t = 0$ s, the rocket hits the ground again at $t = 20$ s.

(c) The maximum height should occur halfway between 2.93 and 17.07 s. Therefore,

$$t_{max} = \frac{2.93 + 17.07}{2} = \frac{20}{2} = 10 \ \text{s}.$$

Substituting $t = 10$ s into equation (2.16) yields

$$h_{max} = 98(10) - 4.9(10)^2 = 490 \ \text{m}.$$

The plot of the rocket trajectory is shown in Fig. 2.6. It can be seen from this figure that the rocket is fired from the ground at a height of zero at 0 s, crosses a height of 245 m at 2.93 s, and continues moving up and reaches the maximum height of 490 m at 10 s. At 10 seconds, it starts its downward descent and after crossing the height of 245 m again at 17.07 s, it reaches the ground again at 20 s.

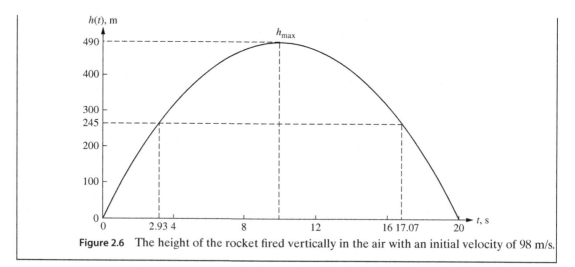

Figure 2.6 The height of the rocket fired vertically in the air with an initial velocity of 98 m/s.

Example 2-2

The equivalent resistance R of two resistors R_1 and R_2 connected in parallel as shown in Fig. 2.4 is given by

$$R = \frac{R_1 R_2}{R_1 + R_2}. \tag{2.19}$$

(a) Suppose $R_2 = 2R_1 + 4\ \Omega$ and the equivalent resistance $R = 8.0\ \Omega$. Substitute these values in equation (2.19) to obtain the following quadratic equation for R_1:

$$2R_1^2 - 20R_1 - 32 = 0.$$

(b) Solve for R_1 by each of the following methods:

   (i) Completing the square.

   (ii) The quadratic formula. Also, determine the value of R_2 corresponding to the only physical solution for R_1.

Solution

(a) Substituting $R_2 = 2R_1 + 4$ and $R = 8.0$ in equation (2.19) gives

$$8.0 = \frac{R_1(2R_1 + 4)}{R_1 + (2R_1 + 4)} = \frac{2R_1^2 + 4R_1}{3R_1 + 4}. \tag{2.20}$$

Multiplying both sides of equation (2.20) by $(3R_1 + 4)$ yields

$$8.0(3R_1 + 4) = 2R_1^2 + 4R_1,$$

or

$$24.0R_1 + 32.0 = 2R_1^2 + 4R_1. \tag{2.21}$$

Rearranging terms in equation (2.21) gives

$$2R_1^2 - 20R_1 - 32 = 0. \tag{2.22}$$

(b) The quadratic equation given in equation (2.22) can be now be solved to find the values of R_1.

(i) Method 1: Completing the square:

Dividing both sides of equation (2.22) by 2 gives

$$R_1^2 - 10\,R_1 - 16 = 0. \tag{2.23}$$

Taking 16 on the other side of equation (2.23) and adding $\left(\dfrac{-10}{2}\right)^2 = 25$ to both sides yields

$$R_1^2 - 10\,R_1 + 25 = 16 + 25. \tag{2.24}$$

Now, writing both sides of equation (2.24) as squares yields

$$(R_1 - 5)^2 = (\pm\sqrt{41})^2 = (\pm\,6.4)^2.$$

Therefore,

$$R_1 - 5 = \pm\,6.4,$$

which gives the values of R_1 as $5 + 6.4 = 11.4\ \Omega$ and $5 - 6.4 = -1.4\ \Omega$. Since the value of R_1 cannot be negative, $R_1 = 11.4\ \Omega$ and $R_2 = 2R_1 + 4 = 2(11.4) + 4 = 26.8\ \Omega$.

(ii) Method 2: Solving equation (2.22) using the quadratic formula:

$$R_1 = \frac{20 \pm \sqrt{(-20)^2 - 4(2)(-32)}}{4}$$

$$= \frac{20 \pm \sqrt{656}}{4} = \frac{20 \pm 25.6}{4} = 11.4, -1.4.$$

Since R_1 cannot be negative, $R_1 = 11.4\ \Omega$. Substituting $R_1 = 11.4\ \Omega$ in $R_2 = 2\,R_1 + 4$ gives

$$R_2 = 2(11.4) + 4 = 26.8\ \Omega.$$

Example 2-3

An assembly of springs shown in Fig. 2.7 has an equivalent stiffness k, given by

$$k = k_1 + \frac{k_1 k_2}{k_1 + k_2}. \tag{2.25}$$

If $k_2 = 2k_1 + 4$ lb/in. and the equivalent stiffness is $k = 3.6$ lb/in., find k_1 and k_2 as follows:

(a) Substitute the values of k and k_2 into equation (2.25) to obtain the following quadratic equation for k_1:

$$5k_1^2 - 2.8\,k_1 - 14.4 = 0. \tag{2.26}$$

(b) Using the method of your choice, solve equation (2.26) and determine the values of both k_1 and k_2.

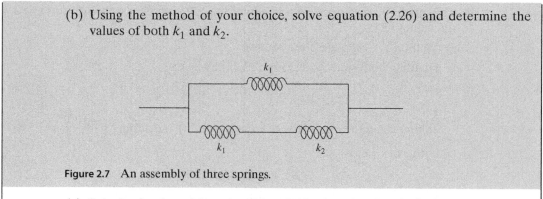

Figure 2.7   An assembly of three springs.

Solution   (a) Substituting $k_2 = 2k_1 + 4$ and $k = 3.6$ in equation (2.25) yields

$$3.6 = k_1 + \frac{k_1(2k_1 + 4)}{k_1 + (2k_1 + 4)} = k_1 + \frac{2k_1^2 + 4k_1}{3k_1 + 4}. \tag{2.27}$$

Multiplying both sides of equation (2.27) by $(3k_1 + 4)$ gives

$$3.6(3k_1 + 4) = k_1(3k_1 + 4) + 2k_1^2 + 4k_1$$

$$10.8k_1 + 14.4 = 3k_1^2 + 4k_1 + 2k_1^2 + 4k_1$$

$$10.8k_1 + 14.4 = 5k_1^2 + 8k_1. \tag{2.28}$$

Rearranging terms in equation (2.28) gives

$$5k_1^2 - 2.8k_1 - 14.4 = 0. \tag{2.29}$$

(b) The quadratic equation (2.29) can be solved using the quadratic formula as

$$k_1 = \frac{2.8 \pm \sqrt{(-2.8)^2 - 4(5)(-14.4)}}{10}$$

$$= \frac{2.8 \pm 17.2}{10}$$

$$= 2.0, \; -1.44.$$

Since k_1 cannot be negative, $k_1 = 2.0$ lb/in. Now, substituting $k_1 = 2.0$ in $k_2 = 2k_1 + 4$ yields

$$k_2 = 2(2) + 4 = 8.0.$$

Therefore,

$$k_2 = 8.0 \text{ lb/in.}$$

Example 2-4

A capacitor C and an inductor L are connected in series as shown in Fig. 2.8. The total reactance X in ohms is given by $X = \omega L - \dfrac{1}{\omega C}$, where ω is the angular frequency in rad/s.

(a) Suppose $L = 1.0$ H and $C = 0.25$ F. If the total reactance is $X = 3.0\ \Omega$, show that the angular frequency ω satisfies the quadratic equation $\omega^2 - 3\omega - 4 = 0$.

(b) Solve the quadratic equation for ω by each of the following methods: factoring, completing the square, and the quadratic formula.

Figure 2.8   Series connection of L and C.

Solution

(a) The total reactance of the series combination of L and C shown in Fig. 2.8 is given by

$$X = \omega L - \frac{1}{\omega C}. \tag{2.30}$$

Substituting $L = 1.0$ H, $C = 0.25$ F, and $X = 3.0\ \Omega$ in equation (2.30) yields

$$3.0 = \omega(1) - \frac{1}{\omega(0.25)}. \tag{2.31}$$

Multiplying both sides of equation (2.31) by ω gives

$$3\omega = \omega^2 - 4. \tag{2.32}$$

Rearranging terms in equation (2.32) yields

$$\omega^2 - 3\omega - 4 = 0. \tag{2.33}$$

(b) The quadratic equation (2.33) can be solved by three different methods: factoring, completing the squares, and the quadratic formula.

(i) Method 1: Factoring:

The quadratic equation (2.33) can be factored as

$$(\omega - 4)(\omega + 1) = 0,$$

which gives $\omega - 4 = 0$ or $\omega + 1 = 0$. Therefore, $\omega = 4$ rad/s or $\omega = -1$ rad/s. Since ω cannot be negative, $\omega = 4$ rad/s.

(ii) Method 2: Completing the squares:

The quadratic equation (2.33) can be written as

$$\omega^2 - 3\omega = 4. \tag{2.34}$$

Adding $\left(-\dfrac{3}{2}\right)^2 = \dfrac{9}{4}$ to both sides of equation (2.34) gives

$$\omega^2 - 3\omega + \left(\frac{9}{4}\right) = 4 + \left(\frac{9}{4}\right).$$

Therefore,

$$\omega^2 - 3\omega + \frac{9}{4} = \frac{25}{4}. \tag{2.35}$$

Writing both sides of equation (2.35) as a square gives

$$\left(\omega - \frac{3}{2}\right)^2 = \left(\pm\frac{5}{2}\right)^2. \tag{2.36}$$

Taking the square root of both sides of equation (2.36) yields

$$\omega - \frac{3}{2} = \pm\frac{5}{2}.$$

Therefore,

$$\omega = \frac{3}{2} \pm \frac{5}{2},$$

which gives $\omega = \dfrac{3}{2} + \dfrac{5}{2} = 4$ rad/s or $\omega = \dfrac{3}{2} - \dfrac{5}{2} = -1$ rad/s. Since ω cannot be negative, $\omega = 4$ rad/s.

(iii) Method 3: Quadratic formula:

Solving the quadratic equation (2.33) using the quadratic formula gives

$$\omega = \frac{3 \pm \sqrt{(-3)^2 - 4(1)(-4)}}{2}. \tag{2.37}$$

Equation (2.37) can be written as

$$\omega = \frac{3 \pm \sqrt{25}}{2} = \frac{3 \pm 5}{2},$$

which gives $\omega = 4, -1$. Since ω cannot be negative, $\omega = 4$ rad/s.

Example 2-5

For the circuit shown in Fig. 2.3, the power P delivered by the voltage source V_s is given by the equation $P = I^2 R + I\,V_L$.

(a) Suppose that $P = 96$ W, $V_L = 32$ V, and $R = 8\ \Omega$. Show that the current I satisfies the quadratic equation $I^2 + 4I - 12 = 0$.

(b) Solve the quadratic equation for I by each of the following methods: factoring, completing the square, and the quadratic formula.

Solution   (a) Substituting $P = 96$ W, $V_L = 32$ V, and $R = 8\,\Omega$ into the power delivered $P = I^2 R + I\,V_L$ yields

$$96 = I^2(8) + I(32). \tag{2.38}$$

Dividing both sides of equation (2.38) by 8 gives

$$12 = I^2 + 4I. \tag{2.39}$$

Rearranging terms in equation (2.39) yields

$$I^2 + 4I - 12 = 0. \tag{2.40}$$

(b) The quadratic equation given in equation (2.40) can be solved by three different methods: factoring, completing the squares, and the quadratic formula.

   (i) Method 1: Factoring:

   The quadratic equation (2.40) can be factored as

$$(I + 6)(I - 2) = 0,$$

   which gives $I + 6 = 0$ or $I - 2 = 0$. Therefore, $I = -6$ A or $I = 2$ A.

   (ii) Method 2: Completing the square:

   The quadratic equation (2.40) can be written as

$$I^2 + 4I = 12. \tag{2.41}$$

   Adding $\left(\dfrac{4}{2}\right)^2 = 4$ to both sides of equation (2.41),

$$I^2 + 4I + 4 = 12 + 4. \tag{2.42}$$

   Writing both sides of equation (2.42) as a square yields

$$(I + 2)^2 = (\pm 4)^2. \tag{2.43}$$

   Taking the square root of both sides of equation (2.43) gives

$$I + 2 = \pm 4.$$

   Therefore,

$$I = -2 \pm 4,$$

   which gives $I = -2 - 4 = -6$ A or $I = -2 + 4 = 2$ A.

   (iii) Method 3: Quadratic formula:

   Solving the quadratic equation (2.40) using the quadratic formula gives

$$I = \frac{-4 \pm \sqrt{(4)^2 - 4(1)(-12)}}{2}. \tag{2.44}$$

   Equation (2.44) can be written as

$$I = \frac{-4 \pm \sqrt{64}}{2} = \frac{-4 \pm 8}{2} = -2 \pm 4,$$

   which gives $I = -2 - 4 = -6$ A or $I = -2 + 4 = 2$ A.

Case I: For $I = -6$ A, the power absorbed by the lamp is $-6 \times 32 = -192$ W. Since the power absorbed by the lamp cannot be negative, $I = -6$ A is not one of the solutions of the quadratic equation given by (2.40).

Case II: For $I = 2$ A, the power absorbed by the lamp is $2 \times 32 = 64$ W and the power dissipated by the resistor is $96 - 64 = 32$ W. The voltage across the resistor $V_R = 2 \times 8 = 16$ V and using KVL, $V_s = 16 + 32 = 48$ V. Therefore, for the applied power of 96 W (source voltage $= 48$ V), $I = 2$ A is the solution of the quadratic equation given by (2.40).

Example 2-6

A diver jumps off a diving board 1.5 m above the water with an initial vertical velocity of 0.6 m/s as shown in Fig. 2.9. The diver's height above the water is given by

$$h(t) = -4.905\,t^2 + 0.6\,t + 1.5 \text{ m.} \tag{2.45}$$

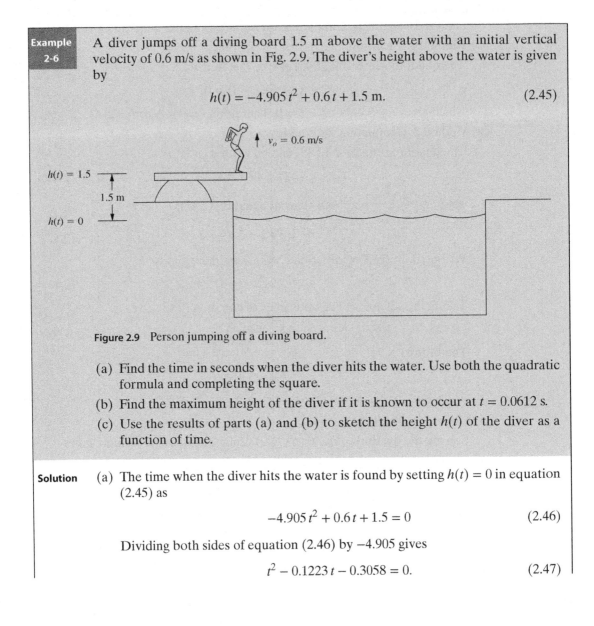

Figure 2.9   Person jumping off a diving board.

(a) Find the time in seconds when the diver hits the water. Use both the quadratic formula and completing the square.

(b) Find the maximum height of the diver if it is known to occur at $t = 0.0612$ s.

(c) Use the results of parts (a) and (b) to sketch the height $h(t)$ of the diver as a function of time.

Solution   (a) The time when the diver hits the water is found by setting $h(t) = 0$ in equation (2.45) as

$$-4.905\,t^2 + 0.6\,t + 1.5 = 0 \tag{2.46}$$

Dividing both sides of equation (2.46) by -4.905 gives

$$t^2 - 0.1223\,t - 0.3058 = 0. \tag{2.47}$$

The quadratic equation given in equation (2.47) can be solved using the quadratic formula and completing the square as outlined below.

(i) Method 1: Quadratic formula:

Solving the quadratic equation (2.47) using the quadratic formula gives

$$t = \frac{0.1223 \pm \sqrt{(0.1223)^2 - 4(1)(-0.3058)}}{2}. \tag{2.48}$$

Equation (2.48) can be written as

$$t = \frac{0.1223 \pm \sqrt{1.2382}}{2} = \frac{0.1223 \pm 1.1127}{2} = 0.0612 \pm 0.5563,$$

which gives $t = 0.0612 - 0.5563 = -0.495$ s or $t = 0.0612 + 0.5564 = 0.617$ s. Since the time cannot be negative, it takes 0.617 s for the diver to hit the water.

(ii) Method 2: Completing the square:

The quadratic equation (2.47) can be written as

$$t^2 - 0.1223\,t = 0.3058. \tag{2.49}$$

Adding $\left(\dfrac{-0.1223}{2}\right)^2 = 0.0037$ to both sides gives

$$t^2 - 0.1223\,t + 0.015 = 0.3058 + 0.0037. \tag{2.50}$$

Writing both sides of equation (2.50) as a perfect square yields

$$(t - 0.0612)^2 = (\pm\sqrt{0.3095})^2. \tag{2.51}$$

Taking the square root of both sides gives

$$t - 0.0612 = \pm 0.5563.$$

Therefore,

$$t = 0.0612 \pm 0.5563,$$

which gives $t = 0.0612 - 0.5563 = -0.495$ s or $t = 0.0612 + 0.5563 = 0.617$ s. Since the time cannot be negative, it takes 0.617 s for the diver to hit the water.

(b) The maximum height of the diver is found by substituting $t = 0.0612$ in equation (2.45) as

$$h_{\max} = h(0.0612) = -4.905(0.0612)^2 + 0.6(0.0612) + 1.5 = 1.518 \text{ m}.$$

(c) Using the results of parts (a) and (b), the diver's height after jumping from the diving board is plotted as shown in Fig. 2.10.

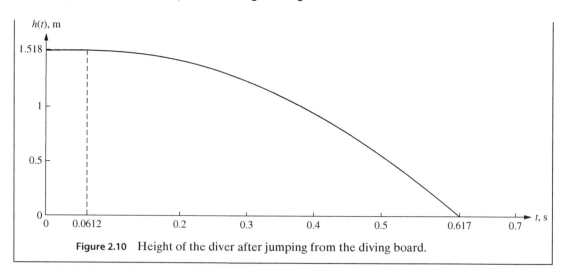

Figure 2.10 Height of the diver after jumping from the diving board.

<table>
</table>

Example 2-7

Pipeline Through Parabolic Hill: A level pipeline is required to pass through a hill having the parabolic profile

$$y = -0.004\,x^2 + 0.3\,x. \tag{2.52}$$

The origin of the x- and y-coordinates is fixed at elevation zero near the base of the hill, as shown in Fig. 2.11.

Figure 2.11 Pipeline path through a parabolic hill.

(a) Write the quadratic equation for a pipeline elevation of $y = 2$ m.

(b) Solve the quadratic equation found in part (a) to determine the positions of the tunnel entry x_A and exit x_B using both the quadratic formula and completing the square.

(c) Find the length of the tunnel.

Solution (a) Since the height y of the tunnel is 2 m, equation (2.52) can be written as

$$2 = -0.004\,x^2 + 0.3\,x$$

which gives

$$0.004\,x^2 - 0.3\,x + 2 = 0$$

or

$$x^2 - 75\,x + 500 = 0. \tag{2.53}$$

(b) The quadratic equation given in equation (2.53) can be solved using the quadratic formula and completing the square as outlined below.

  (i) Method 1: Quadratic formula:

   Solving the quadratic equation (2.53) using the quadratic formula gives

$$x = \frac{75 \pm \sqrt{(-75)^2 - 4(1)(500)}}{2}. \tag{2.54}$$

   Equation (2.54) can be written as

$$x = \frac{75 \pm \sqrt{3625}}{2} = \frac{75 \pm 60.2}{2} = 37.5 \pm 30.1,$$

   which gives $x = 37.5 - 30.1 = 7.4$ m or $x = 37.5 + 30.1 = 67.6$ m. Therefore, the tunnel entry position is 7.4 m and the tunnel exit position is 67.6 m.

  (ii) Method 2: Completing the square:

   The quadratic equation (2.53) can be written as

$$x^2 - 75\,x = -500. \tag{2.55}$$

   Adding $\left(-\dfrac{75}{2}\right)^2 = 1406.25$ to both sides gives

$$x^2 - 75\,x + 1406.25 = -500 + 1406.25. \tag{2.56}$$

   Writing both sides of equation (2.56) as a perfect square yields

$$(x - 37.5)^2 = (\pm \sqrt{906.25})^2. \tag{2.57}$$

   Taking the square root of both sides gives

$$x - 37.5 = \pm 30.1.$$

   Therefore,

$$x = 37.5 \pm 30.1,$$

   which gives $x = 37.5 - 30.1 = 7.4$ m or $x = 37.5 + 30.1 = 67.6$ m. To check if the answer is correct, substitute $x = 7.4$ and $x = 67.6$ into equation (2.53).

Substituting $x = 7.4$ m gives

$$(7.4^2) - 75(7.4) + 500 = 0$$
$$55 - 555 + 500 = 0$$
$$0 = 0$$

Now, substituting $x = 67.6$ m into equation (2.53) gives

$$(67.6^2) - 75(67.6) + 500 = 0$$
$$4570 - 5070 + 500 = 0$$
$$0 = 0$$

Therefore, $x_A = 7.4$ m and $x_B = 67.6$ m are the correct positions of the tunnel entry and exit, respectively.

(c) The tunnel length can be found by subtracting the position x_A from x_B as

$$\text{Tunnel length} = x_B - x_A = 67.6 - 7.4 = 60.2 \text{ m}$$

PROBLEMS

2-1. An analysis of a circuit shown in Fig. P2.1 yields the quadratic equation for the current I as $3I^2 + 6I = 45$, where I is in amps.

   (a) Rewrite the above equation in the form $I^2 + aI + b = 0$, where a and b are constants.

   (b) Solve the equation in part (a) by each of the following methods: factoring, completing the square, and the quadratic formula.

Figure P2.1   Resistive circuit for problem P2-1.

2-2. Repeat problem P2-1 for the circuit shown in Fig. P2.2, which yields the quadratic equation for the current I as $3I^2 - 6I = 45$, where I is in amps.

Figure P2.2   Resistive circuit for problem P2-2.

2-3. Repeat problem P2-1 for the circuit shown in Fig. P2.3, which yields the quadratic equation for the current I as $210 = 10I^2 + 40I$, where I is in amps.

Figure P2.3   Resistive circuit for problem P2-3.

2-4. The current flowing through the inductor shown in Fig. P2.4 is given by the quadratic equation $i(t) = t^2 - 8t$. Find t when
  (a) $i(t) = 9$ A (use the quadratic formula), and
  (b) $i(t) = 84$ A (use completing the square).

$$v(t) = 2t - 8$$

Figure P2.4   Current flowing through an inductor.

2-5. The voltage across the capacitor shown in Fig. P2.5 is given by the quadratic equation $v(t) = t^2 - 6t$. Find t when
  (a) $v(t) = 16$ V (use the quadratic formula), and
  (b) $v(t) = 27$ V (use completing the square).

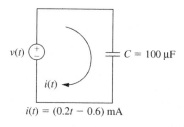

$$i(t) = (0.2t - 0.6) \text{ mA}$$

Figure P2.5   Voltage across a capacitor.

2-6. In the purely resistive circuit shown in Fig. P2.6, the total resistance R of the circuit is given by

$$R = R_1 + \frac{R_1 R_2}{R_1 + R_2}. \qquad (2.58)$$

If the total resistance of the circuit is $R = 100 \; \Omega$ and $R_2 = 2R_1 + 100 \; \Omega$, find R_1 and R_2 as follows:
  (a) Substitute the values of R and R_2 into equation (2.58), and simplify the resulting expression to obtain a single quadratic equation for R_1.
  (b) Using the method of your choice, solve the quadratic equation for R_1 and compute the corresponding value of R_2.

Figure P2.6   Series parallel combination of resistors.

2-7. Repeat problem P2-6 if $R = 1600 \; \Omega$ and $R_2 = R_1 + 500 \; \Omega$.

2-8. The energy dissipated by a resistor shown in Fig. P2.8 varies with time t in s according to the quadratic equation $W = 3t^2 + 6t$. Solve for t if
  (a) $W = 3$ joules
  (b) $W = 9$ joules
  (c) $W = 45$ joules

Figure P2.8   Resistive circuit for problem P2-8.

2-9. The equivalent capacitance C of two capacitors connected in **series** as shown in Fig. P2.9 is given by

$$C = \frac{C_1 C_2}{C_1 + C_2}. \qquad (2.59)$$

If the total capacitance is $C = 120$ μF and $C_2 = C_1 + 100$ μF, find C_1 and C_2 as follows:

(a) Substitute the values of C and C_2 in equation (2.59) and obtain the quadratic equation for C_1.

(b) Solve the quadratic equation for C_1 obtained in part (a) by each of the following methods: factoring, completing the square, and the quadratic formula. Also, compute the corresponding values of C_2.

Figure P2.9 Series combination of two capacitors.

2-10. Repeat problem P2-9 if $C = 75$ μF and $C_2 = C_1 + 200$ μF.

2-11. The equivalent capacitance C of three capacitors connected in series-parallel as shown in Fig. P2.11 is given by

$$C = 25 + \frac{C_1 C_2}{C_1 + C_2}. \tag{2.60}$$

If the total capacitance is $C = 125$ μF and $C_2 = C_1 + 100$ μF, find C_1 and C_2 as follows:

(a) Substitute the values of C and C_2 in equation (2.60) and obtain the quadratic equation for C_1.

(b) Solve the quadratic equation for C_1 obtained in part (a) by completing the square and the quadratic formula. Also, compute the corresponding values of C_2.

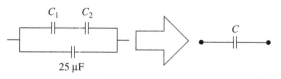

Figure P2.11 Series combination of two capacitors.

2-12. The equivalent inductance L of two inductors connected in **parallel** as shown in Fig. P2.12 is given by

$$L = \frac{L_1 L_2}{L_1 + L_2}. \tag{2.61}$$

If the total inductance $L = 80$ mH and $L_1 = L_2 + 300$ mH, find L_1 and L_2 as follows:

(a) Substitute the values of L and L_1 in equation (2.61) and obtain the quadratic equation for L_2.

(b) Solve the quadratic equation for L_2 obtained in part (a) by completing the square, and the quadratic formula. Also, compute the corresponding values of L_1.

Figure P2.12 Parallel combination of two inductors.

2-13. Repeat problem P2-12 if $L = 150$ mH and $L_1 = L_2 + 400$ mH.

2-14. The equivalent inductance L of three inductors connected in series-parallel as shown in Fig. P2.14 is given by

$$L = 125 + \frac{L_1 L_2}{L_1 + L_2}. \tag{2.62}$$

(a) Suppose $L_2 = L_1 + 200$ mH and that the equivalent inductance is $L = 200$ mH. Substitute these values in equation (2.62) and obtain the following quadratic equation:

$$L_1^2 + 50\,L_1 - 15,000 = 0. \quad (2.63)$$

(b) Solve the quadratic equation (2.63) for L_1 by each of the following methods: factoring, completing the square, and the quadratic formula.

Figure P2.14   Series-parallel combination of three inductors.

2-15. A model rocket is launched in the vertical plane at time $t = 0$ s as shown in Fig. P2.15. The height of the rocket (in feet) satisfies the quadratic equation $h(t) = 64\,t - 16\,t^2$.

(a) Find the value(s) of the time t when $h(t) = 48$ ft.

(b) Find the value(s) of the time t when $h(t) = 60$ ft.

(c) Find the time required for the rocket to hit the ground.

(d) Based on your solution to parts (a) through (c), determine the maximum height of the rocket and sketch the height $h(t)$.

Figure P2.15   A model rocket for problem P2-15.

2-16. The ball shown in Fig. P2.16 is dropped from a height of 1000 meters. The ball falls according to the quadratic equation $h(t) = 1000 - 4.905t^2$. Find the time t in seconds for the ball to reach a height $h(t)$ of

(a) 921.52 m   (b) 686.08 m

(c) 509.5 m   (d) 0 m

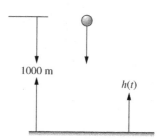

Figure P2.16   A ball dropped from a height of 1000 m.

2-17. At time $t = 0$, a ball is thrown vertically from the top of the building at a speed of 56 ft/s, as shown in Fig. P2.17. The height of the ball at time t is given by

$$h(t) = 32 + 56t - 16t^2 \text{ ft.}$$

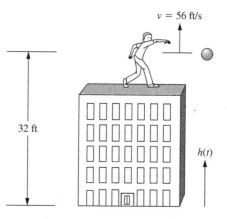

Figure P2.17   A ball thrown vertically from the top of a building.

(a) Find the value(s) of the time t when $h(t) = 32$ ft.

(b) Find the time required for the ball to hit the ground.

(c) Use the results to determine the maximum height, and sketch the height $h(t)$ of the ball.

2-18. Two springs connected in **series** shown in Fig. P2.18 can be represented by a single equivalent spring. The stiffness of the equivalent spring is given by

$$k_{eq} = \frac{k_1 k_2}{k_1 + k_2}, \qquad (2.64)$$

where k_1 and k_2 are the spring constants of the two springs. If $k_{eq} = 1.2$ N/m and $k_2 = 2k_1 - 1$ N/m, find k_1 and k_2 as follows:

(a) Substitute the values of k_{eq} and k_2 in equation (2.64) and obtain the quadratic equation for k_1.

(b) Solve the quadratic equation for k_1 obtained in part (a) by completing the square, and the quadratic formula. Also, compute the corresponding values of k_2.

Figure P2.18   Series combination of two springs.

2-19. The equivalent stiffness of a series-parallel combination of three springs shown in Fig. P2.19 is given by

$$k = k_2 + \frac{k_1 k_2}{k_1 + k_2}. \qquad (2.65)$$

If $k_1 = k_2 + 2$ lb/in and the equivalent stiffness is $k = 1.75$ lb/in, find k_1 and k_2 as follows:

(a) Substitute the values of k and k_1 into equation (2.65) and obtain the quadratic equation for k_2.

(b) Using the method of your choice, solve the quadratic equation

obtained in part (a) and determine the values of k_2. Also, find the values of k_1.

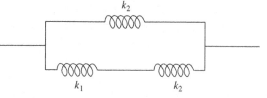

Figure P2.19   Series-parallel combination of three springs.

2-20. An assembly of three springs connected in series as shown in Fig. P2.20 has an equivalent stiffness k given by

$$k = \frac{k_1 k_2 k_3}{k_2 k_3 + k_1 k_3 + k_1 k_2}. \qquad (2.66)$$

(a) Suppose $k_2 = 6$ lb/in., $k_3 = k_1 + 8$ lb/in., and the equivalent stiffness is $k = 2$ lb/in. Substitute these values into equation (2.66) to obtain the following quadratic equation:

$$4 k_1^2 + 8 k_1 - 96 = 0. \qquad (2.67)$$

(b) Solve equation (2.67) for k_1 by each of the following methods: (i) factoring, (ii) quadratic formula, and (iii) completing the square. For each case, determine the value of k_3 corresponding to the only physical solution for k_1.

Figure P2.20   Series combination of three springs.

2-21. Consider a capacitor C and an inductor L connected in parallel, as shown in Fig. P2.21. The total reactance X in ohms is given by $X = \dfrac{\omega L}{1 - \omega^2 LC}$, where ω is the angular frequency in rad/s.

(a) Suppose $L = 1.0$ mH and $C = 1$ F. If the total reactance is $X = 1.0\,\Omega$, show that the angular frequency ω satisfies the quadratic equation $\omega^2 + \omega - 1000 = 0$.

(b) Solve the quadratic equation for ω by the methods of completing the square and the quadratic formula.

Figure P2.21   Parallel connection of L and C.

2-22. Assume that the total reactance in problem P2-21 is $X = -1\,\Omega$. Show that the angular frequency ω satisfies the quadratic equation $\omega^2 - \omega - 1000 = 0$, and find the value of ω using both the quadratic formula and completing the square.

2-23. If $L = 0.5$ H, $C = 0.005$ F, and the total reactance in problem P2-21 is $X = -\frac{20}{3}\,\Omega$, show that the angular frequency ω satisfies the quadratic equation $\omega^2 - 30\omega - 400 = 0$. Also, find the value of ω using both the quadratic formula and completing the square.

2-24. When converting resistances connected in a Δ formation to a Y formation as shown in Fig. P2.24, the resistance R_1 is obtained as

$$R_1 = \frac{R_a R_b}{R_a + R_b + R_c} \qquad (2.68)$$

(a) Suppose $R_1 = 100$, $R_a = R_b = R$, and $R_c = 100 + R$, all measured in ohms. Substitute these values into equation (2.68) to obtain the following quadratic equation for R.

$$R^2 - 300\,R - 10,000 = 0$$

(b) Solve the quadratic equation for R by the methods of completing the square and the quadratic formula.

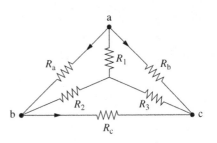

Figure P2.24   Delta to Y conversion circuit.

2-25. When converting resistances connected in a Δ formation to a Y formation as shown in Fig. P2.24, the resistance R_1 is given by equation (2.68).

(a) Suppose $R_1 = 6$, $R_a = R$, $R_b = 3R$, and $R_c = 100 + R$, all measured in ohms. Substitute these values into equation (2.68) to obtain the following quadratic equation for R.

$$R^2 - 10\,R - 200 = 0$$

(b) Solve the quadratic equation for R by the methods of completing the square and the quadratic formula.

2-26. When converting resistances connected in a Δ formation to a Y formation as shown in Fig. P2.24, the resistance R_2 is obtained as

$$R_2 = \frac{R_a R_c}{R_a + R_b + R_c} \qquad (2.69)$$

(a) Suppose $R_2 = 12, R_a = R, R_b = 3R$, and $R_c = 100 + R$, all measured in ohms. Substitute these values into equation (2.69) to obtain the following quadratic equation for R.

$$R^2 + 40R - 1200 = 0$$

(b) Solve the quadratic equation for R by the methods of completing the square and the quadratic formula.

2-27. When converting resistances connected in a Δ formation to a Y formation as shown in Fig. P2.24, the resistance R_3 is obtained as

$$R_3 = \frac{R_b R_c}{R_a + R_b + R_c} \qquad (2.70)$$

(a) Suppose $R_3 = 36, R_a = R, R_b = 3R$, and $R_c = 100 + R$, all measured in ohms. Substitute these values into equation (2.70) to obtain the following quadratic equation for R.

$$R^2 + 40R - 1200 = 0$$

(b) Solve the quadratic equation for R by the methods of completing the square and the quadratic formula.

2-28. The characteristic equation of a series RLC circuit shown in Fig. P2.28 is given as

$$s^2 + \frac{R}{L}s + \frac{1}{LC} = 0. \qquad (2.71)$$

(a) If $R = 7\,\Omega, L = 1\,H$, and $C = 0.1\,F$, solve the quadratic equation (2.71) for the values of s (called the eigenvalues of the system) using the methods of completing the square and the quadratic formula.

(b) Repeat part (a) if $R = 10\,\Omega, L = 1$ H, and $C = \dfrac{1}{25}\,F$.

Figure P2.28 Series RLC circuit for problem P2-28.

2-29. The characteristic equation of a parallel RLC circuit shown in Fig. P2.29 is given as

$$s^2 + \frac{1}{RC}s + \frac{1}{LC} = 0. \qquad (2.72)$$

If $R = 200\,\Omega, L = 50$ mH, and $C = 0.2\,\mu F$, solve the quadratic equation (2.72) for the values of s by the methods of completing the square and the quadratic formula.

Figure P2.29 Parallel RLC circuit for problem P2-29.

2-30. The characteristic equation of a mass, spring, and damper system shown in Fig. P2.30 is given by

$$ms^2 + cs + k = 0. \qquad (2.73)$$

(a) If $m = 1$ kg, $c = 3$ Ns/m, and $k = 2$ N/m, solve the quadratic equation (2.73) for the values of s using the methods of completing the square and the quadratic formula.

(b) Repeat part (a) if $m = 1$ kg, $c = 2$ Ns/m, and $k = 1$ N/m.

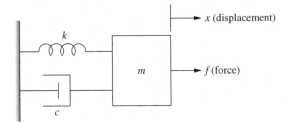

Figure P2.30   Mass, spring, and damper system for problem P2-30.

2-31. The perimeter of a rectangle shown in Fig. P2.31 is given by

$$P = 2\left(\frac{A}{L} + L\right). \qquad (2.74)$$

W	Area = 36 m² Perimeter = 30 m

L

Figure P2.31   A rectangle of length L and width W.

If the perimeter $P = 30$ m and the area $A = W \times L = 36$ m², find the length L and width W as follows:

(a) Substitute the values of P and A in equation (2.74) and obtain the quadratic equation for L.

(b) Solve the quadratic equation for L obtained in part (a) by completing the square, and the quadratic formula. Also, compute the corresponding values of W.

2-32. A diver jumps off a diving board 2.0 m above the water with an initial vertical velocity of 0.981 m/s as shown in Fig. P2.32. The height $h(t)$ above the water is given by

$$h(t) = -4.905\,t^2 + 0.981t + 2.0 \text{ m}.$$

(a) Find the time in seconds when the diver hits the water. Use both the quadratic formula and completing the square.

(b) Find the maximum height of the diver if it is known to occur at $t = 0.1$ s.

(c) Use the results of parts (a) and (b) to sketch the height $h(t)$ of the diver.

2-33. A level pipeline is required to pass through a hill having a parabolic profile

$$y = -0.005\,x^2 + 0.35\,x. \qquad (2.75)$$

The origin of the x- and y-coordinates is fixed at elevation zero near the base of the hill, as shown in Fig. P2.33.

Figure P2.32   Diver jumping off a diving board.

Figure P2.33   Pipeline path through a parabolic hill.

(a) Write the quadratic equation for a pipeline elevation of $y = 3$ m.
(b) Solve the quadratic equation found in part (a) to determine the positions of the tunnel entry x_A and exit x_B using both the quadratic formula and completing the square.
(c) Find the length of the tunnel.

2-34. A research group is using a drop test to measure the force of attenuation of a helmet liner they designed to reduce the occurance of brain injuries for soldiers and athletes. The helmet attached to a weight is propelled downward with an initial velocity v_i of 3 m/s from an initial height of 30 m. The behavior of the falling helmet is characterized by a quadratic equation $h(t) = 30 - 3t - 4.9t^2$.

(a) Write the quadratic equation for time t when the helmet and weight hit the ground, i.e., $h(t) = 0$ m.
(b) Solve the quadratic equation for t obtained in part (a) by completing the square, and the quadratic formula.

2-35. The modulus of elasticity (E) is a measure of a material's resistance to deformation; the larger the modulus, the stiffer the material. During fabrication of a ceramic material from a powder form, pores were generated that affect the strength of the material. The magnitude of elasticity is related to volume fraction porosity P by

$$E = 304\,(1 - 1.9\,P + 0.9\,P^2). \qquad (2.76)$$

(a) If a porous sample of silicon nitride has a modulus of elasticity of $E = 209$ GPa, obtain the quadratic equation for volume fraction porosity P.
(b) Solve the quadratic equation found in part (a) using both the methods of completing the square and the quadratic formula.
(c) Repeat parts (a) and (b) to find the volume fraction porosity of a silicon nitride sample with a modulus of elasticity of 25 GPa.

2-36. Consider the following reaction having an equilibrium constant of 4.66×10^{-3} at a certain temperature:

$$A(g) + B(g) \rightleftharpoons C(g)$$

If 0.300 mol of A and 0.100 mol of B are mixed in a container and allowed to reach equilibrium, the concentrations of $A = 0.300 - x$ and $B = 0.100 - x$ reaction that form the concentration of $C = 2x$ are related to the equilibrium constant by the expression

$$4.66 \times 10^{-3} = \frac{(2x)^2}{(0.300 - x)\,(0.100 - x)}.$$

(a) Write the quadratic equation for x.
(b) Solve the quadratic equation found in part (a) by completing the square and the quadratic formula.
(c) Find the equilibrium concentration A, B, and C.

2-37. Consider the following reaction having an equilibrium constant of 1.00 at $1105°$K temperature:

$$CO(g) + H_2O(g) \rightleftharpoons CO_2(g) + H_2(g)$$

Suppose the feed to a reactor contains 1.000 mol of CO and 2.000 mol of $H_2O(g)$, and the reaction mixture comes to equilibrium at 1105 K. The concentrations of $CO = 1.000 - x$ and $H_2O = 2.000 - x$ reaction that form the concentration of $CO_2 = x$ and $H_2 = x$ are related to the equilibrium constant as

$$1.00 = \frac{2x}{(1.000 - x)(2.000 - x)}$$

(a) Write the quadratic equation for x.
(b) Solve the quadratic equation found in part (a) by completing the square and the quadratic formula.
(c) Find the equilibrium concentration of CO, H_2O, CO_2, and H_2.

2-38. An engineering co-op wants to hire an asphalt contractor to widen the truck entrance to the corporate headquarters as shown in Fig. P2.38.

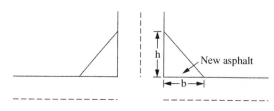

Figure P2.38 New asphalt dimensions of the corporate headquarters driveway.

The height h is required to be 10 ft. more than the width b, and the total area of the new asphalt is given by

$$A = \frac{1}{2}b(10 + b). \qquad (2.77)$$

(a) If $A = 200$ sq. ft., obtain the quadratic equation for b.
(b) Solve the quadratic equation found in part (a) by completing the square and the quadratic formula.
(c) Find the dimensions of the width and height.

2-39. Repeat problem P2-38 if the total area of the new asphalt needs to be 300 sq. ft.

2-40. A city wants to hire a contractor to build a walkway around the swimming pool in one of its parks. The dimensions of the walkway along with the dimensions of the pool are shown in Fig. P2.40. The area of the walkway is given by

$$A = (50 + 5x)(30 + 2x) - 1500$$

(a) If $A = 4500$ sq. ft., obtain the quadratic equation for x.
(b) Solve the quadratic equation found in part (a) by completing the square and the quadratic formula.

Figure P2.40 Walkway around the swimming pool.

CHAPTER 3
Trigonometry in Engineering

3.1 | INTRODUCTION

In this chapter, the direct (forward) and inverse (reverse) kinematics of one-link and two-link planar robots are considered to explain the trigonometric functions and their identities. Kinematics is the branch of mechanics that studies the motion of an object. The direct or forward kinematics is the static geometric problem of determining the position and orientation of the end-effector (hand) of the robot from the knowledge of the joint displacement. In general, the joint displacement can be linear or rotational (angular). But in this chapter, only rotational motion is considered. Furthermore, it is assumed that the planar robot is wristless (i.e., it has no end-effector or hand) and that only the position but not the orientation of the tip of the robot can be changed.

Going in the other direction, the inverse or reverse kinematics is the problem of determining all possible joint variables (angles) that lead to the given Cartesian position and orientation of the end-effector. Since no end-effector is considered in this chapter, the inverse kinematics will determine the joint angle(s) from the Cartesian position of the tip.

3.2 | ONE-LINK PLANAR ROBOT

Consider a one-link planar robot of length l (Fig. 3.1) that is being rotated in the x-y plane by a motor mounted at the center of the table, which is also the location of the robot's joint. The robot has a position sensor installed at the joint that gives the value of the angle θ of the robot measured from the positive x-axis. The angle θ is positive in the counterclockwise direction (0° to 180°) and it is negative in the clockwise direction (0° to −180°). Therefore, as the joint rotates from 0° to 180° and 0° to −180°, the tip of the robot moves on a circle of radius l (the length of the link of the robot) as shown in Fig. 3.2. Note that 180° = π radians.

3.2.1 Kinematics of One-Link Robot

In Fig. 3.2, the point P (tip of the robot) can be represented in rectangular or Cartesian coordinates by a pair (x, y) or in polar coordinates by the pair (l, θ). Assuming

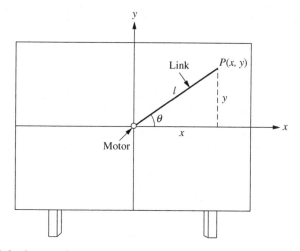

Figure 3.1   One-link planar robot.

that the length of the link l is fixed, a change in the angle θ of the robot changes the position of the tip of the robot. This is known as the direct or forward kinematics of the robot. The position of the tip of the robot (x, y) in terms of l and θ can be found using the right-angle triangle OAP in Fig. 3.2 as

$$\mathbf{cos}(\theta) = \frac{\textbf{Adjacent side}}{\textbf{Hypotenuse}} = \frac{x}{l} \quad \Rightarrow \quad x = l \ \cos(\theta) \tag{3.1}$$

$$\mathbf{sin}(\theta) = \frac{\textbf{Opposite side}}{\textbf{Hypotenuse}} = \frac{y}{l} \quad \Rightarrow \quad y = l \ \sin(\theta) \tag{3.2}$$

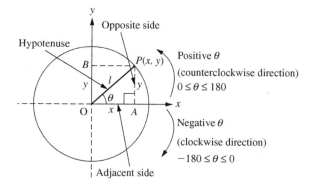

Figure 3.2   Circular path of the one-link robot tip.

| Example 3-1 | Use the one-link robot to find the values of $\cos(\theta)$ and $\sin(\theta)$ for $\theta = 0°$, $90°$, $-90°$, and $180°$. Also, find the values of x and y. |

Solution   **Case I:** $\theta = 0°$

By inspection,

$x = l\cos(0°) = l \;\Rightarrow\; \cos(0°) = 1$

$y = l\sin(0°) = 0 \;\Rightarrow\; \sin(0°) = 0$

Case II: $\theta = 90°$

By inspection,

$x = l\cos(90°) = 0 \;\Rightarrow\; \cos(90°) = 0$

$y = l\sin(90°) = l \;\Rightarrow\; \sin(90°) = 1$

Case III: $\theta = -90°$

By inspection,

$x = l\cos(-90°) = 0 \;\Rightarrow\; \cos(-90°) = 0$

$y = l\sin(-90°) = -l \;\Rightarrow\; \sin(-90°) = -1$

Case IV: $\theta = 180°$

By inspection,

$x = l\cos(180°) = -l \;\Rightarrow\; \cos(180°) = -1$

$y = l\sin(180°) = 0 \;\Rightarrow\; \sin(180°) = 0$

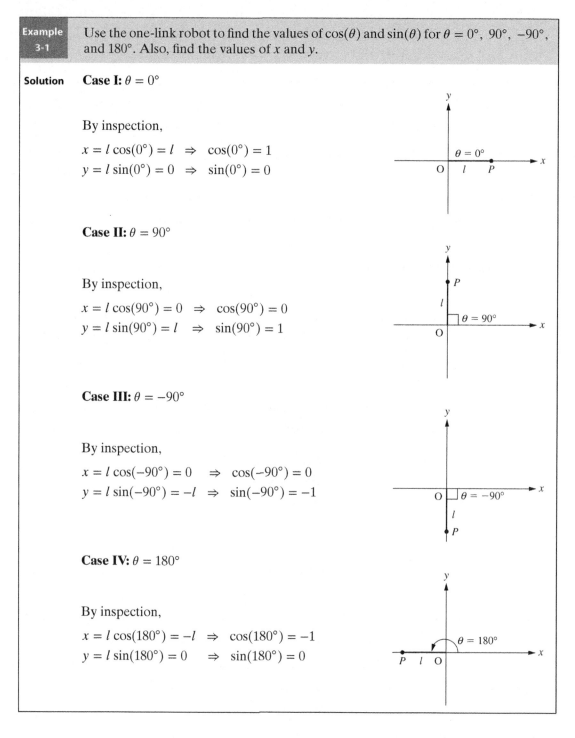

Example 3-2   Find the position $P(x, y)$ of the robot for $\theta = 45°$, $-45°$, $135°$, and $-135°$.

Solution   **Case I:** $\theta = 45°$

$$x = l \cos(45°) = \frac{l}{\sqrt{2}} \quad \Rightarrow \quad \cos(45°) = \frac{1}{\sqrt{2}}$$

$$y = l \sin(45°) = \frac{l}{\sqrt{2}} \quad \Rightarrow \quad \sin(45°) = \frac{1}{\sqrt{2}}$$

Case II: $\theta = -45°$

$$x = l \cos(-45°) = \frac{l}{\sqrt{2}} \quad \Rightarrow \quad \cos(-45°) = \frac{1}{\sqrt{2}}$$

$$y = l \sin(-45°) = -\frac{l}{\sqrt{2}} \quad \Rightarrow \quad \sin(-45°) = -\frac{1}{\sqrt{2}}$$

Case III: $\theta = 135°$

$$x = l \cos(135°) = -\frac{l}{\sqrt{2}} \quad \Rightarrow \quad \cos(135°) = -\frac{1}{\sqrt{2}}$$

$$y = l \sin(135°) = \frac{l}{\sqrt{2}} \quad \Rightarrow \quad \sin(135°) = \frac{1}{\sqrt{2}}$$

Case IV: $\theta = -135°$

$$x = l \cos(-135°) = -\frac{l}{\sqrt{2}}$$

$$\Rightarrow \quad \cos(-135°) = -\frac{1}{\sqrt{2}}$$

$$y = l \sin(-135°) = -\frac{l}{\sqrt{2}}$$

$$\Rightarrow \quad \sin(-135°) = -\frac{1}{\sqrt{2}}$$

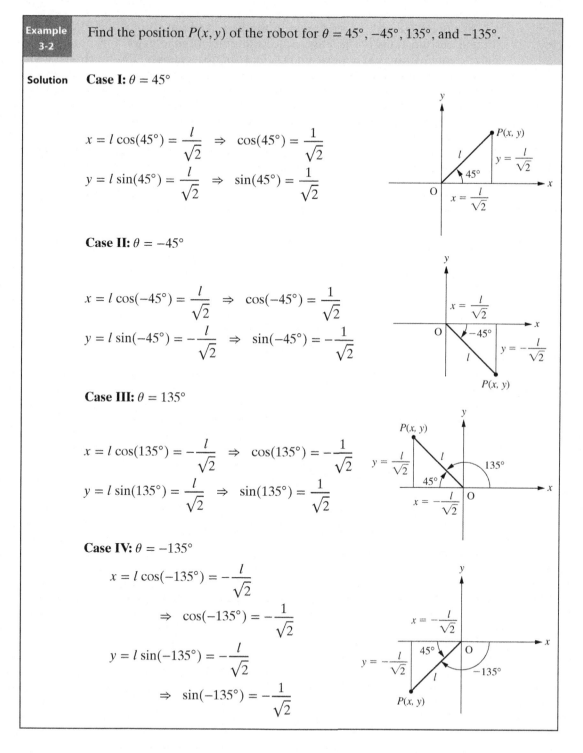

Examples 3-1 and 3-2 show that in the first quadrant ($0° < \theta < 90°$), both the sin and cos functions are positive. Since the other trigonometric functions (tan = sin/cos, cot = 1/tan, sec = 1/cos, and csc = 1/sin, for example) are functions of sin and cos functions, all the trigonometric functions are positive in the first quadrant, as shown in Fig. 3.3. In the second quadrant, sin and csc are positive and all the rest of the trigonometric functions are negative. In the third quadrant, both the sin and cos functions are negative. Therefore, only the tan and cot are positive. Finally, in the fourth quadrant, only the cos and sec are positive. To remember this, one of the phrases commonly used is **"All Sin Tan Cos."** Another is **"All Students Take Calculus,"** which is certainly true of engineering students!

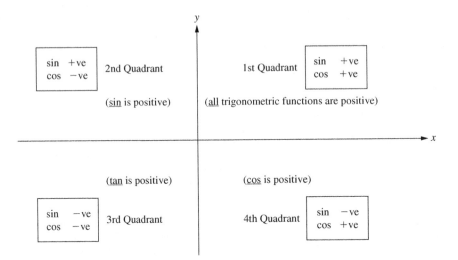

Figure 3.3   Trigonometric functions in the four quadrants.

The values of sin and cos functions for $\theta = 0°$, $30°$, $45°$, $60°$, and $90°$ are given in Table 3.1. The values of sin and cos functions for many other angle can be found using Table 3.1, as explained in the following examples.

TABLE 3.1   Values of sine and cosine functions for common angles.

	Angle				
deg (rad)	$0°$ (0)	$30°$ $(\frac{\pi}{6})$	$45°$ $(\frac{\pi}{4})$	$60°$ $(\frac{\pi}{3})$	$90°$ $(\frac{\pi}{2})$
sin	$\sqrt{\frac{0}{4}} = 0$	$\sqrt{\frac{1}{4}} = \frac{1}{2}$	$\sqrt{\frac{2}{4}} = \frac{1}{\sqrt{2}}$	$\sqrt{\frac{3}{4}} = \frac{\sqrt{3}}{2}$	$\sqrt{\frac{4}{4}} = 1$
cos	$\sqrt{\frac{4}{4}} = 1$	$\sqrt{\frac{3}{4}} = \frac{\sqrt{3}}{2}$	$\sqrt{\frac{2}{4}} = \frac{1}{\sqrt{2}}$	$\sqrt{\frac{1}{4}} = \frac{1}{2}$	$\sqrt{\frac{0}{4}} = 0$

Example 3-3	Find $\sin\theta$ and $\cos\theta$ for $\theta = 120°$. Also, find the position of the tip of the one-link robot for this angle.

Solution   The position of the tip of the robot for $\theta = 120°$ is shown in Fig. 3.4.

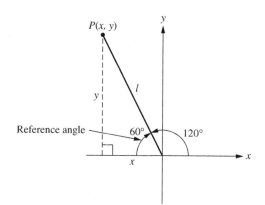

Figure 3.4   One-link planar robot with an angle of $120°$.

Note that the point P is in the second quadrant and, therefore, $\sin(120°)$ should have a positive value and $\cos(120°)$ should be negative. Their values can be found using the reference angle of $\theta = 120°$, which in this case is $60°$. The **reference angle** is always positive, and it is the acute angle formed between the x-axis and the terminal side of the angle ($120°$ in this case).

If the angle θ is in the first quadrant, the reference angle is the same as the angle θ. If the angle θ is in the second quadrant, the reference angle is $180° - \theta$ ($\pi - \theta$, if the angle is in radians). If the angle θ is in the third quadrant, the reference angle is $\theta + 180°$. However, if the angle θ is in the fourth quadrant, the reference angle is the absolute value of θ. Therefore, the values of $\sin(120°)$ and $\cos(120°)$ can be written as

$$x = l\cos(120°) = -l\cos(60°) = -\frac{l}{2}$$

$$y = l\sin(120°) = l\sin(60°) = \frac{\sqrt{3}}{2}l.$$

Note that $\cos(120°) = -\cos(60°) = -\frac{1}{2}$ and $\sin(120°) = \sin(60°) = \frac{\sqrt{3}}{2}$. The values of $\sin(120°)$ and $\cos(120°)$ can also be found using the trigonometric identities

$$\sin(A \pm B) = \sin(A)\cos(B) \pm \cos(A)\sin(B)$$

$$\cos(A \pm B) = \cos(A)\cos(B) \mp \sin(A)\sin(B).$$

Therefore,

$$\sin(120°) = \sin(90° + 30°)$$
$$= \sin(90°)\cos(30°) + \cos(90°)\sin(30°)$$
$$= (1)\left(\frac{\sqrt{3}}{2}\right) + (0)\left(\frac{1}{2}\right)$$
$$= \frac{\sqrt{3}}{2}$$

and

$$\cos(120°) = \cos(90° + 30°)$$
$$= \cos(90°)\cos(30°) - \sin(90°)\sin(30°)$$
$$= (0)\left(\frac{\sqrt{3}}{2}\right) - (1)\left(\frac{1}{2}\right)$$
$$= -\frac{1}{2}.$$

Therefore, the position of the tip of the one-link robot if $\theta = 120°$ is given by $(x, y) = \left(\frac{-l}{2}, \frac{\sqrt{3}l}{2}\right)$.

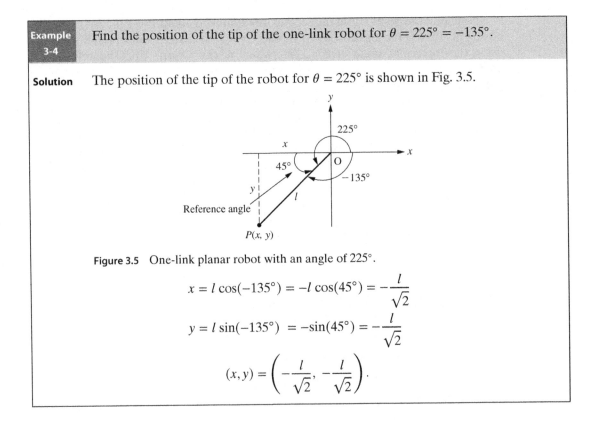

Example 3-4   Find the position of the tip of the one-link robot for $\theta = 225° = -135°$.

Solution   The position of the tip of the robot for $\theta = 225°$ is shown in Fig. 3.5.

Figure 3.5   One-link planar robot with an angle of 225°.

$$x = l\cos(-135°) = -l\cos(45°) = -\frac{l}{\sqrt{2}}$$

$$y = l\sin(-135°) = -\sin(45°) = -\frac{l}{\sqrt{2}}$$

$$(x, y) = \left(-\frac{l}{\sqrt{2}}, -\frac{l}{\sqrt{2}}\right).$$

Example 3-5

Find the position of the tip of the one-link robot for $\theta = 390°$.

Solution   The position of the tip of the robot for $\theta = 390°$ is shown in Fig. 3.6.

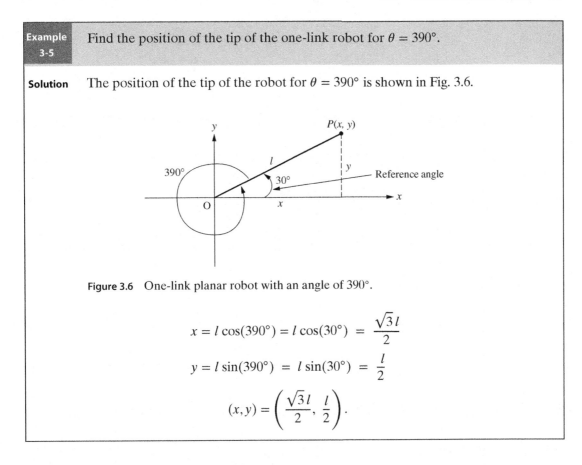

Figure 3.6   One-link planar robot with an angle of 390°.

$$x = l\cos(390°) = l\cos(30°) = \frac{\sqrt{3}\,l}{2}$$

$$y = l\sin(390°) = l\sin(30°) = \frac{l}{2}$$

$$(x, y) = \left(\frac{\sqrt{3}\,l}{2}, \frac{l}{2} \right).$$

Example 3-6

Find the position of the tip of the one-link robot for $\theta = -510°$.

Solution   The position of the tip of the robot for $\theta = -510°$ is shown in Fig. 3.7.

Figure 3.7   One-link planar robot with an angle of −510°.

$$x = l \cos(-510°) = -l \cos(30°) = -\frac{\sqrt{3}\,l}{2}$$

$$y = l \sin(-510°) = -l \sin(30°) = -\frac{l}{2}$$

$$(x, y) = \left(-\frac{\sqrt{3}\,l}{2}, \ -\frac{l}{2} \right).$$

3.2.2   Inverse Kinematics of One-Link Robot

In order to move the tip of the robot to a given position $P(x, y)$, it is required to find the joint angle θ by which the motor needs to move. This is called the inverse problem; for example, given x and y, find the angle θ and length l. Equations (3.1) and (3.2) give the relationship between the tip position and the angle θ. Squaring and adding x and y in these equations gives

$$x^2 + y^2 = (l \cos \theta)^2 + (l \sin \theta)^2$$
$$= l^2 (\sin^2\theta + \cos^2\theta).$$

Using the trigonometric identity $\sin^2\theta + \cos^2\theta = 1$,

$$x^2 + y^2 = l^2.$$

Therefore, $l = \pm\sqrt{x^2 + y^2}$. Since the distance cannot be negative, $l = \sqrt{x^2 + y^2}$. Now dividing y in (3.2) by x in (3.1),

$$\frac{y}{x} = \frac{l \sin\theta}{l \cos\theta} = \tan(\theta). \tag{3.3}$$

Therefore, the angle θ can be determined from the position of the tip of the robot using equation (3.3) as

$$\theta = \tan^{-1}\left(\frac{y}{x}\right) = \text{atan}\left(\frac{y}{x}\right). \tag{3.4}$$

In equation (3.4), y is divided by x before the inverse tangent (arctangent or atan) is calculated and therefore, $\left(\frac{y}{x}\right)$ is either positive or negative. If $\left(\frac{y}{x}\right)$ is positive, the angle obtained from the atan function is between 0 and 90° (first quadrant) and if $\left(\frac{y}{x}\right)$ is negative, the angle obtained from the atan function is between 0 and −90° (fourth quadrant). This is why the atan function is called the two-quadrant arctangent function. However, if both x and y are negative (third quadrant), or x is negative and y is positive (second quadrant), the angles obtained from the atan function will be wrong since the angles should lie in the third or second quadrant, respectively. Therefore, it is important to keep track of the signs of x and y. This can be done by locating the point P in the proper quadrant or using the four-quadrant arctangent function (atan2) as explained in the following examples.

Example 3-7	Find l and θ for the following points (x, y):

Case I: $(x, y) = (1, 0)$:

By inspection, $l = 1$ and $\theta = 0°$.

Also, $l = \sqrt{x^2 + y^2} = \sqrt{1^2 + 0^2} = 1$,

$\theta = \tan^{-1}\left(\frac{0}{1}\right) = \tan^{-1}(0) = 0°$.

Case II: $(x, y) = (0, 1)$:

By inspection, $l = 1$ and $\theta = 90°$.

Also, $l = \sqrt{x^2 + y^2} = \sqrt{0^2 + 1^2} = 1$,

$\theta = \tan^{-1}\left(\frac{1}{0}\right) = \tan^{-1}(\infty) = 90°$.

Case III: $(x, y) = (0, -1)$:

By inspection, $l = 1$ and $\theta = -90°$.

Also, $l = \sqrt{x^2 + y^2} = \sqrt{0^2 + (-1)^2} = 1$,

$\theta = \tan^{-1}(\frac{-1}{0}) = \tan^{-1}(-\infty) = -90°$.

Case IV: $(x, y) = (-1, 0)$:

By inspection, $l = 1$ and $\theta = 180°$.

Also, $l = \sqrt{x^2 + y^2} = \sqrt{(-1)^2 + 0^2} = 1$,

$\theta = \tan^{-1}\left(\frac{0}{-1}\right) = \tan^{-1}(-0) = 180°$.

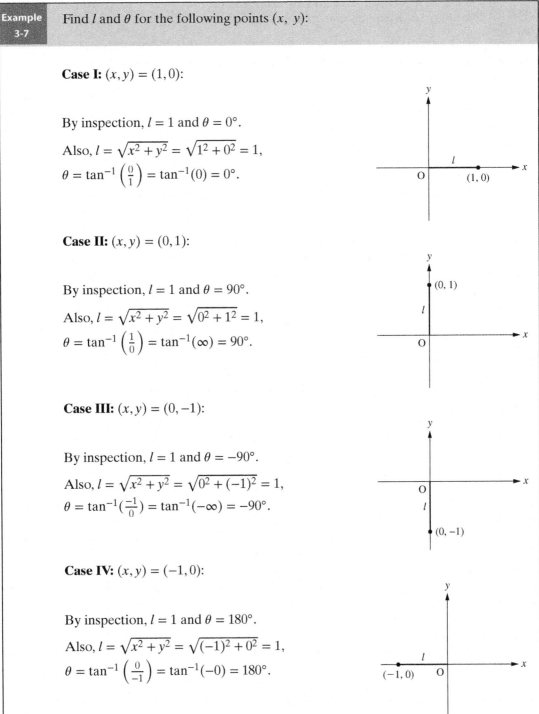

But a calculator will give an answer of 0°. For this case, the calculator answer must be adjusted as explained in example 3-10.

Example 3-8

Find the values of l and θ if $(x, y) = \left(\frac{1}{\sqrt{2}}, \frac{1}{\sqrt{2}}\right)$.

$$l = \sqrt{x^2 + y^2} = \sqrt{\left(\frac{1}{\sqrt{2}}\right)^2 + \left(\frac{1}{\sqrt{2}}\right)^2} = 1,$$

$$\theta = \tan^{-1}\left(\frac{\frac{1}{\sqrt{2}}}{\frac{1}{\sqrt{2}}}\right) = \tan^{-1}(1) = 45°.$$

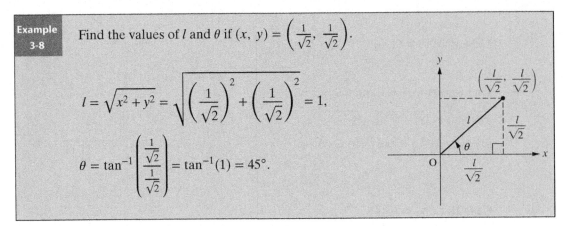

Example 3-9

Find the values of l and θ if $(x, y) = \left(\frac{1}{\sqrt{2}}, -\frac{1}{\sqrt{2}}\right)$.

$$l = \sqrt{x^2 + y^2} = \sqrt{\left(\frac{1}{\sqrt{2}}\right)^2 + \left(-\frac{1}{\sqrt{2}}\right)^2} = 1,$$

$$\theta = \tan^{-1}\left(\frac{-\frac{1}{\sqrt{2}}}{\frac{1}{\sqrt{2}}}\right) = \tan^{-1}(-1) = -45°.$$

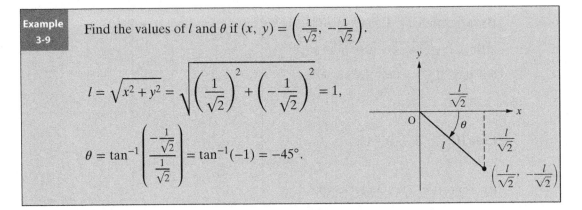

Example 3-10

Find the values of l and θ if $(x, y) = \left(-\frac{1}{\sqrt{2}}, -\frac{1}{\sqrt{2}}\right)$.

$$l = \sqrt{x^2 + y^2} = \sqrt{\left(-\frac{1}{\sqrt{2}}\right)^2 + \left(-\frac{1}{\sqrt{2}}\right)^2} = 1,$$

$$\theta = \tan^{-1}\left(\frac{-\frac{1}{\sqrt{2}}}{-\frac{1}{\sqrt{2}}}\right) = \tan^{-1}(1) = 45°.$$

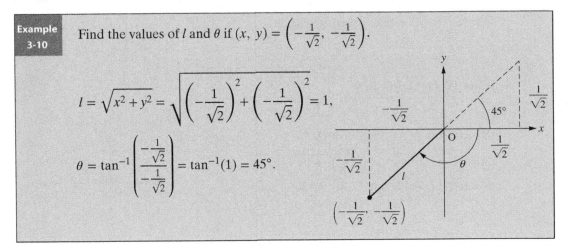

The answer $\theta = 45°$ obtained in example 3-10 is incorrect, and it is the same value obtained in example 3-8 where $(x, y) = \left(\frac{1}{\sqrt{2}}, \frac{1}{\sqrt{2}} \right)$. Remember that the calculator function $\tan^{-1}\left(\frac{y}{x}\right)$ always returns a value in the range of $-90° \le \theta \le 90°$. To obtain the correct answer, it is best to find the quadrant the point lies in and then correct the problem accordingly. Since, in this case, the point lies in the third quadrant, the angle should lie between $-90°$ and $-180°$. The correct answer, therefore, can be obtained by subtracting $180°$ from angle obtain using $\tan^{-1}\left(\frac{y}{x}\right)$. The other method is to obtain the reference angle and then add the reference angle to $-180°$. Therefore, the correct answer is $\theta = 45° - 180° = -135°$.

The correct answer can also be obtained using the atan2(y, x) function. The atan2(y, x) function computes the $\tan^{-1}\left(\frac{y}{x}\right)$ function but uses the sign of both x and y to determine the quadrant in which the resulting angle lies. The atan2(y, x) function is sometimes called a four-quadrant arctangent function and returns a value in the range $-\pi \le \theta \le \pi$ $(-180° \le \theta \le 180°)$. Most of the programming languages including MATLAB have the atan2(y, x) function predefined in their libraries. (Note that the atan2 function requires both x and y values separately instead of $\left(\frac{y}{x}\right)$.) Therefore, using MATLAB gives

$$\text{atan2}\left(-\frac{1}{\sqrt{2}}, -\frac{1}{\sqrt{2}} \right) = -2.3562 \text{ rad}$$

$$= -135°. \tag{3.5}$$

Example 3-11   Find the values of l and θ if $(x, y) = (-0.5, 0.25)$.

$l = \sqrt{x^2 + y^2} = \sqrt{(-0.5)^2 + (0.25)^2} = 0.559$

Using your calculator,

$\theta = \tan^{-1}\left(\frac{0.25}{-0.5} \right) = \tan^{-1}(-0.5) = -26.57°.$

The answer $\theta = -26.57°$ obtained in example 3-11 is clearly incorrect. The correct angle can be obtained using one of the following three methods.

Method 1:   Obtain the reference angle and then subtract the reference angle from $180°$.

$$\theta = 180° - \text{reference angle}$$

$$= 180° - \tan^{-1}\left(\frac{0.25}{0.5} \right)$$

$$= 180° - 26.57°$$

$$= 153.4°. \tag{3.6}$$

Method 2:   Use the $\tan^{-1}\left(\frac{y}{x}\right)$ function and add 180° to the result.

$$\theta = 180° + \tan^{-1}\left(\frac{y}{x}\right)$$

$$= 180° + \tan^{-1}\left(\frac{0.25}{-0.5}\right)$$

$$= 180° + (-26.57°)$$

$$= 153.4°. \tag{3.7}$$

Method 3:   Use the atan2(y,x) function in MATLAB.

$$\theta = \text{atan2}(0.25,\ -0.5)$$

$$= 2.6779 \text{ rad}$$

$$= (2.6779 \text{ rad})\left(\frac{180°}{\pi \text{ rad}}\right)$$

$$= 153.4°. \tag{3.8}$$

3.3   TWO-LINK PLANAR ROBOT

Figure 3.8 shows a two-link planar robot moving in the *x-y* plane. The upper arm of length l_1 is rotated by the shoulder motor, and the lower arm of length l_2 is rotated by the elbow motor. Position sensors are installed at the joints that give the value of the angle θ_1 measured from the positive real axis (*x*-axis) to the upper arm, and the relative angle θ_2 measured from the upper arm to the lower arm of the robot. These angles are **positive** in the **counterclockwise direction** and **negative** in the **clockwise**

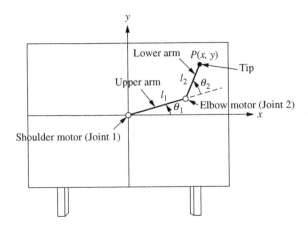

Figure 3.8   Two-link planar robot.

direction. In this section, both the direct and inverse kinematics of the two-link robot are derived.

3.3.1 Direct Kinematics of Two-Link Robot

The direct kinematics of the two-link planar robot is the problem of finding the position of the tip of the robot $P(x, y)$ if the joint angles θ_1 and θ_2 are known. As illustrated in Figs. 3.8 and 3.9,

$$x = x_1 + x_2 \tag{3.9}$$

$$y = y_1 + y_2. \tag{3.10}$$

From the right-angle triangle OAP_1,

$$x_1 = l_1 \cos \theta_1 \tag{3.11}$$

$$y_1 = l_1 \sin \theta_1. \tag{3.12}$$

Similarly, using right-angle triangle P_1BP,

$$x_2 = l_2 \cos(\theta_1 + \theta_2) \tag{3.13}$$

$$y_2 = l_2 \sin(\theta_1 + \theta_2). \tag{3.14}$$

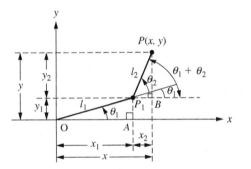

Figure 3.9 Two-link planar robot.

Substituting equations (3.11) and (3.13) into equation (3.9) gives

$$x = l_1 \cos \theta_1 + l_2 \cos(\theta_1 + \theta_2). \tag{3.15}$$

Similarly, substituting equations (3.12) and (3.14) into equation (3.10) yields

$$y = l_1 \sin \theta_1 + l_2 \sin(\theta_1 + \theta_2). \tag{3.16}$$

Equations (3.15) and (3.16) give the position of the tip of the robot in terms of joint angles θ_1 and θ_2.

Example 3-12

Find the position, $P(x, y)$, of the tip of the robot for the following configurations. Also, sketch the orientation of the robot in the x-y plane.

Solution

Case I: $\theta_1 = \theta_2 = 0°$

By inspection:

$x = l_1 + l_2$ and $y = 0$.

Using equations (3.15) and (3.16):

$x = l_1 \cos(0°) + l_2 \cos(0° + 0°) = l_1 + l_2$

$y = l_1 \sin(0°) + l_2 \sin(0° + 0°) = 0$.

Case II: $\theta_1 = 180°, \theta_2 = 0°$

By inspection:

$x = -(l_1 + l_2)$ and $y = 0$.

Using equations (3.15) and (3.16):

$x = l_1 \cos(180°) + l_2 \cos(180° + 0°)$
$= l_1(-1) + l_2(-1) = -(l_1 + l_2)$

$y = l_1 \sin(180°) + l_2 \sin(180° + 0°)$
$= l_1(0) + l_2(0) = 0$.

Case III: $\theta_1 = 90°, \theta_2 = -90°$

By inspection:

$x = l_2$ and $y = l_1$.

Using equations (3.15) and (3.16):

$x = l_1 \cos(90°) + l_2 \cos(90° - 90°)$
$= l_1(0) + l_2(1) = l_2$

$y = l_1 \sin(90°) + l_2 \sin(90° - 90°)$
$= l_1(1) + l_2(0) = l_1$.

Case IV: $\theta_1 = 45°, \theta_2 = -45°$

Using equations (3.15) and (3.16):

$x = l_1 \cos(45°) + l_2 \cos(45° - 45°)$

$= l_1 \left(\dfrac{1}{\sqrt{2}} \right) + l_2(1) = \dfrac{l_1}{\sqrt{2}} + l_2$

$y = l_1 \sin(45°) + l_2 \sin(45° - 45°)$

$= l_1 \left(\dfrac{1}{\sqrt{2}} \right) + l_2(0) = \dfrac{l_1}{\sqrt{2}}$.

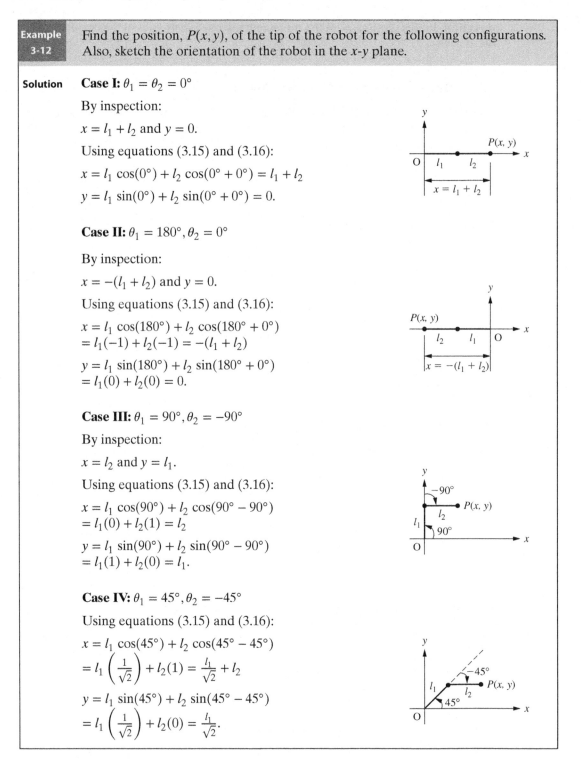

3.3.2 Inverse Kinematics of Two-Link Robot

The inverse kinematics of the two-link planar robot is the problem of finding the joint angles θ_1 and θ_2 if the position of the tip of the robot $P(x,y)$ is known. This problem can be solved using a geometric solution or an algebraic solution. In this chapter, only the algebraic solution will be carried out.

Example 3-13	Find the joint angles θ_1 and θ_2 if the position of the tip of the robot is given by $P(x,y) = (12,6)$ as shown in Fig. 3.10. Assume $l_1 = l_2 = 5\sqrt{2}$.

Figure 3.10 Two-link configuration to find θ_1 and θ_2.

Solution

In the algebraic solution, the joint angles θ_1 and θ_2 are determined using the Pascal laws of cosines and sines. The Pascal law of cosines can be used to find the unknown angles of a triangle if the three sides of the triangle are known. For example, if the three sides of the triangle shown in Fig. 3.11 are known, the unknown angle γ can be found using the Law of Cosines as

$$a^2 = b^2 + c^2 - 2\,b\,c\,\cos\gamma \tag{3.17}$$

or

$$\cos\gamma = \frac{b^2 + c^2 - a^2}{2\,b\,c}.$$

Figure 3.11 A triangle with an unknown angle and three known sides.

Similarly, if the two sides (a and c) and the angle (γ) of the triangle shown in Fig. 3.11 are known, the unknown angle α can be found using the Law of Sines as

$$\frac{\sin \alpha}{c} = \frac{\sin \gamma}{a}$$

or

$$\sin \alpha = \frac{c}{a} \sin \gamma.$$

Solution for θ_2: In Fig. 3.10, the angle θ_2 can be obtained from triangle OPP_1 formed by joining points O and P. In this triangle (Fig. 3.12), three sides are known and one of the angles $180 - \theta_2$ is unknown. Applying the law of cosines to the triangle OP_1P gives

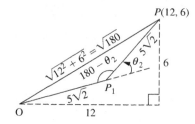

Figure 3.12   Using the law of cosines to find θ_2.

$$\left(\sqrt{180}\right)^2 = \left(5\sqrt{2}\right)^2 + \left(5\sqrt{2}\right)^2 - 2\left(5\sqrt{2}\right)\left(5\sqrt{2}\right) \cos(180° - \theta_2)$$
$$180 = 50 + 50 - 100 \cos(180° - \theta_2)$$
$$80 = -100 \cos(180° - \theta_2)$$
$$-0.8 = \cos(180° - \theta_2). \tag{3.18}$$

Since $\cos(180° - \theta_2) = -\cos\theta_2$, equation (3.18) can be written as $\cos\theta_2 = 0.8$. For the positive value of $\cos\theta_2$, θ_2 lies either in the first or fourth quadrant based on the values of $\sin\theta_2$ as shown in Fig. 3.13. If the value of $\sin\theta_2$ is positive, angle θ_2 is positive. However, if $\sin\theta_2$ is negative, angle θ_2 is negative.

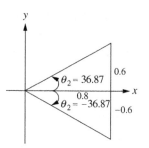

Figure 3.13   Two solutions of θ_2.

Therefore, there are two possible solutions of θ_2, $\theta_2 = 36.87°$ and $\theta_2 = -36.87°$. In Fig. 3.14, the positive solution $\theta_2 = 36.87°$ is called the elbow-up solution and the negative solution $\theta_2 = -36.87°$ is called the elbow-down solution.

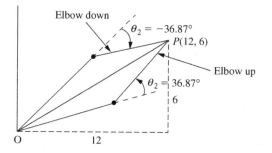

Figure 3.14   Elbow-up and elbow-down solutions of θ_2.

Elbow-Up Solution for θ_1: The angle θ_1 for the elbow-up solution is shown in Fig. 3.15. The angle $\theta_1 + \alpha$ can be obtained from Fig. 3.15 as

$$\tan(\theta_1 + \alpha) = \frac{6}{12}$$

$$\theta_1 + \alpha = \tan^{-1}\left(\frac{6}{12}\right)$$

$$\theta_1 + \alpha = 26.57°$$

$$\theta_1 = 26.57° - \alpha. \tag{3.19}$$

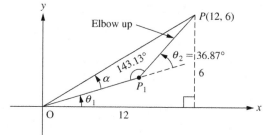

Figure 3.15   Elbow-up configuration to find angle θ_1.

The angle α needed to find θ_1 in equation (3.19) can be obtained using the law of sines or cosines from the triangle OP_1P shown in Fig. 3.16. Using the law of sines gives

$$\frac{\sin \alpha}{5\sqrt{2}} = \frac{\sin 143.13°}{\sqrt{180}}.$$

Therefore,

$$\sin \alpha = \frac{5\sqrt{2}}{\sqrt{180}} \sin 143.13°$$
$$= 0.3164.$$

Since the robot is in the elbow-up configuration, the angle α is positive. Therefore,

$$\alpha = \sin^{-1}(0.3164)$$
$$\alpha = 18.45°.$$

Figure 3.16   Elbow-up configuration to find angle α.

Substituting $\alpha = 18.45°$ into equation (3.19) yields

$$\theta_1 = 26.57 - 18.45 = 8.12°.$$

Therefore, the inverse kinematic solution for the tip position $P(12, 6)$ when the elbow is up is given by

$$\theta_1 = 8.12° \quad \text{and} \quad \theta_2 = 36.87°.$$

Elbow-Down Solution for θ_1: The angle θ_1 for the elbow-down solution is shown in Fig. 3.17. The angle $\theta_1 - \alpha$ can be obtained from Fig. 3.17 as

$$\tan(\theta_1 - \alpha) = \frac{6}{12}$$
$$\theta_1 - \alpha = \tan^{-1}\left(\frac{6}{12}\right)$$
$$\theta_1 - \alpha = 26.57°$$
$$\theta_1 = 26.57° + \alpha. \tag{3.20}$$

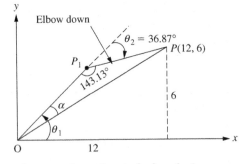

Figure 3.17   Elbow-down configuration to find angle θ_1.

The angle α needed to find θ_1 in equation (3.20) can be obtained using either the law of sines or cosines for the triangle OP_1P shown in Fig. 3.18. Using the law of cosines gives

$$\left(5\sqrt{2}\right)^2 = \left(5\sqrt{2}\right)^2 + \left(\sqrt{180}\right)^2 - 2\left(5\sqrt{2}\right)\left(\sqrt{180}\right)\cos\alpha.$$

Therefore,

$$0 = 180 - 2\left(5\sqrt{2}\right)\left(\sqrt{180}\right)\cos\alpha$$

$$\cos\alpha = \frac{180}{2\times\sqrt{180}\times 5\sqrt{2}}$$

$$\cos\alpha = 0.9487$$

$$\alpha = \cos^{-1}(0.9487)$$

$$\alpha = 18.43°.$$

Figure 3.18   Elbow-down configuration to find angle α.

Substituting $\alpha = 18.43°$ into equation (3.20) yields

$$\theta_1 = 26.57 + 18.43 = 45°.$$

Therefore, the inverse kinematic solution for the tip position $P(12, 6)$ when the elbow is down is given by

$$\theta_1 = 45° \quad \text{and} \quad \theta_2 = -36.87°.$$

3.3.3   Further Examples of Two-Link Planar Robot

Example 3-14

Consider a two-link planar robot, with positive orientations of θ_1 and θ_2 as shown in Fig. 3.19.

(a) Suppose $\theta_1 = \frac{2\pi}{3}$ rad, $\theta_2 = \frac{5\pi}{6}$ rad, $l_1 = 10$ in., and $l_2 = 12$ in. Sketch the orientation of the robot in the x-y plane, and determine the x and y coordinates of point $P(x, y)$.

(b) Suppose now that the same robot is located in the first quadrant and oriented in the elbow-up position, as shown in Fig. 3.19. If the tip of the robot is located at the point $P(x, y) = (12, 16)$, determine the values of θ_1 and θ_2.

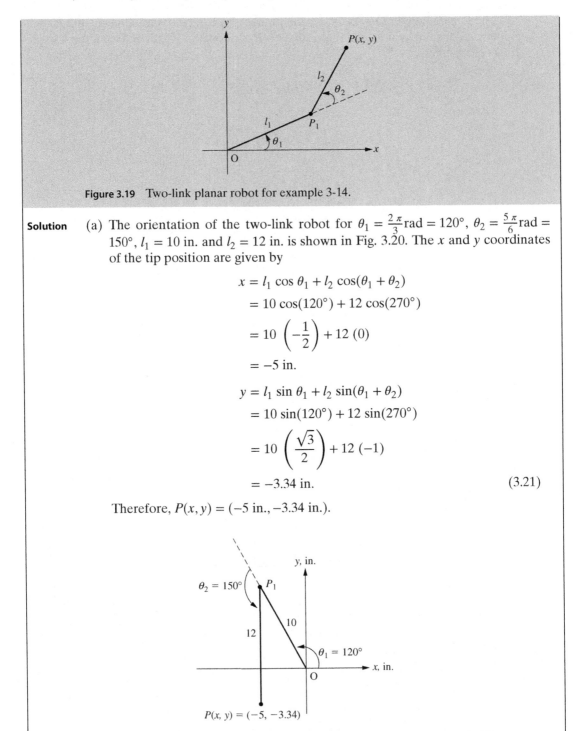

Figure 3.19 Two-link planar robot for example 3-14.

Solution (a) The orientation of the two-link robot for $\theta_1 = \frac{2\pi}{3}$ rad $= 120°$, $\theta_2 = \frac{5\pi}{6}$ rad $= 150°$, $l_1 = 10$ in. and $l_2 = 12$ in. is shown in Fig. 3.20. The x and y coordinates of the tip position are given by

$$x = l_1 \cos \theta_1 + l_2 \cos(\theta_1 + \theta_2)$$
$$= 10 \cos(120°) + 12 \cos(270°)$$
$$= 10 \left(-\frac{1}{2}\right) + 12\,(0)$$
$$= -5 \text{ in.}$$

$$y = l_1 \sin \theta_1 + l_2 \sin(\theta_1 + \theta_2)$$
$$= 10 \sin(120°) + 12 \sin(270°)$$
$$= 10 \left(\frac{\sqrt{3}}{2}\right) + 12\,(-1)$$
$$= -3.34 \text{ in.} \tag{3.21}$$

Therefore, $P(x, y) = (-5 \text{ in.}, -3.34 \text{ in.})$.

Figure 3.20 Orientation of the two-link planar robot for $\theta_1 = 120°$ and $\theta_2 = 150°$.

(b) For the two-link robot located in the first-quadrant as shown in Fig. 3.19, the angle θ_2 can be found using the law of cosines on the triangle OP_1P shown in Fig. 3.21. The unknown angle $180° - \theta_2$ and the three sides of the triangle OP_1P are shown in Fig. 3.21. Using the law of cosines gives

$$20^2 = 10^2 + 12^2 - 2(10)(12) \cos(180° - \theta_2)$$
$$400 = 244 + 240 \cos \theta_2$$
$$156 = 240 \cos \theta_2 \quad \Rightarrow \cos \theta_2 = 0.65. \tag{3.22}$$

Since the robot is in the elbow-up configuration, the angle θ_2 is positive and is given by $\theta_2 = \cos^{-1}(0.65) = 49.46°$. Also, from Fig. 3.21, the angle $\theta_1 + \alpha$ can be determined using the right-angled triangle OAP as

$$\tan(\theta_1 + \alpha) = \frac{16}{12} \quad \Rightarrow \theta_1 = \tan^{-1}\left(\frac{16}{12}\right) - \alpha.$$

Therefore,

$$\theta_1 = 53.13° - \alpha. \tag{3.23}$$

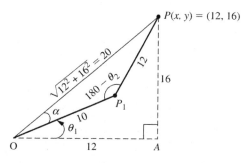

Figure 3.21 The triangle OP_1P to find the angles θ_1 and θ_2.

The angle α can be found from the triangle OP_1P using either the law of cosines or the law of sines. Using the law of sines gives

$$\frac{\sin \alpha}{12} = \frac{\sin(180° - \theta_2)}{20}.$$

Therefore,

$$\sin \alpha = \frac{12}{20} \sin(180° - 49.46°)$$
$$= 0.4560$$
$$\alpha = \sin^{-1}(0.4560)$$
$$\alpha = 27.13°.$$

Substituting $\alpha = 27.13°$ in equation (3.23) yields

$$\theta_1 = 53.13 - 27.13 = 26.0°.$$

Example 3-15	Consider a two-link planar robot with positive orientations of θ_1 and θ_2 as shown in Fig. 3.19. Suppose $\theta_1 = 120°$, $\theta_2 = -30°$, $l_1 = 8$ cm, and $l_2 = 4$ cm.

(a) Sketch the orientation of the robot in the x-y plane.
(b) Determine the x and y coordinates of point $P(x, y)$.
(c) Determine the distance from point P to the origin.

Solution   (a) The orientation of the two-link robot for $\theta_1 = 120°$, $\theta_2 = -30°$, $l_1 = 8$ cm, and $l_2 = 4$ cm is shown in Fig. 3.22.

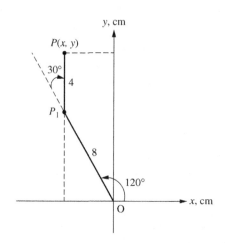

Figure 3.22   Orientation of the two-link planar robot for $\theta_1 = 120°$ and $\theta_2 = -30°$.

(b) The x- and y-coordinates of the tip position are given by

$$x = l_1 \cos \theta_1 + l_2 \cos(\theta_1 + \theta_2)$$
$$= 8 \cos(120°) + 4 \cos(90°)$$
$$= 8 \left(-\frac{1}{2}\right) + 4\,(0)$$
$$= -4 \text{ cm}.$$

$$y = l_1 \sin \theta_1 + l_2 \sin(\theta_1 + \theta_2)$$
$$= 8 \sin(120°) + 4 \sin(90°)$$
$$= 8 \left(\frac{\sqrt{3}}{2}\right) + 4\,(1)$$
$$= 10.93 \text{ cm}.$$

Therefore, $P(x, y) = (-4 \text{ cm}, 10.93 \text{ cm})$.

(c) The distance from the tip $P(x, y)$ to the origin is given by

$$d = \sqrt{x^2 + y^2}$$
$$= \sqrt{(-4)^2 + (10.93)^2}$$
$$= 11.64 \text{ cm}.$$

Therefore, the distance from the tip of the robot to the origin is 11.64 cm.

Example 3-16

Consider a two-link planar robot with positve orientations of θ_1 and θ_2 as shown in Fig. 3.19.

(a) Suppose $\theta_1 = 135°$, $\theta_2 = 45°$ and $l_1 = l_2 = 10$ in. Sketch the orientation of the robot in the x-y plane, and determine the x and y coordinates of point $P(x, y)$.

(b) Suppose now that the tip of the same robot is located in the second quadrant and oriented in the elbow-up position as shown in Fig. 3.23. If the tip of the robot is located at the point $P(x, y) = (-17.07 \text{ in.}, 7.07 \text{ in.})$, determine the values of θ_1 and θ_2.

Figure 3.23   Two-link planar robot in the elbow-up position with $P(x, y) = (-17.07 \text{ in.}, 7.07 \text{ in.})$.

Solution

(a) The orientation of the two-link robot for $\theta_1 = 135°$, $\theta_2 = 45°$, and $l_1 = l_2 = 10$ in. is shown in Fig. 3.24. The x and y coordinates of the tip position are given by

$$x = l_1 \cos \theta_1 + l_2 \cos(\theta_1 + \theta_2)$$
$$= 10 \cos 135° + 10 \cos 180°$$
$$= 10 \left(-\frac{\sqrt{2}}{2} \right) + 10 \, (-1)$$
$$= -17.07 \text{ in.}$$
$$y = l_1 \sin \theta_1 + l_2 \sin(\theta_1 + \theta_2)$$
$$= 10 \sin 135° + 10 \sin 180°$$
$$= 10 \left(\frac{\sqrt{2}}{2} \right) + 10(0)$$
$$= 7.07 \text{ in.} \tag{3.24}$$

Therefore, $P(x, y) = (-17.07 \text{ in.}, \ 7.07 \text{ in.})$.

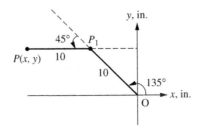

Figure 3.24   The orientation of the two-link planar robot with $\theta_1 = 135°$, and $\theta_2 = 45°$.

(b) The angle θ_2 can be found using the law of cosines on the triangle OP_1P in Fig. 3.25. The unknown angle $180° - \theta_2$ and the three sides of the triangle OP_1P are shown in Fig. 3.25. Using the law of cosines gives

$$(18.48)^2 = 10^2 + 10^2 - 2(10)(10)\cos(180° - \theta_2)$$
$$341.4 = 200 + 200\cos\theta_2$$
$$141.4 = 200\cos\theta_2 \quad \Rightarrow \quad \cos\theta_2 = 0.707.$$

Since the robot is in the elbow-up configuration, angle θ_2 is positive and is given by $\theta_2 = \cos^{-1}(0.707) = 45.0°$. Also, from Fig. 3.25, the angle $\theta_1 + \alpha$ can be determined using the right-angled triangle OAP as

$$\theta_1 + \alpha = \text{atan2}(7.07, -17.07)$$
$$\theta_1 + \alpha = 157.5°.$$

Therefore,

$$\theta_1 = 157.5° - \alpha. \tag{3.25}$$

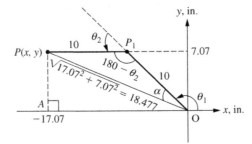

Figure 3.25   The triangle OP_1P to find the angles θ_1 and θ_2.

The angle α can be found from the triangle OP_1P using either the law of cosines or the law of sines. Using the law of cosines gives

$$10^2 = (18.48)^2 + 10^2 - 2(10)(18.48)\cos\alpha$$
$$-341.4 = -369.54\cos\alpha$$
$$0.9239 = \cos\alpha.$$

Therefore, $\alpha = 22.5°$. Substituting $\alpha = 22.5°$ in equation (3.25) yields

$$\theta_1 = 157.5 - 22.5° = 135.0°.$$

Example 3-17

Consider a two-link planar robot with positive orientations of θ_1 and θ_2 as shown in Fig. 3.19.

(a) Suppose $\theta_1 = -135°$, $\theta_2 = -45°$, and $l_1 = l_2 = 10$ in. Sketch the orientation of the robot in the x-y plane, and determine the x- and y-coordinates of point $P(x, y)$.

(b) Suppose now that the tip of same robot is located in the third quadrant and oriented in the elbow-down position (clockwise direction), as shown in Fig. 3.26. If the tip of the robot is located at point $P(x, y) = (-17.07$ in., -7.07 in.$)$, determine the values of θ_1 and θ_2.

Figure 3.26 Two-link planar robot in the elbow-down position with $P(x, y) = (-17.07$ in., -7.07 in.$)$.

Solution

(a) The orientation of the two-link robot for $\theta_1 = -135°$, $\theta_2 = -45°$, and $l_1 = l_2 = 10$ in. is shown in Fig. 3.27. The x- and y-coordinates of the tip position are given by

$$x = l_1 \cos \theta_1 + l_2 \cos(\theta_1 + \theta_2)$$
$$= 10 \cos(-135°) + 10 \cos(-180°)$$
$$= 10\left(-\frac{\sqrt{2}}{2}\right) + 10 \, (-1)$$
$$= -17.07 \text{ in.}$$
$$y = l_1 \sin \theta_1 + l_2 \sin(\theta_1 + \theta_2)$$
$$= 10 \sin(-135°) + 10 \sin(-180°)$$
$$= 10\left(-\frac{\sqrt{2}}{2}\right) + 10(0)$$
$$= -7.07 \text{ in.}$$

Therefore, $P(x, y) = (-17.07$ in., -7.07 in.$)$.

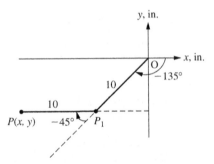

Figure 3.27   The orientation of the two-link planar robot with $\theta_1 = -135°$ and $\theta_2 = -45°$.

(b) The angle θ_2 can be found using the law of cosines on the triangle OP_1P shown in Fig. 3.28. The unknown angle $180° - \theta_2$ and the three sides of the triangle OP_1P are shown in Fig. 3.28. Using the law of cosines gives

$$(18.48)^2 = 10^2 + 10^2 - 2(10)(10)\cos(180° - \theta_2)$$

$$341.4 = 200 + 200 \cos \theta_2$$

$$141.4 = 200 \cos \theta_2 \quad \Rightarrow \quad \cos \theta_2 = 0.707.$$

Therefore, $\theta_2 = \cos^{-1}(0.707) = 45.0°$ or $-45°$. Since the angle θ_2 is in the clockwise direction, $\theta_2 = -45°$. Also, from Fig. 3.28, the angle $\theta_1 + \alpha$ can be determined using the right-angled triangle OAP as

$$\theta_1 + \alpha = \text{atan2}(-7.07, 17.07)$$

$$\theta_1 + \alpha = -157.5°.$$

Therefore,

$$\theta_1 = -157.5° - \alpha. \tag{3.26}$$

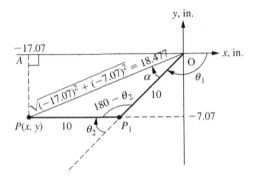

Figure 3.28   The triangle OP_1P used to find the angles θ_1 and θ_2.

The angle α can be found from the triangle OP_1P using either the law of cosines or the law of sines. Using the law of cosines yields

$$10^2 = (18.477)^2 + 10^2 - 2(10)(18.477)\cos\alpha$$

$$-341.4 = -369.54\cos\alpha$$

$$0.9239 = \cos\alpha.$$

Since angle α is in the clockwise direction, $\alpha = -22.5°$. Substituting $\alpha = -22.5°$ in equation (3.26) gives

$$\theta_1 = -157.5° - (-22.5°) = -135.0°.$$

Therefore, $\theta_1 = -135°$.

Example 3-18

Consider a two-link planar robot with positive orientations of θ_1 and θ_2 as shown in Fig. 3.19.

(a) Suppose $\theta_1 = -45°$, $\theta_2 = 45°$, and $l_1 = l_2 = 10$ in. Sketch the orientation of the robot in the x-y plane, and determine the x- and y-coordinates of point $P(x, y)$.

(b) Suppose now that the tip of same robot is located in the fourth quadrant and oriented in the elbow-up position (counterclockwise direction), as shown in Fig. 3.29. If the tip of the robot is located at the point $P(x, y) = (17.07$ in., -7.07 in.$)$, determine the values of θ_1 and θ_2.

Figure 3.29 Two-link planar robot in the elbow-up position with $P(x, y) = (17.07$ in., -7.07 in.$)$

Solution

(a) The orientation of the two-link robot for $\theta_1 = -45°$, $\theta_2 = 45°$, and $l_1 = l_2 = 10$ in. is shown in Fig. 3.30. The x- and y-coordinates of the tip position are given by

$$x = l_1\cos\theta_1 + l_2\cos(\theta_1 + \theta_2)$$

$$= 10\cos(-45°) + 10\cos(0°)$$

$$= 10\left(\frac{\sqrt{2}}{2}\right) + 10(1)$$

$$= 17.07 \text{ in.}$$

$$y = l_1 \sin \theta_1 + l_2 \sin(\theta_1 + \theta_2)$$
$$= 10 \sin(-45°) + 10 \sin(0°)$$
$$= 10 \left(-\frac{\sqrt{2}}{2} \right) + 10(0)$$
$$= -7.07 \text{ in.}$$

Therefore, $P(x, y) = (17.07 \text{ in.}, -7.07 \text{ in.})$.

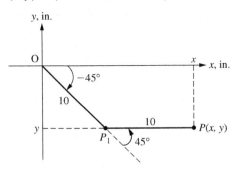

Figure 3.30   The orientation of the two-link planar robot with $\theta_1 = -45°$ and $\theta_2 = 45°$.

(b) The angle θ_2 can be found using the law of cosines on the triangle OP_1P shown in Fig. 3.31. The unknown angle $180° - \theta_2$ and the three sides of the triangle OP_1P are shown in Fig. 3.31. Using the law of cosines gives

$$(18.48)^2 = 10^2 + 10^2 - 2(10)(10) \cos(180° - \theta_2)$$
$$341.4 = 200 + 200 \cos \theta_2$$
$$141.4 = 200 \cos \theta_2 \Rightarrow \cos \theta_2 = 0.707.$$

Since the robot is in the elbow-up configuration, angle $\theta_2 = \cos^{-1}(0.707) = 45°$. Also, from Fig. 3.31, the angle $\theta_1 - \alpha$ can be determined using the right-angled triangle OAP as

$$\theta_1 - \alpha = \text{atan2}(-7.07, 17.07)$$
$$\theta_1 - \alpha = -22.5°.$$

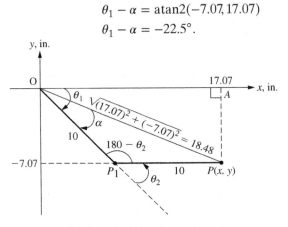

Figure 3.31   The triangle OP_1P to find the angles θ_1 and θ_2.

Therefore,

$$\theta_1 = -22.5° + \alpha. \qquad (3.27)$$

The angle α can be found from the triangle OP_1P using either the law of cosines or the law of sines. Using the law of cosines yields

$$10^2 = (18.48)^2 + 10^2 - 2(10)(18.48)\cos\alpha$$

$$-341.4 = -369.5\cos\alpha$$

$$0.9239 = \cos\alpha.$$

Since angle α is in the clockwise direction, $\alpha = -22.5°$. Substituting $\alpha = -22.5°$ in equation (3.27) gives

$$\theta_1 = -22.5 + (-22.5)° = -45.0°.$$

Therefore, $\theta_1 = -45°$.

3.4 FURTHER EXAMPLES OF TRIGONOMETRY IN ENGINEERING

Example 3-19 A one-link planar robot of length $l = 1.5$ m is moving in the x-y plane. If the joint angle $\theta = -165°$, locate the tip $P(x,y)$ of the robot in the x-y plane.

Solution The tip of the one-link robot for $\theta = -165°$ is shown in Fig. 3.32. It can be seen from this figure that the tip is located in the third quadrant and the reference angle is 15°. Since both the sin and cos functions are negative in the third quadrant, the position $P(x,y)$ of the tip is given by

$$x = 1.5\cos(165°) = -1.5\cos(15°) = -1.5 \times 0.9659 = -1.449 \text{ m}$$
$$y = 1.5\sin(165°) = -1.5\sin(15°) = -1.5 \times 0.2588 = -0.388 \text{ m}$$

Figure 3.32 One-link planar robot for Example 3-19.

Example 3-20

The x- and y-components of the tip of a one-link planar robot are given as -10 cm and 5 cm, respectively. Locate the tip of the robot in the x-y plane. Also, find the length l of the link and the angle θ.

Solution

The tip of the one-link robot with $x = -10$ cm and $y = 5$ cm is shown in Fig. 3.33. The length l is given by

$$l = \sqrt{(-10)^2 + (5)^2} = \sqrt{100 + 25} = \sqrt{125} = 11.18 \text{ cm}.$$

Figure 3.33 One-link planar robot for example 3-20.

Since the tip of the robot is in the second quadrant, the angle θ is given by

$$\theta = 180° - \operatorname{atan}\left(\frac{5}{10}\right)$$

$$= 180° - \operatorname{atan}(0.5)$$

$$= 180° - 26.57°$$

$$= 153.4°.$$

Using the $\operatorname{atan2}(y, x)$ function in MATLAB, the angle θ is given by

$$\theta = \operatorname{atan2}(5, -10)$$

$$= 2.6779 \text{ rad}$$

$$= (2.6779 \text{ rad}) \left(\frac{180°}{\pi \text{ rad}}\right)$$

$$= 153.4°.$$

Example 3-21

In Example 1-8, the civil engineer calculated the elevation of the building cornerstone located between two benchmarks shown in Fig. 1.19. The same engineer is now required to calculate the angle of inclination and horizontal distance between the two benchmarks, as shown in Fig. 3.34.

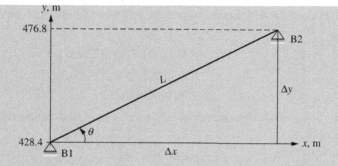

Figure 3.34   Elevation between the two benchmarks.

The distance L between the two benchmarks B1 and B2 is 1001.2 m, and their elevations are 428.4 m and 476.8 m, respectively.

(a) Find the angle of inclination θ of the grade. Also, calculate the percent grade, $\dfrac{\Delta y}{\Delta x} \times 100$.

(b) Calculate the horizontal distance between the two benchmarks.

(c) Check the results of part (b) using the Pythagorean theorem.

Solution   (a) The angle of inclination θ can be determined from the right-angle triangle shown in Fig. 3.34 as

$$\sin \theta = \frac{\Delta y}{L}$$

$$= \frac{E_2 - E_1}{L}$$

$$= \frac{476.8 - 428.4}{1001.2}$$

$$= 0.0483.$$

Therefore, the angle of inclination $\theta = \sin^{-1}(0.0483) = 2.768°$. The percent grade can now be calculated as

$$\text{Percent grade} = \frac{\Delta y}{\Delta x} \times 100$$

$$= 100 \times \tan \theta$$

$$= 100 \times \tan(2.768°)$$

$$= 4.84\%.$$

(b) The horizontal distance between the two benchmarks can be calculated as

$$\Delta x = L \cos\theta$$
$$= (1001.2) \cos(2.468°)$$
$$= (1001.2) (0.99883)$$
$$= 1000.0 \text{ m.} \tag{3.28}$$

(c) The horizontal distance can also be calculated using the Pythagorean theorem as

$$\Delta x = \sqrt{L^2 - \Delta y^2}$$
$$= \sqrt{(1001.2)^2 - (48.4)^2}$$
$$= 1000 \text{ m.}$$

Example 3-22

Consider the position of the toes of a person sitting in a chair, as shown in Fig. 3.35.

Figure 3.35 Toes position of a person sitting on a chair.

(a) Suppose $\theta_1 = -30°$, $\theta_2 = 45°$, $l_1 = 20$ in., and $l_2 = 5$ in. Determine the x- and y-coordinates of the position of the toes $P(x, y)$.

(b) Now suppose that the same leg is positioned such that the tip of the toes is located in the first quadrant and oriented in the ankle-up position (counterclockwise direction) as shown Fig. 3.36. If the end of the toes are located at $P(x, y) = (19.5$ in., 2.5 in.$)$, determine the values of θ_1 and θ_2.

Figure 3.36 Ankle-up position.

Solution   (a) The x- and y-coordinates of the toes position can be calculated as

$$x = l_1 \cos \theta_1 + l_2 \cos(\theta_1 + \theta_2)$$
$$= 20 \cos(-30°) + 5 \cos(-30° + 45°)$$
$$= 20 \left(\frac{\sqrt{3}}{2} \right) + 5(0.9659)$$
$$= 22.15 \text{ in.}$$
$$y = l_1 \sin \theta_1 + l_2 \sin(\theta_1 + \theta_2)$$
$$= 20 \sin(-30°) + 5 \sin(-30° + 45°)$$
$$= 20 \left(-\frac{1}{2} \right) + 5(0.2588)$$
$$= -18.71 \text{ in.}$$

Therefore, $P(x, y) = (22.15 \text{ in.}, -18.71 \text{ in.})$.

(b) The angle θ_2 can be found using the law of cosines on the triangle OP_1P shown in Fig. 3.37. The unknown angle $180° - \theta_2$ and the three sides of the triangle OP_1P are shown in Fig. 3.37. Using the law of cosines gives

$$(19.66)^2 = 20^2 + 5^2 - 2(20)(5) \cos(180° - \theta_2)$$
$$386.5 = 425 - 200 \cos(180° - \theta_2)$$
$$-38.5 = -200 \cos(180° - \theta_2) \quad \Rightarrow \cos(180° - \theta_2) = 0.1925$$

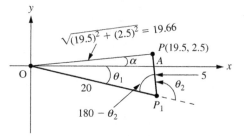

Figure 3.37   The triangle OP_1P to find the angles θ_1 and θ_2.

Since the leg is in the ankle-up configuration, $(180° - \theta_2) = \cos^{-1}(0.1925) = 78.9°$. Therefore, $\theta_2 = 101.1°$. Also, from Fig. 3.37, the angle $\theta_1 + \alpha$ can be determined using from the triangle OP_1P using either the law of cosines or the law of sines. Using the law of sines yields

$$\frac{\sin(\theta_1 + \alpha)}{5} = \frac{\sin(180° - \theta_2)}{19.66}.$$

Therefore,

$$\sin(\theta_1 + \alpha) = \frac{5}{19.66} \sin 78.9°$$

$$\theta_1 + \alpha = 14.45°. \tag{3.29}$$

The angle α can be found from the right-triangle OAP shown in Fig. 3.37 as

$$\alpha = \text{atan2}(2.5, 19.5) = 7.31°.$$

Substituting the value of α in equation (3.29) gives

$$\theta_1 = 14.45 - \alpha$$

$$= 14.45° - 7.31°$$

or

$$\theta_1 = 7.14°.$$

Example 3-23

In a motion capture study of a runner, one frame shows the subject supporting her weight on one leg, as shown in Fig. 3.38. The length of the foot segment (from ankle to toe) is 7.9 in. and the length of the lower leg (from ankle to knee) is 17.1 in.

(a) Given the angles shown in Fig. 3.38, find the position of the knee if the runner's toes touch the ground at the point $x = y = 0$.

Figure 3.38   Position of the runner's leg during motion capture study.

(b) Now suppose that the same leg is positioned such that the knee is located in the second quadrant and oriented in the knee-down position (clockwise direction) as shown Fig. 3.39. If the end of the toes are located at $P(x, y) = (-4 \text{ in.}, 24 \text{ in.})$, determine the values of θ_1 and θ_2.

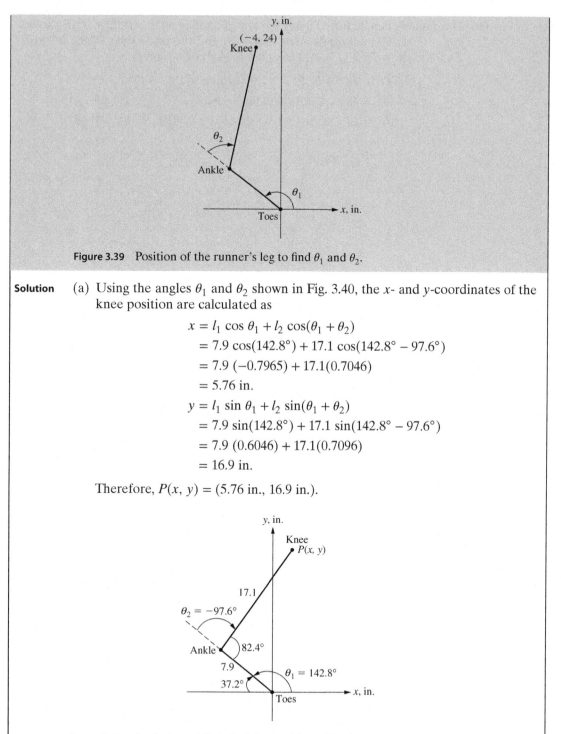

Figure 3.39   Position of the runner's leg to find θ_1 and θ_2.

Solution   (a) Using the angles θ_1 and θ_2 shown in Fig. 3.40, the x- and y-coordinates of the knee position are calculated as

$$x = l_1 \cos \theta_1 + l_2 \cos(\theta_1 + \theta_2)$$
$$= 7.9 \cos(142.8°) + 17.1 \cos(142.8° - 97.6°)$$
$$= 7.9 \, (-0.7965) + 17.1(0.7046)$$
$$= 5.76 \text{ in.}$$

$$y = l_1 \sin \theta_1 + l_2 \sin(\theta_1 + \theta_2)$$
$$= 7.9 \sin(142.8°) + 17.1 \sin(142.8° - 97.6°)$$
$$= 7.9 \, (0.6046) + 17.1(0.7096)$$
$$= 16.9 \text{ in.}$$

Therefore, $P(x, y) = (5.76 \text{ in.}, 16.9 \text{ in.})$.

Figure 3.40   Angle θ_1 and θ_2 to find the position of the knee.

(b) The angle θ_2 can be found using the law of cosines on the triangle TAK shown in Fig. 3.41. The unknown angle $180° - \theta_2$ and the three sides of the triangle TAK are shown in Fig. 3.41. Using the law of cosines gives

$$(24.33)^2 = 7.9^2 + 17.1^2 - 2(7.9)(17.1)\cos(180° - \theta_2)$$
$$592 = 425 - 270.18\cos(180° - \theta_2)$$
$$167 = 270.18\cos(\theta_2) \Rightarrow \cos(\theta_2) = 0.6181 \Rightarrow \theta_2 = 51.82°.$$

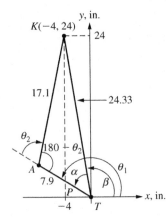

Figure 3.41 The triangle TAK to find the angles θ_2.

Also from Fig. 3.41, the angle α can be determined from the triangle TAK using either the law of cosines or the law of sines. Using the law of sines yields

$$\frac{\sin(\alpha)}{17.1} = \frac{\sin(180° - \theta_2)}{24.33}.$$

Therefore,

$$\sin(\alpha) = \frac{17.1}{24.33}\sin 129.2°$$

which gives

$$\alpha = 33°.$$

The angle β can be found from the right-triangle TKP shown in Fig. 3.41 as

$$\beta = \text{atan2}(24, -4) \Rightarrow \beta = 99.46°.$$

Angle θ_1 can now be found by adding angles α and β as shown in Fig. 3.41 as

$$\theta_1 = \alpha + \beta$$
$$= 51.82° + 99.46°$$
$$= 151.28°.$$

Therefore, $\theta_1 = 151.8°$. Also, since the robot is in the knee-down configuration, the angle $\theta_2 = -51.82°$.

PROBLEMS

3-1. A laser range finder records the distance from the laser to the base and from the laser to the top of a building as shown in Fig. P3.1. Find the angle θ and the height of the building.

Figure P3.1 Using a range finder to find the height of a building.

3-2. The eyes of a 7 ft 4 in. player are 82 in. from the floor, as shown in Fig. P3.2. If the height of the basketball hoop is 10 ft from the floor, find the distance l and angle θ from the player's eye to the hoop.

Figure P3.2 A basketball player in front of the basketball hoop.

3-3. Repeat problem P3-2 if the player's height is 72 in.

3-4. To calculate the property tax, a city hires a coop student to determine the area of different lots in a new subdivision. The coop student calculates the area of lot 1 shown in Fig. P3.4 as 94,640 m². Is this the correct answer? If not, find the correct answer.

Figure P3.4 Dimension of lot 1 in the new subdivision.

3-5. The same coop student calculates the area of lot 2 shown in Fig. P3.5 as 50,000 m². Is this the correct answer? If not, find the correct answer.

Figure P3.5 Dimension of lot 2 in the new subdivision.

3-6. A laser beam is directed through a small hole in the center of a circle of radius 1.73 m. The origin of the beam is 5 m from the circle as shown in Fig. P3.6.

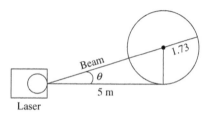

Figure P3.6 Laser beam for problem P3-6.

What should be the angle θ of the beam for the beam to go through the hole? Use the law of sines.

3-7. A truss structure consists of three isosceles triangles as shown in Fig. P3.7. Determine the angle θ using the laws of cosines or sines.

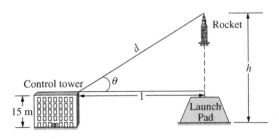

Figure P3.7   Truss structure for problem P3-7.

3-8. A rocket takes off from a launch pad located $l = 500$ m from the control tower as shown in Fig. P3.8. If the control tower is 15 m tall, determine the height h of the rocket from the ground when it is located at a distance $d = 575$ m from the top of the control tower. Also, determine the angle θ.

Figure P3.8   A rocket taking off from a launch pad for problem P3-8.

3-9. Repeat problem P3-8 if $l = 300$ m and $d = 500$ m.

3-10. A one-link planar robot moves in the x-y plane as shown in Fig. P3.10. For the given l and θ, find the position $P(x, y)$ of the tip.

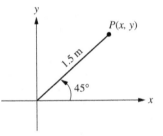

Figure P3.10   A one-link planar robot for problem P3-10.

3-11. A one-link planar robot moves in the x-y plane as shown in Fig. P3.11. For the given l and θ, find the position $P(x, y)$ of the tip.

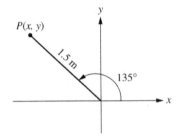

Figure P3.11   A one-link planar robot for problem P3-11.

3-12. A one-link planar robot moves in the x-y plane as shown in Fig. P3.12. For the given l and θ, find the position $P(x, y)$ of the tip.

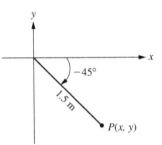

Figure P3.12   A one-link planar robot for problem P3-12.

3-13. A one-link planar robot moves in the *x-y* plane as shown in Fig. P3.13. For the given *l* and θ, find the position $P(x, y)$ of the tip.

Figure P3.13 A one-link planar robot for problem P3-13.

3-14. Consider the one-link planar robot shown in Fig. P3.14.

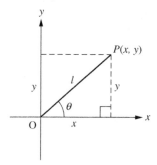

Figure P3.14 A one-link planar robot for problem P3-14.

If $l = 5$ cm, sketch the position of the tip of the robot and determine the (x, y) coordinates of position P for

(a) $\theta = \dfrac{\pi}{4}$ rad

(b) $\theta = \dfrac{4\pi}{4}$ rad

(c) $\theta = -135°$

(d) $\theta = -\dfrac{\pi}{4}$ rad

3-15. Repeat problem P3-14 if $l = 10$ in. and

(a) $\theta = 150°$

(b) $\theta = -\dfrac{2\pi}{3}$ rad

(c) $\theta = 420°$

(d) $\theta = -\dfrac{9\pi}{4}$ rad

3-16. Consider again the one-link planar robot shown in Fig. P3.14. Determine the length *l* and angle θ if the tip of the robot is located at the following points $P(x, y)$.

(a) $P(x, y) = (3, 4)$ cm

(b) $P(x, y) = (-4, 3)$ cm

(c) $P(x, y) = (-3, -3)$ cm

(d) $P(x, y) = (5, -4)$ cm

3-17. Repeat problem P3-16 if

(a) $P(x, y) = (4, 2)$ in.

(b) $P(x, y) = (-2, 4)$ in.

(c) $P(x, y) = (-5, -7.5)$ in.

(d) $P(x, y) = (6, -6)$ in.

3-18. Consider the two-link planar robot shown in Fig. P3.18.

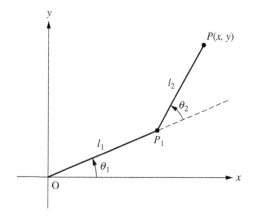

Figure P3.18 A two-link planar robot for problem P3-18.

Sketch the orientation of the robot and determine the (x, y) coordinates of point P for

(a) $\theta_1 = 30°$, $\theta_2 = 45°$, $l_1 = l_2 = 5$ cm

(b) $\theta_1 = 30°$, $\theta_2 = -45°$, $l_1 = l_2 = 5$ cm

(c) $\theta_1 = \dfrac{3\pi}{4}$ rad, $\theta_2 = \dfrac{\pi}{2}$ rad, $l_1 = l_2 = 5$ cm

(d) $\theta_1 = \dfrac{3\pi}{4}$ rad, $\theta_2 = -\dfrac{\pi}{2}$ rad, $l_1 = l_2 = 5$ cm

(e) $\theta_1 = -30°$, $\theta_2 = 45°$, $l_1 = l_2 = 5$ cm

(f) $\theta_1 = -30°$, $\theta_2 = -45°$, $l_1 = l_2 =$ 5 cm

(g) $\theta_1 = -\dfrac{3\pi}{4}$ rad, $\theta_2 = \dfrac{\pi}{2}$ rad, $l_1 = l_2 =$ 5 cm

(h) $\theta_1 = -\dfrac{3\pi}{4}$ rad, $\theta_2 = -\dfrac{\pi}{2}$ rad, $l_1 = l_2 = 5$ cm

3-19. Suppose that the two-link planar robot shown in Fig. P3.18 is located in the first quadrant and is oriented in the elbow-up position. If the tip of the robot is located at the point $P(x, y) = (9, 9)$, determine the values of θ_1 and θ_2. Assume $l_1 = 6$ in. and $l_2 = 8$ in.

3-20. Suppose that the two-link planar robot shown in Fig. P3.18 is located in the first quadrant and is oriented in the elbow-down position. If the tip of the robot is located at the point $P(x, y) = (10, 5)$, determine the values of θ_1 and θ_2. Assume $l_1 = 6$ in. and $l_2 = 8$ in.

3-21. Consider the two-link planar robot with $l_1 = l_2 = 5$ in. and oriented in the elbow-up position, as shown in Fig. P3.21. If the tip of the robot is located at the point $P(x, y) = (7.5, -2.8)$, determine the values of θ_1 and θ_2.

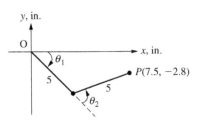

Figure P3.21 A two-link planar robot for problem P3-21.

3-22. Consider the two-link planar robot with $l_1 = l_2 = 5$ in. and oriented in the elbow-down position as shown in Fig. P3.22. If the tip of the robot is located at the point $P(x, y) = (4.83, -8.36)$, determine the values of θ_1 and θ_2.

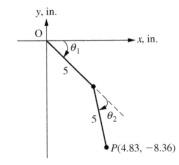

Figure P3.22 A two-link planar robot for problem P3-22.

3-23. Consider a two-link planar robot with $l_1 = l_2 = 10$ cm and oriented in the elbow-down position as shown in Fig. P3.23. If the tip of the robot is located at the point $P(x, y) = (-17, -1)$, determine the values of θ_1 and θ_2.

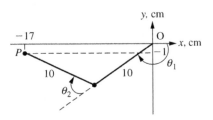

Figure P3.23 A two-link planar robot for problem P3-23.

3-24. Consider a two-link planar robot with $l_1 = l_2 = 10$ cm and oriented in the elbow-up position as shown in Fig. P3.24. If the tip of the robot is located

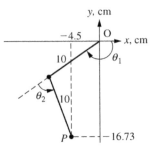

Figure P3.24 A two-link planar robot for problem P3-24.

at the point $P(x, y) = (-4.5, -16.73)$, determine the values of θ_1 and θ_2.

3-25. Consider a two-link planar robot oriented in the elbow-up position as shown in Fig. P3.25. If the tip of the robot is located at the point $P(x, y) = (-13, 12)$, determine the values of θ_1 and θ_2. Assume $l_1 = 10$ in. and $l_2 = 8$ in.

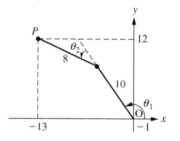

Figure P3.25 A two-link planar robot for problem P3-25.

3-26. Consider a two-link planar robot oriented in the elbow-down position as shown in Fig. P3.26. If the tip of the robot is located at the point $P(x, y) = (-1, 15)$, determine the values of θ_1 and θ_2. Assume $l_1 = 10$ in. and $l_2 = 8$ in.

Figure P3.26 A two-link planar robot for problem P3-26.

3-27. Consider a two-link planar robot oriented in the elbow-up position as shown in Fig. P3.27. If the tip of the robot is located at the point $P(x, y) = (13.66 \text{ cm}, -3.66 \text{ cm})$, determine the

values of θ_1 and θ_2 using the laws of cosines and sines.

Figure P3.27 A two-link planar robot for problem P3-27.

3-28. An airplane travels at a heading of 60° northwest with an air speed of 500 mph as shown in Fig. P3.28. The wind is blowing at 30° southwest at a speed of 50 mph. Find the magnitude of the velocity V and the angle θ of the plane relative to the ground using the laws of sines and cosines.

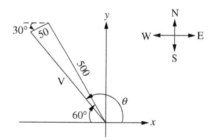

Figure P3.28 Velocity of an airplane for problem P3-28.

3-29. A large barge is crossing a river at a heading of 30° northwest with a speed of 12 mph against the water as shown in Fig. P3.29. The river flows due east at a speed of 4 mph. Find the magnitude of the velocity V and the angle θ of the barge using the laws of sines and cosines.

Figure P3.29 A barge crossing a river against the water current.

3-30. The impedance triangle of a resistor (R) and an inductor (L) connected in series in an AC circuit is shown in Fig. P3.30, where $R = 100\ \Omega$ is the resistance of the resistor and $X_L = 30\ \Omega$ is the inductive reactance of the inductor. Find the impedance Z and the phase angle θ.

Figure P3.30 A series AC circuit containing R and L.

3-31. The impedance triangle of a resistor (R) and a capacitor (C) connected in series in an AC circuit is shown in Fig. P3.31, where $R = 100\ \Omega$ is the resistance of the resistor and $X_C = 30\ \Omega$ is the capacitive reactance of the capacitor. Find the impedance Z and the phase angle θ.

Figure P3.31 A series AC circuit containing R and C.

3-32. The impedance triangle of a resistor (R) and an inductor (L) connected in series in an AC circuit is shown in Fig. P3.32, where $R = 1000\ \Omega$ is the resistance of the resistor and $Z = 1.005\ \text{K}\Omega$ is the total impedance of the circuit. Find the inductive reactance X_L and the phase angle θ.

Figure P3.32 Impedance triangle to find the inductive reactance.

3-33. The impedance triangle of a resistor (R) and a capacitor (C) connected in series in an AC circuit is shown in Fig. P3.33, where $R = 1000\ \Omega$ is the resistance of the resistor and $Z = 1.118\ \text{K}\Omega$ is the total impedance of the circuit. Find the capacitive reactance X_C and the phase angle θ.

Figure P3.33 Impedance triangle to find the capacitive reactance.

3-34. The phasor diagram of a series RL circuit is shown in Fig. P3.34, where V_R is the voltage across the resistor, V_L is the voltage across the inductor, and V is the AC voltage applied to the RL circuit in volts. Find the total voltage V and the phase angle θ.

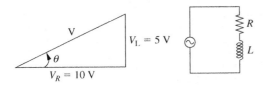

Figure P3.34   Phasor diagram of an RL circuit.

3-35. The phasor diagram of a series RC circuit is shown in Fig. P3.35, where V_R is the voltage across the resistor, V_C is the voltage across the capacitor, and V is the AC voltage applied to the RL circuit in volts. Find the total voltage V and the phase angle θ.

Figure P3.35   Phasor diagram of an RC circuit.

3-36. A three-phase AC system has a trigonal planar configuration as shown in Fig. P3.36. The voltage of each phase is 100 V and the angle between the adjacent phase is 120°. Find the voltage between phase a and b (i.e., find V_{ab}).

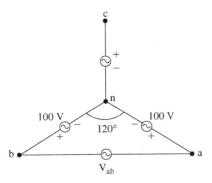

Figure P3.36   Three-phase AC system.

3-37. Consider the elevation between the two benchmarks shown in Fig. P3.37. The distance L between the benchmarks B1 and B2 is 500 m, and their elevations are 500 m and 600 m, respectively.

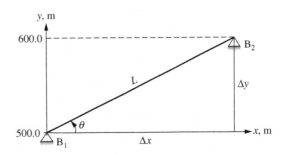

Figure P3.37   Elevation between the two benchmarks.

(a) Find the angle of inclination θ of the grade. Also calculate the percent grade.

(b) Calculate the horizontal distance between the benchmarks.

(c) Check the results in part (b) using the Pythagorean theorem.

3-38. Repeat problem P3-37 if the distance L between the two benchmarks is 200 m.

3-39. Consider the elevation between the two benchmarks shown in Fig. P3.39. The distance L between the benchmarks B1 and B2 is 200 m, and their elevations are 500 m and 400 m, respectively.

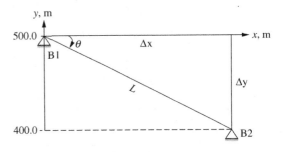

Figure P3.39   Elevation between the two benchmarks for problem P3-39.

(a) Find the angle of inclination θ of the grade. Also calculate the percent grade.

(b) Calculate the horizontal distance between the benchmarks.

(c) Check the results in part (b) using the Pythagorean theorem.

3-40. Repeat problem P3-39 if the distance L between the two benchmarks is 100 m.

3-41. To find the height of a building, a surveyor measures the angle of the building from two different points A and B as shown in Fig. P3.41. The distance between the two points is 10 m. Find the height h of the building.

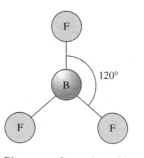

Figure P3.41   Survey set up to find the height of a building.

3-42. The gas boron trifluoride (BF_3) has a trigonal planar configuration as shown in Fig. P3.42. The B-F bond length is 1.3 Angstrom. Adjacent fluoride molecules form a 120° angle. Find the distance between adjacent fluoride molecules.

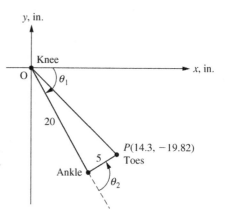

Figure P3.42   Planar configuration of boron trifluoride.

3-43. The force exerted on the kneecap by the quadriceps and patellar tendons are shown in Fig. P3.43. The force exerted by the quadriceps is 38 N and the force by the patellar is 52 N.

(a) Find the magnitude of the resultant force.

(b) What is the angle the patellar forms with the horizontal, as indicated in the Fig. P3.43?

Figure P3.43   Forces exerted on the knee.

3-44. Repeat P3-43 if the force exerted by the quadriceps is 30 N and the force by the patellar is 40 N.

3-45. Consider the position of the toes of a person sitting in a chair as shown in Fig. P3.45.

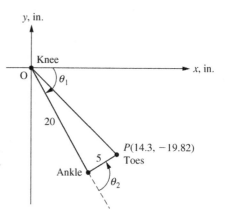

Figure P3.45   Ankle-up position of the leg.

(a) Suppose $\theta_1 = -45°$, $\theta_2 = 30°$, $l_1 = 20$ in., and $l_2 = 5$ in. Determine the

x and y coordinates of the position of the toes $P(x,y)$.

(b) Now suppose that the leg is positioned such that the tip of the toes is located in the fourth quadrant and oriented in the ankle-up position (counterclockwise direction) as shown Fig. P3.45. If the end of the toes are located at $P(x,y) = (14.33$ in., -19.82 in.$)$, determine the values of θ_1 and θ_2.

3-46. In a motion capture study of a runner, one frame shows the subject supporting her weight on one leg, as shown in Fig. P3.46. The length of the foot segment (from ankle to toe) is 8 in. and the length of the lower leg (from ankle to knee) is 18 in.

(a) Given the angles shown in Fig. P3.46(a), find the position of the knee if the runner's toes touch the ground at the point $x = y = 0$.

(b) Now suppose that the same leg is positioned such that the knee is located in the second quadrant and oriented in the knee-down position (clockwise direction), as shown in Fig. P3.46. If the toes are located at $P(x,y) = (-6.25$ in., 25 in.$)$, determine the values of θ_1 and θ_2.

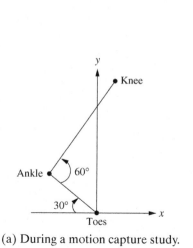

(a) During a motion capture study.

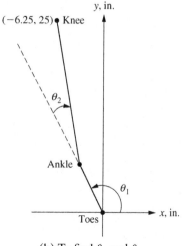

(b) To find θ_1 and θ_2.

Figure P3.46 Position of the runner's leg.

CHAPTER 4

Two-Dimensional Vectors in Engineering

The applications of two-dimensional vectors in engineering are introduced in this chapter. Vectors play a very important role in engineering. The quantities such as displacement (position), velocity, acceleration, forces, electric and magnetic fields, and momentum have not only a magnitude but also a direction associated with them. To describe the displacement of an object from its initial point, both the distance and direction are needed. A vector is a convenient way to represent both magnitude and direction and can be described in either a Cartesian or a polar coordinate system (rectangular or polar forms).

For example, an automobile traveling north at 65 mph can be represented by a two-dimensional vector in polar coordinates with a magnitude (speed) of 65 mph and a direction along the positive y-axis. It can also be represented by a vector in Cartesian coordinates with an x-component of zero and a y-component of 65 mph. The tip of the one-link and two-links planar robots introduced in Chapter 3 will be represented in this chapter using vectors both in Cartesian and polar coordinates. The concepts of unit vectors, magnitude, and direction of a vector will be introduced.

4.1   INTRODUCTION

Graphically, a vector $\overrightarrow{OP}$ or simply $\vec{P}$ with the initial point O and the final point P can be drawn as shown in Fig. 4.1. The magnitude of the vector is the distance between points O and P (magnitude $= P$) and the direction is given by the direction of the

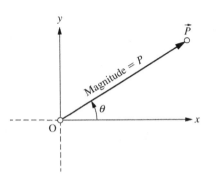

Figure 4.1   A representation of a vector.

arrow or the angle θ in the counterclockwise direction from the positive x-axis as shown in Fig. 4.1. The arrow above P indicates that P is a vector. In many engineering books, the vectors are also written as a boldface **P**.

4.2   POSITION VECTOR IN RECTANGULAR FORM

The position of the tip of a one-link robot represented as a 2-D vector $\vec{P}$ (Fig. 4.2) can be written in rectangular form as

$$\vec{P} = P_x\,\hat{i} + P_y\,\hat{j},$$

where $\hat{i}$ is the unit vector in the x-direction and $\hat{j}$ is the unit vector in the y-direction as shown in Fig. 4.2. Note that the magnitude of the unit vectors is equal to 1. The x- and y-components, P_x and P_y, of the vector $\vec{P}$ are given by

$$P_x = P\cos\theta$$
$$P_y = P\sin\theta.$$

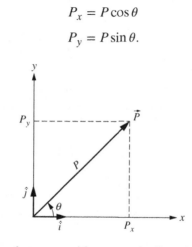

Figure 4.2   One-link planar robot as a position vector in Cartesian coordinates.

4.3   POSITION VECTOR IN POLAR FORM

The position of the tip of a one-link robot represented as a 2-D vector $\vec{P}$ (Fig. 4.2) can be also be written in polar form as

$$\vec{P} = P\angle\theta.$$

where P is the magnitude and θ is the angle or direction of the position vector $\vec{P}$, and can be obtained from the Cartesian components P_x and P_y as

$$P = \sqrt{P_x^2 + P_y^2}$$
$$\theta = \operatorname{atan2}(P_y,\ P_x).$$

Example 4-1

The length (magnitude) of a one-link robot shown in Fig. 4.2 is given as $P = 0.5$ m and the direction is $\theta = 30°$. Find the x- and y-components P_x and P_y and write $\vec{P}$ in rectangular vector notation.

Solution

The x- and y-components P_x and P_y are given by

$$P_x = 0.5 \cos 30°$$

$$= 0.5 \left(\frac{\sqrt{3}}{2} \right) = 0.433 \text{ m}$$

$$P_y = 0.5 \sin 30°$$

$$= 0.5 \left(\frac{1}{2} \right) = 0.25 \text{ m}.$$

Therefore, the position of the tip of the one-link robot can be written in vector form as

$$\vec{P} = 0.433 \,\hat{i} + 0.25 \,\hat{j} \text{ m}.$$

Example 4-2

The length of a one-link robot shown in Fig. 4.2 is given as $P = \sqrt{2}$ m and the direction is $\theta = 135°$. Find the x- and y-components P_x and P_y and write $\vec{P}$ in vector notation.

Solution

The x- and y-components P_x and P_y are given by

$$P_x = \sqrt{2} \cos 135° = -\sqrt{2} \cos 45°$$

$$= -\sqrt{2} \left(\frac{1}{\sqrt{2}} \right) = -1.0 \text{ m}$$

$$P_y = \sqrt{2} \sin 135° = \sqrt{2} \sin 45°$$

$$= \sqrt{2} \left(\frac{1}{\sqrt{2}} \right) = 1.0 \text{ m}.$$

Therefore, the position of the tip of the one-link robot can be written in vector form as

$$\vec{P} = -1.0 \,\hat{i} + 1.0 \,\hat{j} \text{ m}.$$

Example 4-3

The x- and y-components of the one-link robot are given as $P_x = \dfrac{\sqrt{3}}{4}$ m and $P_y = \dfrac{1}{4}$ m, as shown in Fig. 4.3. Find the magnitude (length) and direction of the robot represented as a position vector $\vec{P}$.

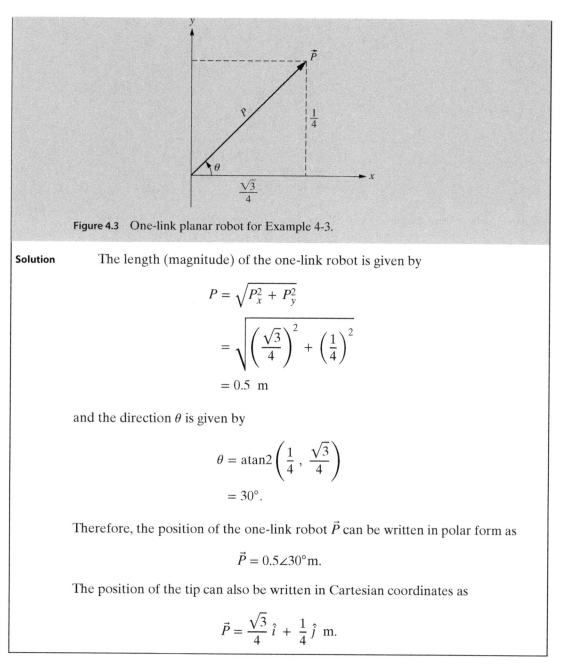

Figure 4.3 One-link planar robot for Example 4-3.

Solution The length (magnitude) of the one-link robot is given by

$$P = \sqrt{P_x^2 + P_y^2}$$

$$= \sqrt{\left(\frac{\sqrt{3}}{4}\right)^2 + \left(\frac{1}{4}\right)^2}$$

$$= 0.5 \ \text{m}$$

and the direction θ is given by

$$\theta = \text{atan2}\left(\frac{1}{4}, \frac{\sqrt{3}}{4}\right)$$

$$= 30°.$$

Therefore, the position of the one-link robot $\vec{P}$ can be written in polar form as

$$\vec{P} = 0.5 \angle 30° \, \text{m}.$$

The position of the tip can also be written in Cartesian coordinates as

$$\vec{P} = \frac{\sqrt{3}}{4} \hat{i} + \frac{1}{4} \hat{j} \ \text{m}.$$

Example 4-4 A person pushes down on a vacuum cleaner with a force of $F = 20$ lb at an angle of $-40°$ relative to ground, as shown in Fig. 4.4. Determine the horizontal and vertical components of the force.

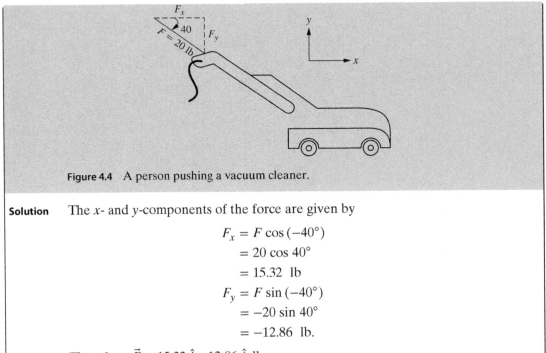

Figure 4.4   A person pushing a vacuum cleaner.

Solution   The x- and y-components of the force are given by

$$F_x = F \cos(-40°)$$
$$= 20 \cos 40°$$
$$= 15.32 \text{ lb}$$
$$F_y = F \sin(-40°)$$
$$= -20 \sin 40°$$
$$= -12.86 \text{ lb}.$$

Therefore, $\vec{F} = 15.32 \, \hat{i} - 12.86 \, \hat{j}$ lb.

4.4   VECTOR ADDITION

The sum of two vectors $\vec{P}_1$ and $\vec{P}_2$ is a vector $\vec{P}$ written as

$$\vec{P} = \vec{P}_1 + \vec{P}_2. \tag{4.1}$$

Vectors can be added graphically or algebraically. Graphically, the addition of two vectors can be obtained by placing the initial point of $\vec{P}_2$ on the final point of $\vec{P}_1$ and then drawing a line from the initial point of $\vec{P}_1$ to the final point of $\vec{P}_2$, forming a triangle as shown in Fig. 4.5.

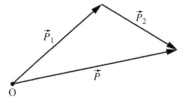

Figure 4.5   Graphical addition of two vectors.

Algebraically, the addition of two vectors given in equation (4.1) can be carried out by adding the x- and y-components of the two vectors. Vectors $\vec{P}_1$ and $\vec{P}_2$ can be written in Cartesian form as

$$\vec{P}_1 = P_{x1} \, \hat{i} + P_{y1} \, \hat{j} \tag{4.2}$$

$$\vec{P}_2 = P_{x2} \, \hat{i} + P_{y2} \, \hat{j}. \tag{4.3}$$

Substituting equations (4.2) and (4.3) into equation (4.1) gives

$$\vec{P} = (P_{x1}\,\hat{i} + P_{y1}\,\hat{j}) + (P_{x2}\,\hat{i} + P_{y2}\,\hat{j})$$

$$= (P_{x1} + P_{x2})\,\hat{i} + (P_{y1} + P_{y2})\,\hat{j}$$

$$= P_x\,\hat{i} + P_y\,\hat{j}$$

where $P_x = P_{x1} + P_{x2}$ and $P_y = P_{y1} + P_{y2}$. Therefore, addition of vectors algebraically amounts to adding their x- and y-components.

4.4.1 Examples of Vector Addition in Engineering

Example 4-5

A two-link planar robot is shown in Fig. 4.6. Find the magnitude and angle of the position of the tip of the robot if the length of the first link $P_1 = \dfrac{1}{\sqrt{2}}$ m, the length of the second link $P_2 = 0.5$ m, $\theta_1 = 45°$, and $\theta_2 = -15°$. In other words, write $\vec{P}$ in polar coordinates.

Figure 4.6 Position of two-link robot using vector addition.

Solution The x- and y-components of the first link of the planar robot $\vec{P}_1$ can be written as

$$P_{x1} = P_1 \cos 45°$$

$$= \left(\frac{1}{\sqrt{2}}\right)\left(\frac{1}{\sqrt{2}}\right)$$

$$= 0.5 \ \text{m}$$

$$P_{y1} = P_1 \sin 45°$$

$$= \left(\frac{1}{\sqrt{2}}\right)\left(\frac{1}{\sqrt{2}}\right)$$

$$= 0.5 \ \text{m}.$$

Therefore, $\vec{P}_1 = 0.5\,\hat{i} + 0.5\,\hat{j}$ m. Similarly, the x- and y-components of the second link $\vec{P}_2$ can be written as

$$P_{x2} = P_2 \cos 30°$$

$$= 0.5 \left(\frac{\sqrt{3}}{2} \right)$$

$$= 0.433 \text{ m}$$

$$P_{y2} = P_2 \sin 30°$$

$$= 0.5 \left(\frac{1}{2} \right)$$

$$= 0.25 \text{ m.}$$

Therefore, $\vec{P}_2 = 0.433\,\hat{i} + 0.25\,\hat{j}$ m. Finally, since $\vec{P} = \vec{P}_1 + \vec{P}_2$,

$$\vec{P} = (0.5\,\hat{i} + 0.5\,\hat{j}) + (0.433\,\hat{i} + 0.25\,\hat{j})$$
$$= 0.933\,\hat{i} + 0.75\,\hat{j}.$$

The magnitude and direction of the vector $\vec{P}$ are given by

$$P = \sqrt{(0.933)^2 + (0.75)^2} = 1.197 \text{ m.}$$

$$\theta = \text{atan2}(0.75, 0.933) = 38.79°.$$

Therefore, $\vec{P} = 1.197\angle 38.79° \text{ m.}$

Example 4-6

A sinusoidal current is flowing through the RL circuit shown in Fig. 4.7. The voltage phasors (vector used to represent voltages and currents in AC circuits) across the resistor $R = 20\,\Omega$ and inductor $L = 100$ mH are given as $\vec{V}_R = 2\,\angle 0°$V and $\vec{V}_L = 3.77\,\angle 90°$V, respectively. If the total voltage phasor $\vec{V}$ across R and L is $\vec{V} = \vec{V}_R + \vec{V}_L$, find the magnitude and phase (angle) of $\vec{V}$.

Figure 4.7 Sum of voltage phasors in an RL circuit.

Solution   The x- and y-components of the voltage phasor $\vec{V}_R$ are given by

$$V_{Rx} = 2 \cos 0°$$
$$= 2.0 \ \text{V}$$
$$V_{Ry} = 2 \sin 0°$$
$$= 0 \ \text{V}.$$

Therefore, $\vec{V}_R = 2.0 \, \hat{\imath} + 0 \, \hat{\jmath}$ V. Similarly, the x- and y-components of the voltage phasor $\vec{V}_L$ are given by

$$V_{Lx} = 3.77 \cos 90°$$
$$= 0 \ \text{V}$$
$$V_{Ly} = 3.77 \sin 90°$$
$$= 3.77 \ \text{V}.$$

Therefore, $\vec{V}_L = 0 \, \hat{\imath} + 3.77 \, \hat{\jmath}$ V. Finally, since $\vec{V} = \vec{V}_R + \vec{V}_L$,

$$\vec{V} = (2.0 \, \hat{\imath} + 0 \, \hat{\jmath}) + (0 \, \hat{\imath} + 3.77 \, \hat{\jmath})$$
$$= 2.0 \, \hat{\imath} + 3.77 \, \hat{\jmath} \ \text{V}.$$

Thus, the magnitude and phase of the total voltage phasor $\vec{V}$ are given by

$$V = \sqrt{(2.0)^2 + (3.77)^2} = 4.27 \ \text{V}$$
$$\theta = \text{atan2}(3.77, 2.0) = 62.05°.$$

Therefore, $\vec{V} = 4.27\angle 62.05°\text{V}$.

Example 4-7   A ship travels 200 miles at 45° northeast, then 300 miles due east as shown in Fig. 4.8. Find the resulting position of the ship.

Figure 4.8   Resulting position of the ship after travel.

Solution   The x- and y-components of the position vector $\vec{P}_1$ are given by

$$P_{x1} = P_1 \cos 45°$$

$$= 200 \left(\frac{1}{\sqrt{2}} \right)$$

$$= 141.4 \text{ mi}$$

$$P_{y1} = P_1 \sin 45°$$

$$= 200 \left(\frac{1}{\sqrt{2}} \right)$$

$$= 141.4 \text{ mi.}$$

Therefore, $\vec{P}_1 = 141.4\,\hat{i} + 141.4\,\hat{j}$ mi. Similarly, the x- and y-components of the position vector $\vec{P}_2$ are given by

$$P_{x2} = P_2 \cos 0°$$

$$= 300\,(1)$$

$$= 300 \text{ mi}$$

$$P_{y2} = P_2 \sin 0°$$

$$= 300\,(0)$$

$$= 0 \text{ mi.}$$

Therefore, $\vec{P}_2 = 300\,\hat{i} + 0\,\hat{j}$ mi. Finally, since $\vec{P} = \vec{P}_1 + \vec{P}_2$,

$$\vec{P} = (141.4\,\hat{i} + 141.4\,\hat{j}) + (300\,\hat{i} + 0\,\hat{j})$$

$$= 441.4\,\hat{i} + 141.4\,\hat{j} \text{ mi.}$$

Thus, the distance and direction of the ship after traveling 200 miles northeast and then 300 miles east are given by

$$P = \sqrt{(441.4)^2 + (141.4)^2} = 463.5 \text{ mi}$$

$$\theta = \text{atan2}(141.4, 441.4) = 17.76°.$$

Therefore, $\vec{P} = 463.5\angle 17.76°$ miles. In other words, the ship is now located at 463.5 miles, 17.76° northeast from its original location.

Example 4-8

Relative Velocity: An airplane is flying at an air speed of 100 mph at a heading of 30° southeast, as shown in Fig. 4.9. If the velocity of the wind is 20 mph due west, determine the resultant velocity of the plane with respect to the ground.

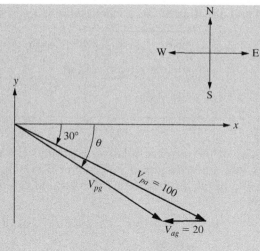

Figure 4.9 Velocity of the plane relative to ground.

Solution The x- and y-components of the velocity of the plane relative to air, $\vec{V}_{pa}$, are given by

$$V_{xpa} = V_{pa} \cos (-30°)$$

$$= 100 \left(\frac{\sqrt{3}}{2} \right)$$

$$= 86.6 \ \text{mph}$$

$$V_{ypa} = V_{pa} \sin (-30°)$$

$$= -100 \left(\frac{1}{2} \right)$$

$$= -50.0 \ \text{mph}.$$

Therefore, $\vec{V}_{pa} = 86.6 \ \hat{i} - 50.0 \ \hat{j}$ mph. Similarly, the x- and y-components of the velocity of air (wind) relative to ground $\vec{V}_{ag}$ are given by

$$V_{xag} = V_{ag} \cos (180°)$$

$$= 20 \ (-1)$$

$$= -20 \ \text{mph}$$

$$V_{yag} = V_{ag} \sin (180°)$$

$$= 20 \ (0)$$

$$= 0 \ \text{mph}.$$

Therefore, $\vec{V}_{ag} = -20 \ \hat{i} + 0 \ \hat{j}$ mph. Finally, the velocity of the plane relative to ground $\vec{V}_{pg} = \vec{V}_{pa} + \vec{V}_{ag}$ is given by

$$\vec{V}_{pg} = (86.6 \ \hat{i} - 50 \ \hat{j}) + (-20 \ \hat{i} + 0 \ \hat{j})$$

$$= 66.6 \ \hat{i} - 50 \ \hat{j} \ \text{mph}.$$

Thus, the speed and direction of the airplane relative to ground are given by

$$V_{pg} = \sqrt{(66.6)^2 + (-50)^2} = 83.3 \text{ mph}$$

$$\theta = \text{atan2}(-50, 66.6) = -36.9°.$$

Therefore, $\vec{V}_{pg} = 83.3\angle{-36.9°}$ mph.

Note: The velocity of the airplane relative to ground can also be found using the laws of cosines and sines discussed in Chapter 3. Using the triangle shown in Fig. 4.10, the speed of the airplane relative to ground can be determined using the law of cosines as

$$V_{pg}^2 = 20^2 + 100^2 - 2\,(20)\,(100)\cos(30°)$$

$$= 6936$$

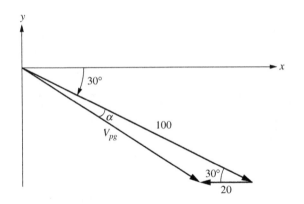

Figure 4.10 The triangle to determine the speed and direction of the plane.

Therefore, $V_{pg} = 83.28$ mph. Also, using the law of sines, the angle α can be found as

$$\frac{\sin 30°}{V_{pg}} = \frac{\sin \alpha}{20}.$$

Therefore, $\sin \alpha = \dfrac{20 \sin 30°}{V_{pg}} = 0.12$ and the value of $\alpha = 6.896°$. The direction of the velocity of the airplane relative to ground can now be found as $\theta = 30 + \alpha = 36.89°$. The velocity of the plane relative to ground is, therefore, given as

$$\vec{V}_{pg} = 83.3\angle{-36.9°} \text{ mph.}$$

Note that while this geometric approach works fine when adding two vectors, it becomes unwieldy when adding three or more vector quantities. In such cases, the algebraic approach is preferable.

Example 4-9

Static Equilibrium: A 100 kg object is hanging from two cables of equal length as shown in Fig. 4.11. Determine the tension in each cable.

Figure 4.11   An object hanging from two cables.

Solution   The free-body diagram (FBD) of the system shown in Fig. 4.11 can be drawn as shown in Fig. 4.12.

Figure 4.12   Free-body diagram of the system shown in Fig. 4.11.

It is assumed that the system shown in Fig. 4.11 is in static equilibrium (not accelerating), and therefore the sum of all the forces are equal to zero **(Newton's first law)**, i.e.,

$$\sum \vec{F} = 0$$

or

$$\vec{T}_1 + \vec{T}_2 + \vec{W} = 0. \qquad (4.4)$$

The x- and y-components of the tension $\vec{T}_1$ are given by

$$T_{x1} = -T_1 \cos 45°$$

$$= -T_1 \left(\frac{1}{\sqrt{2}} \right) \text{ N}$$

$$T_{y1} = T_1 \sin 45°$$

$$= T_1 \left(\frac{1}{\sqrt{2}} \right) \text{ N}.$$

Therefore, $\vec{T}_1 = -\dfrac{T_1}{\sqrt{2}}\,\hat{i} + \dfrac{T_1}{\sqrt{2}}\,\hat{j}$ N. Similarly, the x- and y-components of the tension $\vec{T}_2$ and weight $\vec{W}$ are given by

$$T_{x2} = T_2 \cos 45°$$

$$= T_2 \left(\frac{1}{\sqrt{2}}\right) \text{ N}$$

$$T_{y2} = T_2 \sin 45°$$

$$= T_2 \left(\frac{1}{\sqrt{2}}\right) \text{ N}$$

$$W_x = W \cos(-90°)$$

$$= 0 \text{ N}$$

$$W_y = W \sin(-90°)$$

$$= -981 \text{ N}.$$

Therefore, $\vec{T}_2 = \dfrac{T_2}{\sqrt{2}}\,\hat{i} + \dfrac{T_2}{\sqrt{2}}\,\hat{j}$ N and $\vec{W} = 0\,\hat{i} + -981\,\hat{j}$ N. Substituting $\vec{T}_1, \vec{T}_2$, and $\vec{W}$ into equation (4.4) gives

$$\left(-\frac{T_1}{\sqrt{2}}\hat{i} + \frac{T_1}{\sqrt{2}}\hat{j}\right) + \left(\frac{T_2}{\sqrt{2}}\hat{i} + \frac{T_2}{\sqrt{2}}\hat{j}\right) + (0\hat{i} - 981\hat{j}) = 0$$

$$\left(-\frac{T_1}{\sqrt{2}} + \frac{T_2}{\sqrt{2}}\right)\hat{i} + \left(\frac{T_1}{\sqrt{2}} + \frac{T_2}{\sqrt{2}} - 981\right)\hat{j} = 0. \tag{4.5}$$

In equation (4.5), the x-component is the sum of the forces in the x-direction and the y-component is the sum of forces in the y-direction. Since the right-hand side of equation (4.5) is zero, the sum of forces in the x- and y-directions are zero, or $\sum F_x = 0$ and $\sum F_y = 0$. Therefore,

$$\sum F_x = \left(-\frac{T_1}{\sqrt{2}} + \frac{T_2}{\sqrt{2}}\right) = 0 \tag{4.6}$$

and,

$$\sum F_y = \left(\frac{T_1}{\sqrt{2}} + \frac{T_2}{\sqrt{2}}\right) - 981 = 0. \tag{4.7}$$

Adding equations (4.6) and (4.7) gives $\dfrac{2\,T_2}{\sqrt{2}} = 981$ or $T_2 = 693.7$ N. Also, from equation (4.6), $T_2 = T_1$. Therefore, both cables have the same tension, that is, $T_1 = T_2 = 693.7$ N.

Example **4-10**	**Static Equilibrium:** A 100 kg television is loaded onto a truck using a ramp at a 30° angle. Find the normal and frictional forces on the TV if it is left sitting on the ramp.

Figure 4.13 Loading a TV onto the truck using a ramp.

Solution The free-body diagram (FBD) of the TV sitting on the ramp as shown in Fig. 4.13 is given in Fig. 4.14 where $W = 100 \times 9.81 = 981$ Newton.

Figure 4.14 Free-body diagram of a TV on the 30° ramp.

Note that we are using the rotated axes to simplify the computation below. It is assumed that the system shown in Fig. 4.13 is in static equilibrium; therefore, the sum of all the forces is equal to zero (Newton's first law):

$$\sum \vec{F} = 0$$
$$\vec{F} + \vec{N} + \vec{W} = 0. \tag{4.8}$$

The x- and y-components of the TV weight $\vec{W}$ are given by

$$W_x = -W \sin 30°$$
$$= -981 \left(\frac{1}{2} \right)$$
$$= -490.5 \text{ N}$$
$$W_y = -W \cos 30°$$
$$= -981 \left(\frac{\sqrt{3}}{2} \right)$$
$$= -849.6 \text{ N}.$$

Therefore, $\vec{W} = -490.5\,\hat{i} - 849.6\,\hat{j}$ N. Similarly, the x- and y-components of the frictional force $\vec{F}$ and normal force $\vec{N}$ are given by

$$F_x = F \cos 0°$$
$$= F \text{ N}$$
$$F_y = F \sin 0°$$
$$= 0 \text{ N}$$
$$N_x = N \cos (90°)$$
$$= 0 \text{ N}$$
$$N_y = N \sin (90°)$$
$$= N \text{ N}.$$

Therefore, $\vec{F} = F\,\hat{i} + 0\,\hat{j}$ N and $\vec{N} = 0\,\hat{i} + N\,\hat{j}$ N. Substituting $\vec{F}$, $\vec{N}$, and $\vec{W}$ into equation (4.8) gives

$$(F\,\hat{i} + 0\,\hat{j}) + (0\,\hat{i} + N\,\hat{j}) + (-490.5\,\hat{i} - 849.6\,\hat{j}) = 0$$
$$(F + 0 - 490.5)\,\hat{i} + (0 + N - 849.6)\,\hat{j} = 0. \qquad (4.9)$$

Equating the x- and y-components in equation (4.9) to zero yields

$$F - 490.5 = 0 \quad \Rightarrow \quad F = 490.5 \text{ N}$$
$$N - 849.6 = 0 \quad \Rightarrow \quad N = 849.6 \text{ N}.$$

Example 4-11

A waiter extends his arm to hand a plate of food to his customer. The free-body diagram is shown in Fig. 4.15, where $F_m = 400$ N is the force in the deltoid muscle, $W_a = 40$ N is the weight of the arm, $W_p = 20$ N is the weight of the plate of food, and R_x and R_y are the x- and y-components of the reaction forces at the shoulder.

(a) Using the x-y coordinate system shown in Fig. 4.15, write the muscle force $\vec{F}_m$, the weight of the arm $\vec{W}_a$, and the weight of the plate $\vec{W}_p$ in standard vector notation (i.e., using unit vectors $\hat{i}$ and $\hat{j}$).

(b) Determine the values of R_x and R_y required for static equilibrium: $\vec{R} + \vec{F}_m + \vec{W}_a + \vec{W}_p = 0$. Also compute the magnitude and direction of $\vec{R}$.

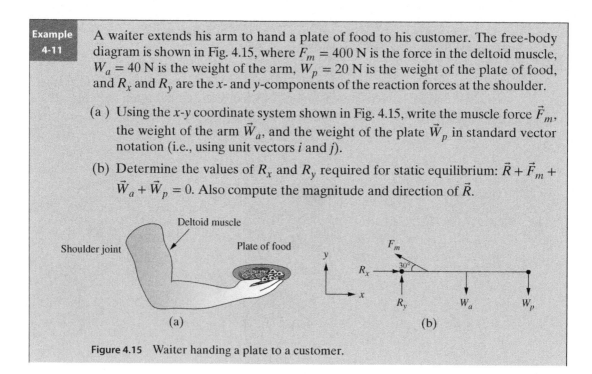

Figure 4.15   Waiter handing a plate to a customer.

Solution   (a) The x- and y-components of the deltoid muscle force $\vec{F}_m$ are given by

$$F_{m,x} = F_m \cos 150°$$
$$= -400 \cos 30°$$
$$= -400 \times 0.866$$
$$= -346.4 \text{ N}$$
$$F_{m,y} = F_m \sin 150°$$
$$= 400 \sin 30°$$
$$= 400 \times 0.5$$
$$= 200 \text{ N}.$$

Therefore, $\vec{F}_m = -346.4\,\hat{i} + 200\,\hat{j}$ N. Similarly, the x- and y-components of the weight of the arm $\vec{W}_a$ and weight of the plate W_p are given by

$$W_{a,x} = W_a \cos (-90)°$$
$$= 0 \text{ N}$$
$$W_{a,y} = W_a \sin (-90)°$$
$$= -40 \text{ N}$$
$$W_{p,x} = W_p \cos (-90)°$$
$$= 0 \text{ N}$$
$$W_{p,y} = W_p \sin (-90)°$$
$$= -20 \text{ N}.$$

Therefore, $\vec{W}_a = 0\,\hat{i} - 40\,\hat{j}$ N and $\vec{W}_p = 0\,\hat{i} - 20\,\hat{j}$ N.

(b) It is assumed that the system shown in Fig. 4.15 is in static equilibrium; therefore, the sum of all the forces is equal to zero (Newton's first law):

$$\vec{F}_m + \vec{W}_a + \vec{W}_p + \vec{R} = 0.$$
$$(-346.4\,\hat{i} + 200\,\hat{j}) + (0\,\hat{i} - 40\,\hat{j}) + (0\,\hat{i} - 20\,\hat{j}) + (R_x\,\hat{i} + R_y\,\hat{j}) = 0$$
$$(R_x - 346.4 + 0 + 0)\,\hat{i} + (R_y + 200 - 40 - 20)\,\hat{j} = 0 \qquad (4.10)$$

Equating the x- and y-components in equation (4.10) to zero yields

$$\sum F_x = 0: \quad \Rightarrow \quad R_x - 346.4 = 0 \quad \Rightarrow \quad R_x = 346.4\,\text{N}$$
$$\sum F_y = 0: \quad \Rightarrow \quad R_y + 200 - 40 - 20 = 0 \quad \Rightarrow \quad R_y = -140\,\text{N}.$$

Therefore, $\vec{R} = 346.4\,\vec{i} - 140\,\vec{j}$ N. The magnitude and direction of $\vec{R}$ can be obtained as

$$\vec{R} = \sqrt{(346.4)^2 + (-140)^2} \; \angle \text{atan2}(-140, 346.4)$$
$$= 373.6 \angle -22°\,\text{N} \qquad (4.11)$$

Example 4-12

Using motion capture, the positions of each arm segment are measured while a person throws a ball. The length from shoulder to elbow (P_1) is 12 in. and the length from elbow to hand (P_2) is 18 in. The angle θ_1 is 45° and θ_2 is 20°.

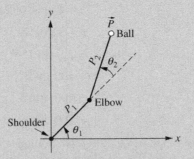

Figure 4.16 Position of the arm throwing a ball.

(a) Using the x-y coordinate system shown in Fig. 4.16, write the position of the ball $\vec{P} = \vec{P}_1 + \vec{P}_2$ in the standard vector notation.

(b) Find the magnitude and direction of $\vec{P}$.

Solution

(a) The x- and y-components of the position $\vec{P}_1$ are given by

$$P_{1x} = P_1 \cos 45°$$

$$= 12 \times \frac{\sqrt{2}}{2}$$

$$= 8.49 \ \text{in.}$$

$$P_{1y} = P_1 \sin 45°$$

$$= 12 \times \frac{\sqrt{2}}{2}$$

$$= 8.49 \ \text{in.}$$

Therefore, $\hat{P}_1 = 8.484\,\hat{i} + 8.484\,\hat{j}$ in. Similarly, the x- and y-components of the position $\vec{P}_2$ are given by

$$P_{2x} = P_2 \cos (45° + 20°)$$

$$= 18 \times (0.4226)$$

$$= 7.61 \ \text{in.}$$

$$P_{2y} = P_2 \sin (45° + 20°)$$

$$= 18 \times (0.9063)$$

$$= 16.31 \ \text{in.}$$

Therefore, $\vec{P}_2 = 7.61\,\hat{i} + 16.31\,\hat{j}$ in. The position of the arm $\vec{P}$ can now be found in standard vector notation by adding vectors $\hat{P}_1$ and $\hat{P}_2$ as

$$\vec{P} = \vec{P}_1 + \vec{P}_2$$
$$= 8.49\,\hat{i} + 8.49\,\hat{j} + 7.61\,\hat{i} + 16.31\,\hat{j}$$
$$= 16.1\,\hat{i} + 25.80\,\hat{j} \ \text{in.} \qquad (4.12)$$

(b) The magnitude of the position $\vec{P}$ is given by

$$P = \sqrt{P_x^2 + P_y^2}$$
$$= \sqrt{16.1^2 + 25.8^2}$$
$$= 29.6 \ \text{in.}$$

The direction of position $\vec{P}$ is given by

$$\theta = \text{atan2}(P_y, P_x)$$
$$= \text{atan2}(25.8, 16.1)$$
$$= 58°.$$

Therefore, the vector $\vec{P}$ can be written in the polar form as $\vec{P} = 29.6\,\angle 58°$ in.

PROBLEMS

4-1. Locate the tip of a one-link robot of 12 in. length as a 2-D position vector with a direction of 60°. Draw the position vector and find its x- and y-components. Also, write $\vec{P}$ in both its rectangular and polar forms.

4-2. Locate the tip of a one-link robot of 1.5 ft length as a 2-D position vector with a direction of −30°. Draw the position vector and find its x- and y-components. Also, write $\vec{P}$ in both its rectangular and polar forms.

4-3. Locate the tip of a one-link robot of 1 m length as a 2-D position vector with a direction of 120°. Draw the position vector and find its x- and y-components. Also, write $\vec{P}$ in both its rectangular and polar forms.

4-4. Locate the tip of a one-link robot of 2 m length as a 2-D position vector with

a direction of −135°. Draw the position vector and find its x- and y-components. Also, write $\vec{P}$ in both its rectangular and polar forms.

4-5. The tip of a one-link robot is represented as a position vector $\vec{P}$ as shown in Fig. P4.5. Find the x- and y-components of the vector if the length of the link is $P = 10$ in. and $\theta = 30°$. Also, write the vector $\vec{P}$ in rectangular and polar form.

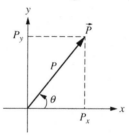

Figure P4.5 A one-link robot represented in polar coordinates.

4-6. Repeat problem P4-5 if $P = 14.42$ cm and $\theta = 123.7°$.

4-7. Repeat problem P4-5 if $P = 15$ cm and $\theta = -120°$.

4-8. Repeat problem P4-5 if $P = 6$ in. and $\theta = -60°$.

4-9. The x- and y-components of a vector $\vec{P}$ shown in Fig. P4.9 are given as $P_x = 2$ cm and $P_y = 3$ cm. Find the magnitude and direction, and write the vector $\vec{P}$ in its rectangular and polar forms.

Figure P4.9   A position vector for problem P4-9.

4-10. The x- and y-components of a vector $\vec{P}$ shown in Fig. P4.10 are given as $P_x = 3$ in. and $P_y = -4$ in. Find the magnitude and direction, and write the vector $\vec{P}$ in its rectangular and polar forms.

Figure P4.10   A position vector for problem P4-10.

4-11. The x- and y-components of a vector $\vec{P}$ shown in Fig. P4.11 are given as $P_x = -1.5$ cm and $P_y = 2$ cm. Find the magnitude and direction, and write the vector $\vec{P}$ in its rectangular and polar forms.

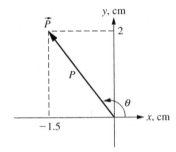

Figure P4.11   A position vector for problem P4-11.

4-12. The x- and y-components of a vector $\vec{P}$ shown in Fig. P4.12 are given as $P_x = -1$ in. and $P_y = -2$ in. Find the magnitude and direction, and write the vector $\vec{P}$ in its rectangular and polar forms.

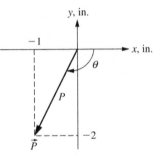

Figure P4.12   A position vector for problem P4-12.

4-13. A state trooper investigating an accident pushes a wheel (shown in Fig. 4.13) to measure skid marks. If a trooper applies a force of 20 lb at an angle of $\theta = 45°$, find the horizontal and vertical forces acting on the wheel.

Figure P4.13   A wheel to measure skid marks.

4-14. Repeat problem P4-13 if the trooper is applying a force of 10 lb at an angle of $\theta = 60°$.

4-15. In a RL circuit, the voltage across the inductor V_L leads the voltage across the resistor by 90° as shown in Fig. P4.15. If $V_R = 10$ V and $V_L = 5$ V, find the total voltage $\vec{V} = \vec{V}_R + \vec{V}_L$.

Figure P4.15   Voltage phasor diagram of RL circuit.

4-16. Repeat problem P4-15 if $V_R = 10$ V and $V_L = 15$ V.

4-17. In an RC circuit, the voltage across the capacitor V_C lags the voltage across the resistor by −90° as shown in Fig. P4.17. If $V_R = 20$ V and $V_C = 5$ V, find the total voltage $\vec{V} = \vec{V}_R + \vec{V}_C$.

Figure P4.17   Voltage phasor diagram of RC circuit.

4-18. Repeat problem P4-17 if $V_R = 10$ V and $V_L = 20$ V.

4-19. In an electrical circuit, voltage $\vec{V}_2$ lags voltage $\vec{V}_1$ by 30° as shown in Fig. P4.19.

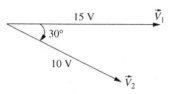

Figure P4.19   Voltages $\vec{V}_1$ and $\vec{V}_2$ for problem P4-19.

Find the sum of the two voltages; in other words, find $\vec{V} = \vec{V}_1 + \vec{V}_2$.

4-20. In an electrical circuit, voltage $\vec{V}_2$ leads voltage $\vec{V}_1$ by 60° as shown in Fig. P4.20. Find the sum of the two voltages; in other words, find $\vec{V} = \vec{V}_1 + \vec{V}_2$.

Figure P4.20   Voltages $\vec{V}_1$ and $\vec{V}_2$ for problem P4-20.

4-21. An airplane travels at a heading of 45° northeast with an air speed of 300 mph. The wind is blowing at 30° southeast at a speed of 40 mph as shown in Fig. P4.21. Find the speed (magnitude of the velocity $\vec{V}$) and the direction θ of the plane relative to the ground using vector addition. Check your answer by finding the magnitude and direction using the laws of sines and cosines.

Figure P4.21   Velocity of airplane for problem P4-21.

4-22. An airplane travels at a heading of −60° with an air speed of 500 mph. The wind is blowing at 30° at a speed of 50 mph as shown in Fig. P4.22. Find the speed (magnitude of the velocity $\vec{V}$) and the direction θ of the plane relative to the ground using vector addition. Check your answer by finding the magnitude and direction using the laws of sines and cosines.

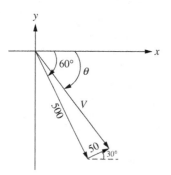

Figure P4.22 Velocity of airplane for problem P4-22.

4-23. A large barge is crossing a river at a heading of 30° northwest with a speed of 15 mph against the water as shown in Fig. P4.23. The river flows due east at a speed of 5 mph.

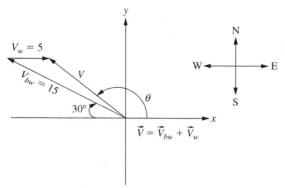

Figure P4.23 A barge crossing a river against the current.

(a) Calculate the resultant velocity $\vec{V}$ of the barge using vector addition (use the $\hat{i}$ and $\hat{j}$ notation).

(b) Determine the magnitude and direction of $\vec{V}$.

(c) Repeat part (b) using the laws of sines and cosines.

4-24. A ship is crossing a river at a heading of −150° with a speed of $V_{SW} = 30$ mph against the water as shown in Fig. P4.24. The river is flowing in the direction of 135° with a speed of $V_W = 10$ mph.

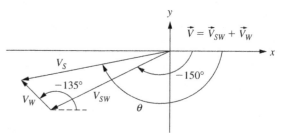

Figure P4.24 A ship crossing a river against the current.

(a) Calculate the resultant velocity $\vec{V}$ of the ship using vector addition (use the $\hat{i}$ and $\hat{j}$ notation).

(b) Determine the magnitude and direction of $\vec{V}$.

(c) Repeat part (b) using the laws of sines and cosines.

4-25. A boat moves diagonally across a river at a heading of 45° southwest and at a water speed of $V_{bw} = 12$ mph as shown in Fig 4.25. The river flows due east at a speed of $V_w = 4$ mph.

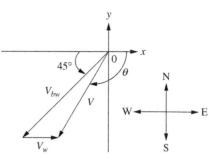

Figure P4.25 A boat moving diagonally across a river.

(a) Determine the resultant velocity $\vec{V}$ of the boat, which is given by $\vec{V} = \vec{V}_{bw} + \vec{V}_w$. In so doing, express all vectors using standard vector notation (i.e., using unit vectors $\hat{i}$ and $\hat{j}$). Given your result for $\vec{V}$, also determine the magnitude V and direction θ.

(b) Repeat part (a) using the laws of sines and cosines.

4-26. A two-link planar robot is shown in Fig. P4.26.

(a) Calculate the position of the tip $\vec{P}$ of the planar robot using vector addition (use the $\hat{i}$ and $\hat{j}$ notation).

(b) Determine the magnitude and direction of the position of the robot tip. In other words, write vector $\vec{P}$ in the polar form.

(c) Repeat part (b) using the laws of sines and cosines.

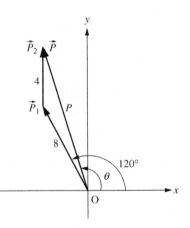

Figure P4.26 A two-link robot located for problem P4-26.

4-27. A two-link planar robot is shown in Fig. P4.27.

(a) Calculate the position of the tip $\vec{P}$ of the planar robot using vector addition (use the $\hat{i}$ and $\hat{j}$ notation).

(b) Determine the magnitude and direction of $\vec{P}$.

(c) Repeat part (b) using the laws of sines and cosines.

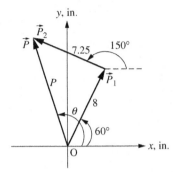

Figure P4.27 A two-link robot for problem P4-27.

4-28. A two-link planar robot is shown in Fig. P4.28.

(a) Calculate the position of the tip $\vec{P}$ of the planar robot using vector addition (use the $\hat{i}$ and $\hat{j}$ notation).

(b) Determine the magnitude and direction of the position of the robot tip. In other words, write vector $\vec{P}$ in the polar form.

(c) Repeat part (b) using the laws of sines and cosines.

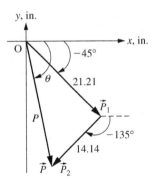

Figure P4.28 A two-link robot for problem P4-28.

4-29. A two-link planar robot is shown in Fig. P4.29.

(a) Calculate the position of the tip $\vec{P}$ of the planar robot using vector addition (use the $\hat{i}$ and $\hat{j}$ notation).

(b) Determine the magnitude and direction of $\vec{P}$.

(c) Repeat part (b) using the laws of sines and cosines.

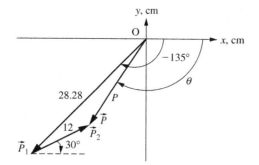

Figure P4.29   A two-link robot for problem P4-29.

4-30. A 200 lb weight is suspended by two cables as shown in Fig. P4.30.

(a) Determine the angle α.

(b) Express $\vec{T}_1$ and $\vec{T}_2$ in rectangular vector notation and determine the values of T_1 and T_2 required for static equilibrium (i.e., $\vec{T}_1 + \vec{T}_2 + \vec{W} = 0$).

4-31. A weight of 100 kg is suspended from the ceiling by cables that make 30° and 60° angles with the ceiling as shown in

Fig. P4.31. Assuming that the weight is not moving ($\vec{T}_1 + \vec{T}_2 + \vec{W} = 0$), find the tensions $\vec{T}_1$ and $\vec{T}_2$.

Figure P4.31   A weight suspended from two cables for problem P4-31.

4-32. A vehicle weighing 10 kN is parked on an inclined driveway, as shown in Fig. P4.32.

(a) Determine the angle θ.

(b) Express the normal force $\vec{N}$, the frictional force $\vec{F}$, and the weight $\vec{W}$ in rectangular vector notation.

(c) Determine the values of F and N required for equilibrium (i.e., $\vec{F} + \vec{N} + \vec{W} = 0$).

Figure P4.32   A vehicle parked on an inclined driveway for problem P4-32.

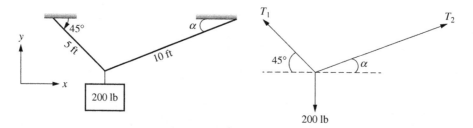

Figure P4.30   A weight suspended from two cables for problem P4-30.

4-33. A vehicle weighing 2000 lb is parked on an inclined driveway, as shown in Fig. P4.33.
(a) Determine the angle θ.
(b) Express the normal force $\vec{N}$, the frictional force $\vec{F}$, and the weight $\vec{W}$ in rectangular vector notation.
(c) Determine the values of F and N required for equilibrium (i.e., $\vec{F} + \vec{N} + \vec{W} = 0$).

Figure P4.33 A vehicle parked on an inclined driveway for problem P4-33.

4-34. A crate of weight $W = 100$ lb sits on a ramp oriented at 27 degrees relative to ground, as shown in Fig. P4.34. The free-body diagram showing the external forces is also shown in Fig. P4.34.
(a) Using the x-y coordinate system shown in Fig. P4.34, write the friction force $\vec{F}$ and the normal force $\vec{N}$ in rectangular vector notation (i.e., in terms of unit vectors $\hat{i}$ and $\hat{j}$).
(b) Determine the values of F and N required for static equilibrium; in other words, find the values of F and N if $\vec{F} + \vec{N} + \vec{W} = 0$.

Figure P4.34 A crate resting on a ramp for problem P4-34.

4-35. A 500 N television sits on an inclined ramp, as shown in Fig. P4.35. The free-body diagram showing the external forces is also shown in Fig. P4.35.
(a) Determine the angle θ.
(b) Using the x-y coordinate system shown in Fig. P4.35, write the friction force $\vec{F}$ and the normal force $\vec{N}$ in rectangular vector notation (i.e., in terms of unit vectors $\hat{i}$ and $\hat{j}$).
(c) Determine the values of F and N required for static equilibrium; in other words, find the values of F and N if $\vec{F} + \vec{N} + \vec{W} = 0$.

Figure P4.35 A crate resting on an inclined ramp for problem P4-35.

4-36. A two-bar truss supports a weight of $W = 750$ lb as shown in Fig. P4.36. The truss is constructed such that $\theta = 38.7°$.
(a) Using the positive x-y coordinate system shown in Fig. P4.36, write the forces $\vec{F}_1$, $\vec{F}_2$, and weight $\vec{W}$ in rectangular vector notation (i.e., in terms of unit vectors $\hat{i}$ and $\hat{j}$).
(b) Determine the values of F_1 and F_2 such that $\vec{F}_1 + \vec{F}_2 + \vec{W} = 0$.

4-37. A two-bar truss is subjected to a vertical load $P = 346$ lb, as shown in Fig. P4.37. The truss is constructed such that $\theta_1 = 39.6°$ and $\theta_2 = 62.4°$.
(a) Using the positive x-y coordinate system shown in Fig. P4.37, write the forces $\vec{F}_1$, $\vec{F}_2$, and $\vec{P}$ in rectangular vector notation (i.e., in terms of unit vectors $\hat{i}$ and $\hat{j}$).
(b) Determine the values of F_1 and F_2 such that $\vec{F}_1 + \vec{F}_2 + \vec{P} = 0$.

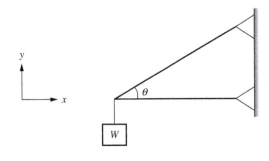

Figure P4.36 A weight supported by a two-bar truss.

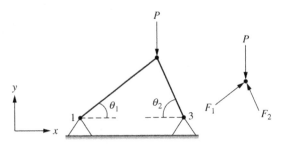

Figure P4.37 A two-bar truss subjected to a vertical load.

4-38. A force $F = 100$ N is applied to a two-bar truss as shown in Fig. P4.38. Express forces $\vec{F}$, $\vec{F}_1$, and $\vec{F}_2$ in terms of unit vectors $\hat{i}$ and $\hat{j}$ and determine the values of F_1 and F_2 such that $\vec{F}_1 + \vec{F}_2 + \vec{F} = 0$.

4-39. A block of weight $W = 125$ lb rests on a hinged shelf as shown in Fig. P4.39. The shelf is constructed such that $\theta = 41.7°$.

(a) Using the positive x-y coordinate system shown in Fig. P4.39, write the forces $\vec{F}_1$, $\vec{F}_2$, and weight $\vec{W}$ in

rectangular vector notation (i.e., in terms of unit vectors $\hat{i}$ and $\hat{j}$).

(b) Determine the values of F_1 and F_2 such that $\vec{F}_1 + \vec{F}_2 + \vec{W} = 0$.

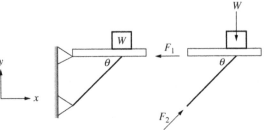

Figure P4.39 A weight resting on a hinged shelf.

4-40. A waiter extends his arm to hand a plate of food to his customer. The free-body diagram is shown in Fig. P4.40 where $F_m = 250\sqrt{2}$ N is the force in the deltoid muscle, $W_a = 35$ N is the weight of the arm, $W_p = 15$ N is the weight of the plate of food, R_x and R_y are the

Figure P4.38 Force applied to joint 2 of a truss.

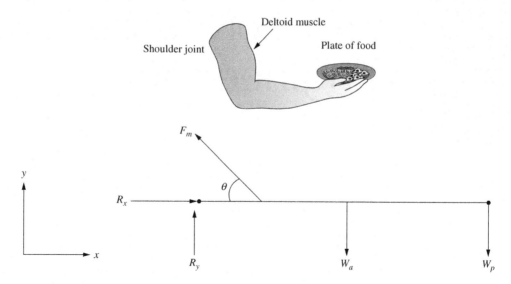

Figure P4.40   Waiter handing a plate to a customer.

x- and y-components of the reaction forces at the shoulder, and $\theta = 45°$.

(a) Using the x-y coordinate system shown in Fig. P4.40, write the deltoid muscle force $\vec{F}_m$, the weight of the arm $\vec{W}_a$, and the weight of the plate $\vec{W}_p$ in the standard vector notation (i.e., using unit vectors i and j).

(b) Determine the values of R_x and R_y required for static equilibrium: $\vec{R} + \vec{F}_m + \vec{W}_a + \vec{W}_p = 0$. Also compute the magnitude and direction of $\vec{R}$.

4-41. Repeat problem P4-40 if $F_m = 50$ lb, $W_a = 7$ lb, $W_p = 3$ lb, and $\theta = 30°$

4-42. Using motion capture, the positions $\vec{P}_1$ and $\vec{P}_2$ of each arm segment are measured while a person throws a ball. The length from shoulder to elbow (P_1) is 10 in. and the length from elbow to the hand holding the ball (P_2) is 13 in. The angle θ_1 is 60° and θ_2 is 65°.

(a) Using the x-y coordinate system shown in Fig. P4.42, write the position of the ball $\vec{P} = \vec{P}_1 + \vec{P}_2$ in standard vector notation.

(b) Determine both the magnitude and direction of $\vec{P}$.

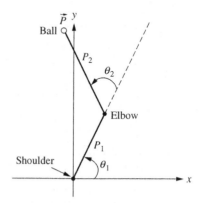

Figure P4.42   Position of the arm throwing a ball.

4-43. Repeat problem P4-42 if $P_1 = 1$ ft, $P_2 = 1.5$ ft, $\theta_1 = 45°$, and θ_2 is 60°.

Complex Numbers in Engineering

5.1   INTRODUCTION

Complex numbers play a significant role in all engineering disciplines, and a good understanding of this topic is necessary. However, it is especially important for the electrical engineer to master this topic. Although imaginary numbers are not commonly used in daily life, in engineering and physics they are in fact used to represent physical quantities such as impedance of RL, RC, or RLC circuit.

Complex numbers are numbers that consist of two parts, one real and one imaginary. An imaginary number is the square root of a negative real number (-1). The square root of a negative real number is said to be imaginary because there is no real number that gives a negative number after it has been squared. The imaginary number $\sqrt{-1}$ is represented by the letter i by mathematicians and by almost all the engineering disciplines except electrical engineering. Electrical engineers use the letter j to represent imaginary number because the letter i is used in electrical engineering to represent current. To remove this confusion, $\sqrt{-1}$ will be represented by the letter j throughout this chapter.

In general, imaginary numbers are used in combination with a real number to form a complex number, $a + bj$, where a is the real part (real number) and bj is the imaginary part (real number times the imaginary unit j). The complex number is useful for representing two-dimensional variables where both dimensions are physically significant and are represented on a complex number plane (which looks very similar to the Cartesian plane discussed in Chapter 4). On this plane, the imaginary part of the complex number is measured by the vertical axis (on the Cartesian plane, this is the y-axis) and the real number part goes on the horizontal axis (x-axis on the Cartesian plane). The one-link robot discussed in Chapter 3 could be described using a complex number where the real part would be its component in the x-direction and the imaginary part would be its component in the y-direction. **Note that the example of one-link planar robot is used only to show similarities between the two-dimensional vector and the complex number. The position of the tip of the robot is generally not described by a complex number**.

In many ways, operations with complex numbers follow the same rules as those for real numbers. However, the two parts of a complex number cannot be combined.

Even though the parts are joined by a plus sign, the addition cannot be performed. The expression must be left as an indicated sum.

5.2 POSITION OF ONE-LINK ROBOT AS A COMPLEX NUMBER

The one-link planar robot shown in Fig. 5.1 can be represented by a complex number as

$$P = P_x + j\ P_y$$

or

$$P = P_x + P_y\ j,$$

where $j = \sqrt{-1}$ is the imaginary number, $P_x = Re(P) = l\cos(\theta)$ is the real part, and $P_y = Im(P) = l\sin(\theta)$ is the imaginary part of the complex number P. The numbers P_x and P_y are like the components of P in the x- and y-directions (analogous to a 2-D vector). Similarly, the one-link planar robot can be represented in polar form as

$$P = |P|\ \angle\theta,$$

where $|P| = l = \sqrt{P_x^2 + P_y^2}$ is the magnitude and $\theta = \mathrm{atan2}(P_y,\ P_x)$ is the angle of the complex number P.

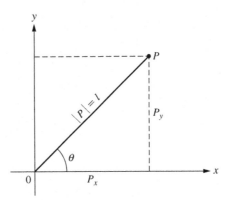

Figure 5.1 Representation of a one-link planar robot as a complex number.

Therefore,

$$
\begin{aligned}
P &= P_x + j\ P_y \\
&= |P|\cos\theta + j\,(|P|\sin\theta) \\
&= |P|\,(\cos\theta + j\sin\theta) \\
&= |P|\,e^{j\theta}
\end{aligned}
\tag{5.1}
$$

where $\cos\theta + j\sin\theta = e^{j\theta}$ is Euler's formula. The complex number written as in equation (5.1) is known as the exponential form of the complex number and can be used to express the cosine and sine functions. The cosine function is the real part of the exponential function and the sine function as the imaginary part of the exponential function; i.e.,

$$\cos\theta = Re(e^{j\,\theta})$$

and

$$\sin\theta = Im(e^{j\,\theta})$$

In summary, a point P in the rectangular plane (x-y plane) can be described as a vector or a complex number as

$$\vec{P} = P_x\,\vec{i} + P_y\,\vec{j} \cdots\cdots \text{vector form}$$
$$P = P_x + j\,P_y \cdot\cdots\cdots \text{complex number in rectangular form}$$
$$P = |P|\ \angle\theta \cdots\cdots\cdots \text{complex number in polar form}$$
$$P = |P|\ e^{j\,\theta} \cdots\cdots\cdots \text{complex number in exponential form}$$

IMPEDANCE OF R, L, AND C AS A COMPLEX NUMBER

5.3.1   Impedance of a Resistor R

The impedance of the resistor shown in Fig. 5.2 can be written as $Z_R = R\ \Omega$, where R is the resistance in ohms (Ω). If $R = 100\ \Omega$, the impedance of the resistor can be written as a complex number in rectangular and polar forms as

$$Z_R = R\ \Omega$$
$$= 100\ \Omega$$
$$= 100 + j\,0\ \Omega$$
$$= 100\ \angle 0°\Omega$$
$$= 100\ e^{j0°}\Omega.$$

Figure 5.2   A resistor.

5.3.2   Impedance of an Inductor L

The impedance of the inductor shown in Fig. 5.3 can be written as $Z_L = j\omega L\ \Omega$, where L is the inductance in henry (H) and $\omega = 2\pi f$ is the angular frequency in rad/s (f is the linear frequency or frequency in hertz (Hz)). If $L = 25$ mH and $f = 60$ Hz, the impedance of the inductor can be written as a complex number in rectangular and polar forms as

$$Z_L = j\,\omega\,L\ \Omega$$
$$= j\,(2\,\pi\,60)\,(0.025)\ \Omega$$
$$= 0\,+\,j\,9.426\ \Omega$$
$$= 9.426\ \angle 90°\,\Omega$$
$$= 9.426\ e^{j\,90°}\,\Omega.$$

Figure 5.3   An inductor.

5.3.3   Impedance of a Capacitor *C*

The impedance of the capacitor shown in Fig. 5.4 can be written as $Z_C = \dfrac{1}{j\omega\,C}\ \Omega$, where *C* is the capacitance in farad (F) and ω is the angular frequency in rad/s. If $C = 20\ \mu$F and $f = 60$ Hz, the impedance of the capacitor can be written as a complex number in rectangular form as

$$Z_C = \frac{1}{j\,\omega\,C}\ \Omega$$
$$= 0 + \frac{1}{j\,(120\,\pi)\,(20 * 10^{-6})}\ \Omega$$
$$= 0 + \frac{132.6}{j}\ \Omega$$
$$= 0 + \frac{132.6}{j}\,(\tfrac{j}{j})\ \Omega$$
$$= 0 + \frac{132.6\,j}{j^2}\ \Omega$$
$$= 0 - 132.6\,j\ \Omega$$

where $j^2 = (\sqrt{-1})^2 = -1$. The impedance of the capacitor can also be written in polar and exponential form as

$$Z_C = 132.6\ \angle{-90°}\,\Omega$$
$$= 132.6\ e^{-j\,90°}\,\Omega.$$

Figure 5.4   A capacitor.

5.4 IMPEDANCE OF A SERIES RLC CIRCUIT

The total impedance of the series RLC circuit shown in Fig. 5.5 is given by

$$Z_T = Z_R + Z_L + Z_C \tag{5.2}$$

where $Z_R = R \ \Omega$, $Z_L = j\omega L \ \Omega$, and $Z_C = \dfrac{1}{j\omega C} \ \Omega$.

Figure 5.5 A series RLC circuit.

For the values of $R = 100 \ \Omega$, $L = 25$ mH, $C = 20 \ \mu F$, and $\omega = 120\pi$ rad/s, the impedance of R, L, and C were calculated in Section 5.3 as

$$Z_R = 100 + j\,0 \ \Omega$$
$$Z_L = 0 + j\,9.426 \ \Omega$$
$$Z_C = 0 - j\,132.6 \ \Omega.$$

Since $Z_T = Z_R + Z_L + Z_C$, the total impedance of the series RLC circuit can be calculated as

$$Z_T = (100 + j\,0) + (0 + j\,9.426) + (0 - j\,132.6) \ \Omega$$
$$= (100 + 0 + 0) + j\,(0 + 9.426 + (-132.6)) \ \Omega$$
$$= 100 - j\,123.174 \ \Omega.$$

Therefore, the total impedance of the series RLC circuit shown in Fig. 5.5 in rectangular form is $Z_T = 100 - j\,123.174 \ \Omega$. The polar and exponential forms of the total impedance can be calculated from the rectangular form as

Polar Form: $Z_T = |Z_T| \angle \theta$, where

$$|Z_T| = \sqrt{100^2 + (-123.174)^2}$$
$$= 158.7 \ \Omega$$
$$\theta = \text{atan2}(-123.174, 100)$$
$$= -50.93°.$$

Therefore, $Z_T = 158.7 \ \angle{-50.93°}\,\Omega$.

Exponential Form: $Z_T = |Z_T|\, e^{j\,\theta} = 158.7 \ e^{-j\,50.93°} \ \Omega$.

Note: The addition and subtraction of complex numbers is best done in the rectangular form. If the complex numbers are given in the polar or exponential form, they should be converted to rectangular form to carry out the addition or subtraction of these complex numbers. However, if the result is needed in the polar or exponential form, the conversion from the rectangular to polar or exponential forms is carried out as a last step.

In summary, the addition and subtraction of two complex numbers can be carried out using the following steps.

Addition of Two Complex Numbers: The addition of two complex numbers $Z_1 = a_1 + jb_1$ and $Z_2 = a_2 + jb_2$ can be obtained as

$$Z = Z_1 + Z_2$$
$$= (a_1 + jb_1) + (a_2 + jb_2)$$
$$= (a_1 + a_2) + j(b_1 + b_2)$$
$$= a + jb.$$

Therefore, the real part, a, of the addition of complex numbers, Z, is obtained by adding the real parts of the complex numbers being added, and the imaginary part of the addition of the two complex numbers is obtained by adding the imaginary parts of complex numbers being added.

Subtraction of Two Complex Numbers: Similarly, the subtraction of the two complex numbers Z_1 and Z_2 can be obtained as

$$Z = Z_1 - Z_2$$
$$= (a_1 + jb_1) - (a_2 + jb_2)$$
$$= (a_1 - a_2) + j(b_1 - b_2)$$
$$= c + jd.$$

Therefore, the real part, c, of the subtraction of two complex numbers Z_1 and Z_2 ($Z_1 - Z_2$) is obtained by subtracting the real part a_2 of complex numbers Z_2 from the real part a_1 of complex number Z_1. Similarly, the imaginary part of the subtraction of the two complex numbers $Z_1 - Z_2$ is obtained by subtracting the imaginary part b_2 of complex number Z_2 from the imaginary part b_1 of complex number Z_1, respectively.

5.5 IMPEDANCE OF R AND L CONNECTED IN PARALLEL

The total impedance Z of a resistor R connected in parallel with an inductor L as shown in Fig. 5.6 is given by

$$Z = \frac{Z_R Z_L}{Z_R + Z_L}$$

where $Z_R = R \ \Omega$ and $Z_L = j\omega L \ \Omega$.

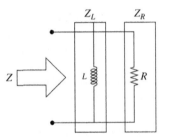

Figure 5.6   A parallel RL circuit.

For $R = 100\ \Omega$, $L = 25\ \text{mH}$ and $\omega = 120\ \pi$ rad/s, the impedances of R and L were calculated in Section 5.3 as

$$Z_R = 100 + j\,0\ \Omega$$
$$= 100\ e^{j0^\circ}\,\Omega$$
$$Z_L = 0\ +\ j\,9.426\ \Omega$$
$$= 9.426\ e^{j\,90^\circ}\,\Omega.$$

Since $Z = \dfrac{Z_R\,Z_L}{Z_R + Z_L}$, the total impedance of the R connected in parallel with L can be calculated as

$$Z_T = \frac{(100\ e^{j0^\circ})(9.426\ e^{j\,90^\circ})}{(100 + j\,0) + (0\ +\ j\,9.426)}$$

$$= \frac{942.6\ e^{j\,90^\circ}}{100\ +\ j\,9.426}$$

$$= \frac{942.6\ e^{j\,90^\circ}}{100.443\ e^{j\,5.384^\circ}}$$

$$= 9.384\ e^{j\,84.62^\circ}$$

$$= 9.384\ \angle 84.62^\circ.$$

Therefore, the total impedance of a $100\ \Omega$ resistor connected in parallel with a 25 mH inductor in the polar form is $Z = 9.384\ \angle 84.62^\circ\Omega$. The rectangular form of the total impedance can be calculated from the polar form as

$$Z = 9.384\cos(84.62^\circ) + j\,9.384\sin(84.62^\circ)$$
$$= 0.88 + j\,9.34\ \Omega.$$

Therefore, $Z = 0.88 + j\,9.34\ \Omega$.

Note that the multiplication and division of complex numbers can be carried out in rectangular or polar forms. However, it will be shown in Section 5.7 that it is best to carry out these operations in polar form. If the complex numbers are given in rectangular form, they should be converted to polar form and the result of the multiplication or division is then obtained in the polar form. However, if the result is needed in rectangular form, the conversion from polar to rectangular form is carried out as a last step. The steps to carry out the multiplication and division in the polar form are explained next.

Multiplication of Complex Number in Polar Form: The multiplication of the two complex numbers $Z_1 = M_1 \angle\theta_1$ and $Z_2 = M_2 \angle\theta_2$ can be carried out as

$$Z = Z_1 * Z_2$$

$$= M_1 \angle\theta_1 * M_2 \angle\theta_2$$

$$= M_1 e^{j\angle\theta_1} * M_2 e^{j\angle\theta_2}$$

$$= (M_1 M_2) e^{j\angle(\theta_1+\theta_2)}$$

$$= (M_1 M_2) \angle(\theta_1 + \theta_2)$$

$$= |Z| \angle\theta.$$

Therefore, the magnitude $|Z|$ of the multiplication of complex numbers given in polar form is obtained by multiplying the magnitudes of complex numbers being multiplied and the angle $\angle\theta$ of the resultant is the sum of the angles of the complex numbers being multiplied. Note that this procedure is not restricted to multiplication of two numbers only; it can be used for multiplying any number of complex numbers.

Division of Complex Number in Polar Form: The division of the two complex numbers Z_1 and Z_2 in the polar form can be carried out as

$$Z = \frac{Z_1}{Z_2}$$

$$= \frac{M_1 \angle\theta_1}{M_2 \angle\theta_2}$$

$$= \frac{M_1 e^{j\angle\theta_1}}{M_2 e^{j\angle\theta_2}}$$

$$= \frac{M_1}{M_2} e^{j(\angle(\theta_1-\theta_2))}$$

$$= \frac{M_1}{M_2} \angle(\theta_1 - \theta_2)$$

$$= |Z| \angle\theta.$$

Therefore, the magnitude $|Z|$ of the division of two complex numbers given in polar form is obtained by dividing the magnitudes of dividend complex number Z_1 by the magnitude of divisor complex number Z_2. The angle $\angle\theta$ of the resultant is obtained by subtracting the angle of the divisor complex numbers Z_2 from the angle of the dividend complex number Z_1. Note that if the dividend or the divisor are the product of complex numbers, the product of all the dividend and the divisor complex numbers should be obtained first, and then the division of the two complex numbers should be carried out.

5.6 ARMATURE CURRENT IN A DC MOTOR

The winding of an electric motor shown in Fig. 5.7 has a resistance of $R = 10\ \Omega$ and an inductance of $L = 25$ mH. If the motor is connected to a 110 V, 60 Hz voltage source as shown, find the current $I = \dfrac{V}{Z}$ A flowing through the winding of the motor, where $Z = Z_R + Z_L$ and $V = 110$ V.

Figure 5.7 Voltage applied to a motor.

The total impedance of the winding of the motor is given as $Z = Z_R + Z_L$, where $Z_R = R = 10 + j\,0\ \Omega$ and $Z_L = j\,\omega\,L = 0 + j\,9.426\ \Omega$. Therefore, $Z = (10 + j\,0) + (0 + j\,9.426) = 10 + j\,9.426\ \Omega$, and the current flowing through the winding of the motor is

$$I = \frac{V}{Z}$$

$$= \frac{110}{10 + j\,9.426}$$

$$= \frac{110 + j\,0}{10 + j\,9.426}\ \text{A.} \tag{5.3}$$

Since it is easier to multiply and divide in exponential or polar forms, the current in equation (5.3) will be calculated in the polar/exponential form. Converting the numerator and denominator in equation (5.3) to exponential form yields

$$I = \frac{110\ e^{j0°}}{\sqrt{10^2 + 9.426^2}\ \ e^{j\,\text{atan2}(9.426,10)}}$$

$$= \frac{110\ e^{j0°}}{13.74\ \ e^{j\,43.3°}}$$

$$= \frac{110\ e^{j(0° - 43.3°)}}{13.74}$$

$$= 8.01\ e^{-j\,43.3°}$$

$$= 8.01\ \angle{-43.3°}\,\text{A.} \tag{5.4}$$

Therefore, the current flowing through the winding of the motor is 8.01 A. The phasor diagram (vector diagram) showing the voltage and current vectors is shown in Fig. 5.8. It can be seen from Fig. 5.8 that the current is lagging (negative angle) the voltage by 43.3°. The polar form of the current given in equation (5.4) can be converted to rectangular form as

$$I = 8.01\, e^{-j\,43.3°}$$

$$= 8.01\,(\cos 43.3° - j \sin 43.3°)$$

$$= 5.83 - j\,5.49 \ \text{A}.$$

Note: In general $e^{\pm\theta} = \cos\theta \pm j \sin\theta$ requires θ in radians. However, converting to radians is unnecessary for the purpose of multiplying and dividing complex numbers.

Figure 5.8   The current and voltage vector.

5.7   FURTHER EXAMPLES OF COMPLEX NUMBERS IN ELECTRIC CIRCUITS

Example 5-1

A current, I, flowing through the RL circuit shown in Fig. 5.9 produces a voltage, $V = I\,Z$, where $Z = R + j\,X_L$. Find V if $I = 0.1\ \angle 30°\,\text{A}$.

$R = 100\ \Omega$

$X_L = 30\ \Omega$

Figure 5.9   Current flowing through an RL circuit.

Solution   The impedance Z of the RL circuit can be calculated as

$$Z = R + j\,X_L \ \Omega$$

$$= 100 + j\,30 \ \Omega.$$

The voltage $V = I\,Z = (0.1\ \angle 30°)(100 + j\,30)$ V will be calculated by multiplying the two complex numbers using their rectangular forms as well as their polar/exponential forms to show that it is much easier to multiply complex numbers using the polar/exponential forms.

Rectangular Form	Polar/Exponential Form
$I = 0.1 \angle 30°$	$I = 0.1 \angle 30° = 0.1\,e^{j\,30°}\,\text{A}$
$= 0.1\,(\cos 30° + j\sin 30°)$	$Z = 100 + j\,30$
$= 0.0866 + j\,0.05\,\text{A}$	$= \sqrt{100^2 + 30^2}\,\angle\text{atan2}\,(30,100)$
$Z = 100 + j\,30\ \Omega$	$= 104.4\,\angle 16.7° = 104.4\,e^{j\,16.7°}\,\Omega$
$V = I\,Z$	$V = I\,Z$
$= (0.0866 + j\,0.05)(100 + j\,30)$	$= (0.1\,e^{j\,30°})(104.4\,e^{j\,16.7°})$
$= 8.66 + j\,2.598 + j\,5 + 1.5\,j^2$	$= (10.44)\,e^{j(30° + 16.7°)}$
$= 8.66 + j\,2.598 + j\,5 + 1.5\,(-1)$	$= (10.44)\,\angle 46.7°\,\text{V}$
$= 7.16 + j\,7.598$	
$= 10.44\,\angle 46.7°\,\text{V}$	

Example 5-2

In the voltage divider circuit shown in Fig. 5.10, the impedance of the resistor is given by $Z_1 = R$. The total impedance of the inductor and capacitor in series is given by $Z_2 = j X_L + \frac{1}{j} X_C$, where $j = \sqrt{-1}$. Suppose $R = 10$, $X_L = 10$, and $X_C = 20$, all measured in ohms:

(a) Express the impedance Z_1 and Z_2 in both rectangular and polar forms.

(b) Suppose the source voltage is $V = 100\,\sqrt{2}\,\angle 45°\,\text{V}$. Compute the voltage V_1 given by

$$V_1 = \frac{Z_2}{Z_1 + Z_2}\,V. \qquad (5.5)$$

Figure 5.10 Voltage divider circuit for Example 5-2.

Solution (a) The impedance Z_1 can be written in rectangular form as

$$Z_1 = R$$
$$= 10 + j\,0\ \Omega.$$

The impedance Z_1 can be written in polar form as

$$Z_1 = \sqrt{10^2 + 0^2} \angle \text{atan2}(0, 10)$$
$$= 10 \angle 0° \Omega.$$

The impedance Z_2 can be written in rectangular form as

$$Z_2 = jX_L + \frac{1}{j}X_C$$
$$= j10 + \frac{1}{j}20 \left(\frac{j}{j}\right)$$
$$= j10 - j20$$
$$= -j10$$
$$= 0 - j10 \ \Omega.$$

The impedance Z_2 can be written in polar form as

$$Z_2 = \sqrt{0^2 + (-10)^2} \angle \text{atan2}(-10, 0)$$
$$= 10 \angle -90° \Omega.$$

(b)

$$V_1 = \frac{Z_2}{Z_1 + Z_2} V$$

$$= \left(\frac{10 \angle -90°}{(10 + j0) + (0 - j10)}\right)(100 \sqrt{2} \angle 45°)$$

$$= \left(\frac{10 \angle -90°}{10 - j10}\right)(100 \sqrt{2} \angle 45°)$$

$$= \left(\frac{10 \angle -90°}{10 \sqrt{2} \angle -45°}\right)(100 \sqrt{2} \angle 45°)$$

$$= \left(\frac{1}{\sqrt{2}} \angle -45°\right)(100 \sqrt{2} \angle 45°)$$

$$= 100 \angle 0° V$$
$$= 100 + j0 V.$$

Example 5-3

In the circuit shown in Fig. 5.11, the impedance of the various components are $Z_R = R$, $Z_L = jX_L$, and $Z_C = \frac{1}{j}X_C$, where $j = \sqrt{-1}$. Suppose $R = 10$, $X_L = 10$, and $X_C = 10$, all measured in ohms.

(a) Express the total impedance $Z = Z_C + \dfrac{Z_R Z_L}{Z_R + Z_L}$ in both rectangular and polar forms.

(b) Suppose a voltage $V = 50\sqrt{2}\,\angle45°\,\mathrm{V}$ is applied to the circuit shown in Fig. 5.11. Find the current I flowing through the circuit if I given by

$$I = \frac{V}{Z} \tag{5.6}$$

Solution   (a) The impedance Z_R can be written in rectangular and polar forms as

$$Z_R = R$$
$$= 10 + j0 \ \Omega$$
$$= 10\,\angle0°\,\Omega.$$

Figure 5.11   Total impedance of the circuit for Example 5-3.

The impedance Z_L can be written in rectangular and polar forms as

$$Z_L = jX_L$$
$$= 0 + j10 \ \Omega$$
$$= 10\,\angle90°\,\Omega.$$

The impedance Z_C can be written in rectangular and polar forms as

$$Z_C = \frac{1}{j}X_C$$
$$= -jX_C$$
$$= 0 - j10 \ \Omega$$
$$= 10\,\angle-90°\,\Omega.$$

The total impedance can now be calculated as

$$Z = Z_C + \frac{Z_R Z_L}{Z_R + Z_L}$$

$$= 0 - j10 + \frac{(10\,\angle0°)\,(10\,\angle90°)}{(10 + j0) + (0 + j10)}$$

$$= 0 - j10 + \frac{100\,\angle90°}{(10 + j10)}$$

$$= 0 - j\,10 + \frac{100\,\angle 90°}{10\,\sqrt{2}\,\angle 45°}$$

$$= 0 - j\,10 + 5\,\sqrt{2}\,\angle 45°$$

$$= 0 - j\,10 + 5 + j\,5$$

$$= 5 - j\,5\ \Omega$$

$$= 5\,\sqrt{2}\,\angle{-45°}\,\Omega.$$

(b)

$$I = \frac{V}{Z}$$

$$= \frac{50\,\sqrt{2}\,\angle 45°}{5\,\sqrt{2}\,\angle{-45°}}$$

$$= 10\,\angle 90°\,\mathrm{A}$$

$$= 0 + j\,10\ \mathrm{A}.$$

Example 5-4	A sinusoidal voltage source $v_s = 100\cos(100\,t + 45°)$ V is applied to a series RLC circuit. Write the voltage source v_s in the exponential form.

Solution   Since $\cos(\theta) = Re(e^{j\,\theta})$, therefore, the voltage source v_s can be written in the exponential form as

$$v_s = Re(100\,e^{j(100\,t + 45°)})\,\mathrm{V}.$$

Example 5-5	A sinusoidal current source $i_s = 100\sin(120\,\pi\,t + 60°)$ mA is applied to a parallel RL circuit. Write the current source i_s in the exponential form.

Solution   Since $\sin(\theta) = Im(e^{j\,\theta})$, therefore, the current source i_s can be written in the exponential form as

$$i_s = Im(100\,e^{j(120\,\pi\,t + 60°)})\,\mathrm{mA}.$$

5.8   COMPLEX CONJUGATE

The complex conjugate of a complex number $z = a + j\,b$ is

$$z^* = a - j\,b.$$

The multiplication of a complex number by its conjugate results in a real number that is the square of the magnitude of the complex number:

$$z\,z^* = (a + j\,(b))(a - j\,(b))$$
$$= a^2 - j\,ab + j\,ab - j^2\,b^2$$
$$= a^2 - (-1)\,b^2$$
$$= a^2 + b^2.$$

Also,

$$|z|^2 = (\sqrt{a^2 + b^2})^2$$
$$= a^2 + b^2.$$

Therefore, $z\,z^* = |z|^2 = a^2 + b^2$.

Example 5-6

If $z = 3 + j\,4$, find $z\,z^*$ using the rectangular and polar forms.

Solution The conjugate of the complex number $z = 3 + j\,4$ is given by

$$z^* = 3 - j\,4.$$

Calculating $z\,z^*$ using the rectangular form yields

$$z\,z^* = (3 + j\,4)(3 - j\,4)$$
$$= 3^2 - j\,(3)(4) + j\,(3)(4) - j^2\,4^2$$
$$= 3^2 - (-1)\,4^2$$
$$= 3^2 + 4^2$$
$$= 25.$$

Therefore, $z\,z^* = 25$. Note: $|z|^2 = (\sqrt{3^2 + 4^2})^2 = 25$. Now, calculating $z\,z^*$ using the polar form, we have

$$z = 3 + j\,4$$

$$z = \sqrt{3^2 + 4^2} \;\angle \text{atan2}(4,\ 3)$$

$$z = 5\ \angle 53.1°$$

$$z^* = 3 - j\,4$$

$$z^* = \sqrt{3^2 + (-4)^2} \;\angle \text{atan2}(-4,\ 3)$$

$$z^* = 5\ \angle{-53.1°}$$

$$z\,z^* = (5\,\angle 53.1°)(5\,\angle -53.1°)$$

$$= (5)(5)\angle(53.1° - 53.1°)$$

$$= 25\,\angle 0°$$

$$= 25.$$

Therefore, $z\,z^* = 25$. Note that the complex conjugate of a complex number in polar form has the same magnitude as the complex number, but the angle of the complex conjugate is the negative of the angle of the complex number.

PROBLEMS

5-1. In the series RL circuit shown in Fig. P5.1, voltage V_L leads voltage V_R by 90° (i.e., if the angle of V_R is 0°, the angle of V_L is 90°). Assume $V_R = 1\,\angle 0°\,\text{V}$ and $V_L = 1\,\angle 90°\,\text{V}$.
  (a) Write V_R and V_L in rectangular form.
  (b) Determine $V = V_R + V_L$ in both its rectangular and polar forms.
  (c) Write the real and imaginary parts of V.

Figure P5.1 A series RL circuit for problem P5-1.

5-2. Repeat problem P5-1 if $V_R = 10\,\angle -45°\,\text{V}$ and $V_L = 5\,\angle 45°\,\text{V}$.

5-3. Repeat problem P5-1 if $V_R = 9\,\angle -26.6°\,\text{V}$ and $V_L = 4.5\,\angle 63.4°\,\text{V}$.

5-4. In the RC circuit shown in Fig. P5.4, voltage V_C lags voltage V_R by 90° (i.e., if the angle of V_R is 0°, the angle of

V_C is −90°). Assume $V_R = 1\,\angle 0°\,\text{V}$ and $V_C = 1\,\angle -90°\,\text{V}$.
  (a) Write V_R and V_C in rectangular form.
  (b) Determine $V = V_R + V_C$ in both its rectangular and polar forms.
  (c) Write the real and imaginary parts of V.

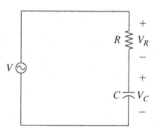

Figure P5.4 RC circuit for problem P5-4.

5-5. Repeat problem P5-4 if $V_R = 9.5\,\angle 18.44°\,\text{V}$ and $V_C = 3.16\,\angle -71.56°\,\text{V}$.

5-6. Repeat problem P5-4 if $V_R = 10\,\angle 60°\,\text{V}$ and $V_C = 17.32\,\angle -30°\,\text{V}$.

5-7. In the parallel RL circuit shown in Fig. P5.7, the total current I is the sum of the currents flowing through the resistor (I_R) and the inductor (I_L). Assume $i_R = 100\,\angle 0°\,\text{mA}$ and $i_L = 200\,\angle -90°\,\text{mA}$.
  (a) Write I_R and I_L in rectangular form.

(b) Determine $I = I_R + I_L$ in both its rectangular and polar forms.

(c) Write the real and imaginary parts of I.

Figure P5.7   A parallel RL circuit for problem P5-7.

5-8. Repeat problem P5-7 if $I_R = 0.707 \angle 45° A$ and $I_L = 0.707 \angle -45° A$.

5-9. Repeat problem P5-7 if $I_R = 86.6 \angle 30° \mu A$ and $I_L = 50 \angle -60° \mu A$.

5-10. In the parallel RC circuit shown in Fig. P5.10, the total current I is the sum of the currents flowing through the resistor (I_R) and the capacitor (I_C). Assume $I_R = 83.2 \angle -33.7° mA$ and $I_C = 55.5 \angle 56.3° mA$.

(a) Write I_R and I_L in rectangular form.

(b) Determine $I = I_R + I_L$ in both its rectangular and polar forms.

(c) Write the real and imaginary parts of I.

Figure P5.10   A parallel RC circuit for problem P5-10.

5-11. Repeat problem P5-10 if $I_R = 0.5 \angle 0° mA$ and $I_C = 0.2 \angle 90° mA$.

5-12. Repeat problem P5-10 if $I_R = 0.929 \angle -21.8° A$ and $I_C = 0.37 \angle 68.2° A$.

5-13. The output voltage across the capacitor in a series RLC circuit is measured by an oscilloscope as $v_o(t) = 15 \cos(120 \pi t - 60°)$ V. Write the voltage $v_o(t)$ in the exponential form.

5-14. The current flowing through the resistor in a parallel RL circuit is given as $i_R(t) = 5 \sin(5 \pi t + 30°)$ mA. Write the current $i_R(t)$ in the exponential form.

5-15. A resistor, capacitor, and an inductor are connected in series as shown in Fig. P5.15. The total impedance of the circuit is $Z = Z_R + Z_L + Z_C$, where $Z_R = R \ \Omega$, $Z_L = j\omega L \ \Omega$, and $Z_C = \dfrac{1}{j\omega C} \ \Omega$. For a particular design $R = 100 \ \Omega$, $L = 500 \ mH$, $C = 25 \ \mu F$, and $\omega = 120 \pi$ rad/s.

(a) Determine the total impedance Z in rectangular form.

(b) Determine the total impedance Z in polar form.

(c) Determine the complex conjugate Z^* and compute the product $Z \ Z^*$.

Figure P5.15   RLC circuit for problem P5-15.

5-16. Two circuit elements are connected in series as shown in Fig. P5.16. The impedance of the first circuit element is $Z_1 = R_1 + j X_{L1}$. The impedance of the second circuit element is $Z_2 = R_2 + j X_{L2}$, where $R_1 = 10 \ \Omega$, $R_2 = 5 \ \Omega$, $X_{L1} = 25 \ \Omega$, and $X_{L2} = 15 \ \Omega$.

(a) Determine the total impedance, $Z = Z_1 + Z_2$.

(b) Determine the magnitude and phase of the total impedance; in other words find $Z = |Z| \angle \theta$.

Figure P5.16   Two circuit elements in series for problem P5-16.

5-17. An RC circuit is subjected to an alternating voltage source V as shown in Fig. P5.17. The relationship between the voltage and current is $V = I\,Z$, where $Z = R - j\,X_C$. For a particular design, $R = 4 \ \Omega$ and $X_C = 2 \ \Omega$.
(a) Find I if $V = 120 \ \angle 30° \text{V}$.
(b) Find V if $I = 7.0 \ \angle 45° \text{A}$.

Figure P5.17   RC circuit subjected to an alternating voltage source for problem P5-17.

5-18. A series-parallel electric circuit consists of the components shown in Fig. P5.18. The values of the impedance of the two components are $Z_1 = \dfrac{-j}{\omega C}$ and $Z_2 = R + j\omega L$, where $C = 5 \ \mu\text{F}$, $R = 100 \ \Omega$, $L = 0.15 \ \text{H}$, $\omega = 120 \pi$ rad/s, and $j = \sqrt{-1}$.
(a) Write Z_1 and Z_2 as complex numbers in both their rectangular and polar forms.

(b) Write down the complex conjugate of Z_2 and calculate the product $Z_2\,Z_2^*$.
(c) Calculate the total impedance $Z = \dfrac{Z_1\,Z_2}{Z_1 + Z_2}$ of the circuit. Write the total impedance in both its rectangular and polar forms.

Figure P5.18   Impedance of series-parallel combination of circuit elements for problem P5-18.

5-19. The circuit shown in Fig. P5.19 consists of a resistor R, an inductor L, and a capacitor C. The impedance of the resistor is $Z_1 = R \ \Omega$, the impedance of the inductor is $Z_2 = j\,\omega\,L \ \Omega$, and the impedance of the capacitor is $Z_3 = \dfrac{1}{j\,\omega\,C} \ \Omega$, where $j = \sqrt{-1}$. For a particular design, $R = 100 \ \Omega$, $L = 15 \ \text{mH}$, $C = 25 \ \mu\text{F}$, and $\omega = 120 \pi$ rad/s.
(a) Compute the quantity $Z_2 + Z_3$, and express the result in both rectangular and polar forms.
(b) Compute the quantity $Z_1 + Z_2 + Z_3$, and express the result in both rectangular and polar forms.
(c) Compute the transfer function $H = \dfrac{Z_2 + Z_3}{Z_1 + Z_2 + Z_3}$ and express the result in both rectangular and polar forms.
(d) Determine the complex conjugate of H and compute the product $H\,H^*$.

Figure P5.19   Transfer function of series-parallel circuit for problem P5-19.

5-20. An electric circuit consists of two components as shown in Fig. P5.20. The values of the impedance of the two components are $Z_1 = R_1 + j X_L$ and $Z_2 = R_2 - j X_C$, where $R_1 = 75\ \Omega, X_L = 100\ \Omega$, $R_2 = 50\ \Omega$, and $X_C = 125\ \Omega$.

(a) Write Z_1 and Z_2 as complex numbers in both their rectangular and polar forms.

(b) Determine the complex conjugate of Z_2 and compute the product $Z_2 Z_2^*$.

(c) Compute the total impedance of the two components $Z = \dfrac{Z_1 Z_2}{Z_1 + Z_2}$ and express the result in both rectangular and polar forms.

Figure P5.20   Impedance of elements connected in parallel for problem P5-20.

5-21. A sinusoidal voltage source $V = 110$ V of frequency 60 Hz ($\omega = 120\,\pi$ rad/s)

is applied to an RLC circuit as shown in Fig. P5.21, where $Z_R = R\ \Omega$, $Z_L = j\omega L\ \Omega$, and $Z_C = \dfrac{1}{j\omega C}\ \Omega$. Assuming that $R = 100\ \Omega$, $L = \dfrac{500}{\pi}$ mH, $C = \dfrac{500}{3\,\pi}\ \mu F$, and $j^2 = -1$,

(a) Write Z_R, Z_L, and Z_C as complex numbers in both their rectangular and polar forms.

(b) If the total impedance is $Z = Z_R + Z_L + Z_C$, write Z in both rectangular and polar forms.

(c) Determine the complex conjugate of Z and compute the product $Z\, Z^*$.

(d) If $V_C = \dfrac{Z_C}{Z_R + Z_L + Z_C} V$, find the voltage in both rectangular and polar forms.

Figure P5.21   Voltage division for problem P5-21.

5-22. A sinusoidal voltage source $V = 110\sqrt{2}\ \angle{-23.2°}\,$V is applied to an circuit shown in Fig. P5.22, where $Z_1 = R_1 - j X_C\ \Omega$ and $Z_2 = R_2 + j X_L\ \Omega$. The voltage V_1 is given by

$$V_1 = \frac{Z_2}{Z_1 + Z_2}\ V.$$

Assuming that $R_1 = 50\ \Omega$, $R_2 = 100\ \Omega$, $X_L = 250\ \Omega$, and $X_C = 100\ \Omega$,

(a) Write Z_1 and Z_2 as complex numbers in polar form.

(b) Determine V_1 in both rectangular and polar forms.

(c) Determine the complex conjugate of Z_1 and compute the product $Z_1 Z_1^*$.

Figure P5.22   Voltage division for problem P5-22.

5-23. An electric circuit consists of a resistor R, an inductor L, and a capacitor C, connected as shown in Fig. P5.23. The impedance of the resistor is $Z_R = R \, \Omega$, the impedance of the inductor is $Z_L = j\omega L \, \Omega$, and the impedance of the capacitor is $Z_C = \dfrac{1}{j\omega C} \, \Omega$, where $j = \sqrt{-1}$. Suppose $R = 9 \, \Omega$, $L = 3$ mH, $C = 250 \, \mu F$, and $\omega = 1000$ rad/s.

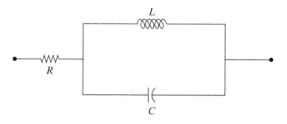

Figure P5.23   Series-parallel combination of R, L, and C for problem P5-23.

(a) Write Z_R, Z_L, and Z_C as complex numbers in both rectangular and polar form.

(b) Calculate the total impedance of the inductor-capacitor parallel combination, which is given as $Z_{LC} =$

$\dfrac{Z_L Z_C}{Z_L + Z_C}$. Express the result in both rectangular and polar forms.

(c) Now suppose that the total impedance of the circuit is $Z = Z_R + Z_{LC}$. Determine the total impedance in both rectangular and polar forms.

(d) Determine the complex conjugate of Z and compute the product $Z \, Z^*$.

5-24. In the circuit shown in Fig. P5.24, the impedances of the various components are $Z_R = R \, \Omega$, $Z_L = j \, X_L \, \Omega$, and $Z_C = \dfrac{1}{j} \, X_C$, where $j = \sqrt{-1}$. Suppose $R = 120 \, \Omega$, $X_L = 120\sqrt{3} \, \Omega$, and $X_C = \dfrac{1}{50\sqrt{3}} \, \Omega$.

(a) Express the impedance Z_R, Z_L, and Z_C as complex numbers in both rectangular and polar forms.

(b) Now, suppose that the total impedance of the circuit is $Z = Z_C + \dfrac{Z_R Z_L}{Z_R + Z_L}$. Determine the total impedance and express the result in both rectangular and polar forms.

(c) Determine the complex conjugate of Z and compute the product $Z \, Z^*$.

Figure P5.24   A capacitor connected in series with a parallel combination of resistor and inductor for problem P5-24.

5-25. In the RC circuit shown in Fig. P5.25, the impedances of R and C are given as $Z_R = R\ \Omega$ and $Z_C = -j\ X_C\ \Omega$, where $j = \sqrt{-1}$. Suppose $R = 100\ \Omega$ and $X_C = 50\ \Omega$.

(a) Express the impedances Z_R and Z_C as complex numbers in both rectangular and polar forms.

(b) Find the total impedance $Z = Z_R + Z_C$ as a complex number in both rectangular and polar forms.

(c) If $V = 100\ \angle 0°\,$V, find the current $I = \dfrac{V}{Z}$ as a complex number in both rectangular and polar forms.

(d) Knowing the current in part (c), find the voltage phasors $V_R = I\,Z_R$ and $V_C = I\,Z_C$ in both rectangular and polar forms.

(e) Show that the KVL is satisfied for the circuit shown in Fig. P5.25; in other words, show $V = V_R + V_C$.

Figure P5.25   An RC circuit for problem P5-25.

5-26. In the RL circuit shown in Fig. P5.26, the impedances of R and L are given as $Z_R = R\ \Omega$ and $Z_C = j\ X_L\ \Omega$, where $j = \sqrt{-1}$. Suppose $R = 100\ \Omega$ and $X_L = 50\ \Omega$.

Figure P5.26   An RL circuit for problem P5-26.

(a) Express the impedances Z_R and Z_L as complex numbers in both rectangular and polar forms.

(b) Find the total impedance $Z = Z_R + Z_L$ as a complex number in both rectangular and polar forms.

(c) If $V = 100\ \angle 0°\,$V, find the current $I = \dfrac{V}{Z}$ as a complex number in both rectangular and polar forms.

(d) Knowing the current in part (c), find the voltage phasors $V_R = I\,Z_R$ and $V_L = I\,Z_L$ in both rectangular and polar forms.

(e) Show that the KVL is satisfied for the circuit shown in Fig. P5.26; in other words, show $V = V_R + V_L$.

5-27. A resistor, capacitor, and inductor are connected in series as shown in Fig. P5.27. The total impedance of the circuit is $Z = Z_R + Z_L + Z_C$, where $Z_R = R\ \Omega$, $Z_L = j\omega L\ \Omega$, and $Z_C = \dfrac{1}{j\omega C}\ \Omega$. For a particular design, $R = 100\ \Omega$, $L = 500$ mH, $C = 25\ \mu$F, and $\omega = 120\ \pi$ rad/s.

Figure P5.27   RLC circuit for problem P5-27.

(a) Express the impedances Z_R, Z_L, and Z_C as complex numbers in both rectangular and polar forms.

(b) Determine the total impedance Z in both its rectangular and polar forms.

(c) If $V = 100 \angle 0°$ V, find the current $I = \dfrac{V}{Z}$ as a complex number in both rectangular and polar forms.

(d) Show that the KVL is satisfied for the circuit shown in Fig. P5.27 (i.e., show $V = V_R + V_L + V_C$).

5-28. In the current divider circuit shown in Fig. P5.28, the sum of the current phasors I_1 and I_2 is equal to the total current phasor I (i.e., $I = I_1 + I_2$). Suppose $I_1 = 1 \angle 0°$ and $I_2 = 1 \angle 90°$, both measured in mA.

Figure P5.28   A current divider circuit for problems P5-28 and P5-29.

(a) Write I_1 and I_2 in rectangular form.

(b) Determine the total current phasor I in both rectangular and polar forms.

(c) Suppose $Z_R = 1000\ \Omega$ and $Z_C = \dfrac{10^3}{j}\ \Omega$. Write V_R and V_C as complex numbers in both rectangular and polar forms if $V_R = I_1 \times Z_R$ and $V_C = I_2 \times Z_C$.

5-29. In the current divider circuit shown in Fig. P5.28, the currents flowing through the resistor I_1 and capacitor I_2 are given by

$$I_1 = \frac{Z_C}{Z_R + Z_C} I$$

$$I_2 = \frac{Z_R}{Z_R + Z_C} I$$

where $Z_R = R\ \Omega$ is the impedance of the resistor and $Z_C = \dfrac{X_C}{j}\ \Omega$ is the impedance of the capacitor.

(a) If $R = 1\ \text{k}\Omega$ and $X_C = 10^3\ \Omega$, express Z_R and Z_C as complex numbers in both rectangular and polar forms.

(b) If $I = 1$ mA, determine I_1 and I_2 in both rectangular and polar forms.

(c) Show $I = I_1 + I_2$.

5-30. In the current divider circuit shown in Fig. P5.30, the sum of the current phasors I_1 and I_2 is equal to the total current phasor I (i.e., $I = I_1 + I_2$).

Figure P5.30   A current divider circuit for problems P5-30 and P5-31.

Suppose $I_1 = \sqrt{2} \angle 45°$ and $I_2 = \sqrt{2} \angle -45°$, both measured in mA.

(a) Write I_1 and I_2 in rectangular form.

(b) Determine the total current phasor I in both rectangular and polar forms.

(c) Suppose $Z_R = 1000\ \Omega$ and $Z_L = j\,1000\ \Omega$. Write V_R and V_L as complex numbers in both rectangular and polar forms if $V_R = I_1 \times Z_R$ and $V_L = I_2 \times Z_L$.

5-31. In the current divider circuit shown in Fig. P5.30, the currents flowing through the resistor I_1 and inductor I_2 are

given by

$$I_1 = \frac{Z_L}{Z_R + Z_L} I$$

$$I_2 = \frac{Z_R}{Z_R + Z_L} I$$

where $Z_R = R\,\Omega$ is the impedance of the resistor and $Z_L = jX_L\,\Omega$ is the impedance of the inductor.

(a) If $R = 1\,k\Omega$ and $X_L = 10^3\,\Omega$, express Z_R and Z_L as complex numbers in both rectangular and polar forms.

(b) If $I = 1\,mA$, determine I_1 and I_2 in both rectangular and polar forms.

(c) Show $I = I_1 + I_2$.

5-32. In the Op–Amp circuit shown in Fig. P5.32, the output voltage V_o is given by

$$V_o = -\frac{Z_C}{Z_R} V_{in}$$

where $Z_R = R\,\Omega$ is the impedance of the resistor and $Z_C = -jX_C\,\Omega$ is the impedance of the capacitor.

(a) If $R = 2\,k\Omega$ and $X_C = 1\,k\Omega$, express Z_R and Z_C as complex numbers in both rectangular and polar forms.

(b) If $V_{in} = 10\,\angle 0°\,V$, determine V_o in both rectangular and polar forms.

Figure P5.32   An Op–Amp circuit for problem P5-32.

5-33. In the Op–Amp circuit shown in Fig. P5.33, the output voltage V_o is given by

$$V_o = -\frac{Z_R}{Z_C} V_{in}$$

where $Z_R = R\,\Omega$ is the impedance of the resistor and $Z_C = -jX_C\,\Omega$ is the impedance of the capacitor.

(a) If $R = 1\,k\Omega$ and $X_C = 2\,k\Omega$, express Z_R and Z_C as complex numbers in both rectangular and polar forms.

(b) If $V_{in} = 5\,\angle 90°\,V$, determine V_o in both rectangular and polar forms.

Figure P5.33   An Op–Amp circuit for problem P5-33.

5-34. In the Op–Amp circuit shown in Fig. P5.34, the output voltage V_o is given by

$$V_o = -\frac{Z_{R2} + Z_C}{Z_{R1}} V_{in}$$

where $Z_{R1} = R1\,\Omega$ is the impedance of the resistor $R1$, $Z_{R2} = R2\,\Omega$ is the impedance of the resistor $R2$, and $Z_C = -jX_C\,\Omega$ is the impedance of the capacitor.

(a) If $R1 = 1\,k\Omega$, $R2 = 2\,k\Omega$, and $X_C = 2\,k\Omega$, express Z_{R1}, Z_{R2}, and Z_C as complex numbers in both rectangular and polar forms.

(b) If $V_{in} = 2\,\angle 45°\,V$, determine V_o in both rectangular and polar forms.

Figure P5.34 An Op–Amp circuit for problem P5-34.

5-35. In the Op–Amp circuit shown in Fig. P5.35, the output voltage V_o is given by

$$V_o = -\frac{Z_{R2}}{Z_{R1} + Z_C} V_{in}$$

where $Z_{R1} = R1\ \Omega$ is the impedance of the resistor $R1$, $Z_{R2} = R2\ \Omega$ is the impedance of the resistor $R2$, and $Z_C = -jX_C\ \Omega$ is the impedance of the capacitor.

(a) If $R1 = 1.5\,\text{k}\Omega$, $R2 = 1\,\text{k}\Omega$ and $X_C = 0.5\ \text{k}\Omega$, express Z_{R1}, Z_{R2}, and Z_C as complex numbers in both rectangular and polar forms.

(b) If $V_{in} = 1\angle-45°\text{V}$, determine V_o in both rectangular and polar forms.

Figure P5.35 An Op–Amp circuit for problem P5-35.

5-36. In the Op–Amp circuit circuit shown in Fig. P5.36, the output voltage V_o is

given by

$$V_o = -\frac{Z_{R2} + Z_{C2}}{Z_{R1} + Z_{C1}} V_{in}$$

where $Z_{R1} = R1\ \Omega$ is the impedance of the resistor $R1$, $Z_{R2} = R2\ \Omega$ is the impedance of the resistor $R2$, $Z_{C1} = -jX_{C1}\ \Omega$ is the impedance of the capacitor $C1$, and $Z_{C2} = -jX_{C2}\ \Omega$ is the impedance of the capacitor $C2$.

(a) If $R1 = 10\,\text{k}\Omega$, $R2 = 5\,\text{k}\Omega$, $X_{C1} = 2.5\ \text{k}\Omega$, and $X_{C2} = 5\ \text{k}\Omega$, express Z_{R1}, Z_{R2}, Z_{C1}, and Z_{C2} as complex numbers in both rectangular and polar forms.

(b) If $V_{in} = 1.5\angle0°\text{V}$, determine V_o in both rectangular and polar forms.

Figure P5.36 An Op–Amp circuit for problem P5-35.

5-37. The Delta-to-Yee (Δ-Y) and Yee-to-Delta (Y-Δ) conversions are used in electrical circuits to find the equivalent impedances of complex circuits. In the circuit shown in Fig. 5.37, the impedances Z_a, Z_b, and Z_c connected in the Delta interconnection can be converted into their equivalent impedances Z_1, Z_2, and Z_3 connected in the Yee interconnection. The impedances Z_1, Z_2, and Z_3 can be written in term of impedances Z_a, Z_b, and Z_c as

$$Z_1 = \frac{Z_a Z_b}{Z_a + Z_b + Z_c}$$

$$Z_2 = \frac{Z_a Z_c}{Z_a + Z_b + Z_c}$$

$$Z_3 = \frac{Z_c Z_b}{Z_a + Z_b + Z_c}. \qquad (5.7)$$

(a) If $Z_a = 20\ \Omega$, $Z_b = -j\,10\ \Omega$, and $Z_c = 20 + j\,50\ \Omega$, express Z_a, Z_b, and Z_c as complex numbers in both rectangular and polar forms.

(b) Determine Z_1, Z_2, and Z_3 in both rectangular and polar forms.

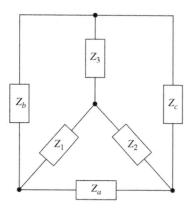

Figure P5.37 Impedances connected in Delta and Yee interconnections.

5-38. Repeat problem P5-37 if $Z_a = 10\ \Omega$, $Z_b = j\,20\ \Omega$, and $Z_c = 20 + j\,10\ \Omega$.

5-39. In the circuit shown in Fig. 5.37, the impedances Z_1, Z_2, and Z_3 connected in the Yee interconnection can be converted into their equivalent impedances Z_a, Z_b, and Z_c connected in the Delta interconnection as

$$Z_a = \frac{Z_1 Z_2 + Z_2 Z_3 + Z_3 Z_1}{Z_3}$$

$$Z_b = \frac{Z_1 Z_2 + Z_2 Z_3 + Z_3 Z_1}{Z_2}$$

$$Z_c = \frac{Z_1 Z_2 + Z_2 Z_3 + Z_3 Z_1}{Z_1}. \qquad (5.8)$$

(a) If $Z_1 = 2.5 - j\,2.5\ \Omega$, $Z_2 = 17.5 + j\,7.5\ \Omega$, and $Z_3 = 3.75 - j\,8.75\ \Omega$, express Z_1, Z_2, and Z_3 as complex numbers in polar form.

(b) Determine Z_a, Z_b, and Z_c in both rectangular and polar forms.

5-40. Repeat problem P5-39 if $Z_1 = 3.33 + j\,3.33\ \Omega$, $Z_2 = 5.0 - j\,1.66\ \Omega$, and $Z_3 = 3.5 + j\,10.6\ \Omega$.

Sinusoids in Engineering

A sinusoid is a signal that describes a smooth repetitive motion of an object that oscillates at a constant rate (frequency) about an equilibrium point. The sinusoid has the form of a sine (sin) or a cosine (cos) function (discussed in Chapter 3) and has applications in all engineering disciplines. These functions are the most important signals because all other signals can be constructed from sine and cosine signals. A few examples of a sinusoid are the motion of a one-link planar robot rotating at a constant rate, the oscillation of an undamped spring-mass system, and the voltage waveform of an electric power source. For example, the frequency of the voltage waveform associated with electrical power in North America is 60 cycles per second (Hz), whereas in many other parts of the world this frequency is 50 Hz. In this chapter, the example of a one-link robot rotating at a constant rate will be used to develop the general form of a sinusoid and explain its amplitude, frequency (both linear and angular), phase angle, and phase shift. The sum of sinusoids of the same frequency will also be explained in the context of both electrical and mechanical systems.

6.1 ONE-LINK PLANAR ROBOT AS A SINUSOID

A one-link planar robot of length l and angle θ is shown in Fig. 6.1. It was shown in Chapter 3 that the tip of the robot has coordinates $x = l \cos \theta$ and $y = l \sin \theta$. Varying θ from 0 to 2π radians and assuming $l = 1$ (l has units of x and y), the plots of $y = l \sin \theta$ and $x = l \cos \theta$ are shown in Figs. 6.2 and 6.3, respectively.

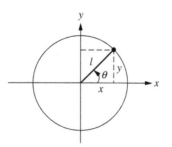

Figure 6.1 A one-link planar robot.

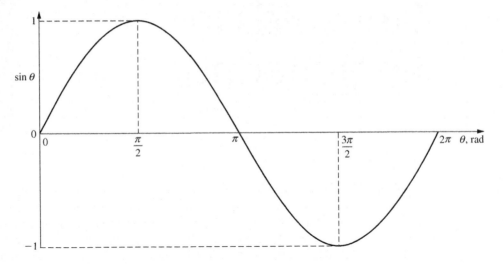

Figure 6.2 y-coordinate of one-link robot with $l = 1$ (sine function).

It can be seen from Fig. 6.2 that $\sin\theta$ goes from 0 (for $\theta = 0$) to 1 (for $\theta = \pi/2$) and back to 0 (for $\theta = \pi$) to -1 (for $\theta = 3\pi/2$) back to 0 (for $\theta = 2\pi$), thus completing one full cycle. From Fig. 6.3, it can be seen that $\cos\theta$ goes from 1 (for $\theta = 0$) to 0 (for $\theta = \pi/2$) to -1 (for $\theta = \pi$) to 0 (for $\theta = 3\pi/2$) and back to 1 (for $\theta = 2\pi$), thus completing one full cycle. Note that the minimum value of the sin and cos functions is -1 and the maximum value of the sin and cos functions is 1.

Figures 6.4 and 6.5 show two cycles of the sine and cosine functions, respectively, and it can be seen from these figures that $\sin(\theta + 2\pi) = \sin(\theta)$ and $\cos(\theta + 2\pi) = \cos(\theta)$, for $0 \le \theta \le 2\pi$. Similarly, the plot of sine and cosine functions will complete

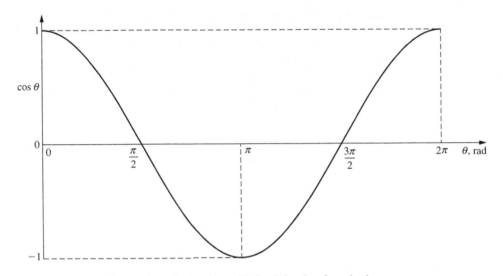

Figure 6.3 x-coordinate of one-link robot with $l = 1$ (cosine function).

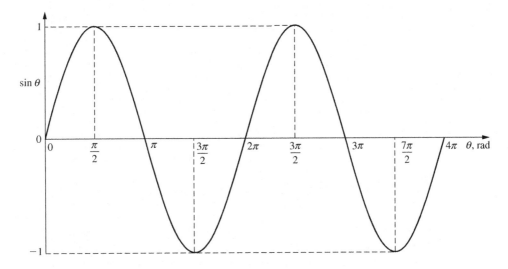

Figure 6.4 Two cycles of the sine function.

another cycle from 4π to 6π and so on for every 2π. Thus, both the sine and cosine functions are called periodic functions with period $T = 2\pi$ rad. Since $\pi = 180°$, the period of the sine and cosine functions can also be written as $360°$.

6.2 ANGULAR MOTION OF THE ONE-LINK PLANAR ROBOT

Suppose now that the one-link planar robot shown in Fig. 6.1 is rotating with an angular frequency ω, as shown in Fig. 6.6.

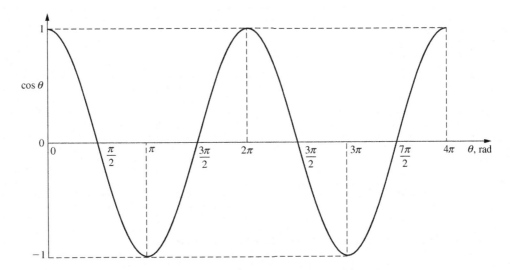

Figure 6.5 Two cycles of the cosine function.

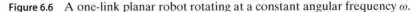

Figure 6.6 A one-link planar robot rotating at a constant angular frequency ω.

The angle traveled in time t is given by $\theta = \omega t$. Therefore, $y(t) = l\sin(\theta) = l\sin(\omega t)$ and $x(t) = l\cos(\theta) = l\cos(\omega t)$. Suppose the robot starts from $\theta = 0$ at time $t = 0$ s and takes $t = 2\pi$s to complete one revolution. Since $\theta = \omega t$, the angular frequency is $\omega = \dfrac{\theta}{t} = \dfrac{2\pi\,\text{rad}}{2\pi\,\text{s}} = 1$ rad/s, and the time period to complete one cycle is $T = 2\pi$s. The resulting plots of $y = l\sin t$ and $x = l\cos t$ are shown in Figs. 6.7 and 6.8, respectively. The x- and y-components oscillate between l and $-l$, where the length l is the *amplitude* of the sinusoids.

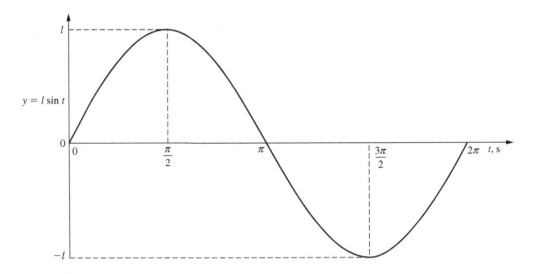

Figure 6.7 The y-component of the one-link planar robot completing one cycle in 2π s.

6.2.1 Relations between Frequency and Period

In Figs. 6.7 and 6.8, it took 2πs to complete one cycle of the sinusoidal signals, and it was found that $\omega = 1$ rad/s (i.e., in 1 s, the robot went through 1 radian of rotation).

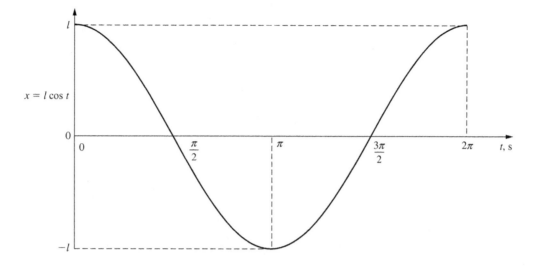

Figure 6.8   The x-component of the one-link planar robot completing one cycle in 2πs.

Since one revolution (cycle) $= 2\pi$rad, a robot rotating at 1 rad/s would go through $1/2\pi$ cycles in 1s. This is called the linear frequency or simply frequency f, with units of cycle/s (s^{-1}). Therefore, the relationship between the **angular frequency** ω and **linear frequency** f is given by

$$\omega = 2\pi f.$$

By definition, the period T is defined as the number of seconds per cycle, which means f is the reciprocal of T. In other words,

$$f = \frac{1}{T}.$$

Since $\omega = 2\pi f$,

$$\omega = \frac{2\pi}{T}.$$

Solving for T gives

$$T = \frac{2\pi}{\omega}.$$

The above relations allow the computation of f, ω, or T when only *one* of the three is given. For example, assume that a one-link robot of length l goes through two revolutions (i.e., 4πrad) in 2πs. The angular frequency is $\omega = \dfrac{4\pi\text{rad}}{2\pi\text{s}} = 2$ rad/s, the period is $T = 2\pi/2 = \pi$s, and the frequency is $f = 2/2\pi = 1/\pi$Hz. Therefore, $y(t) = l\sin(\omega t) = l\sin(2t)$ and $x(t) = l\cos(\omega t) = l\cos(2t)$; and their plots are as shown in Figs. 6.9 and 6.10, respectively.

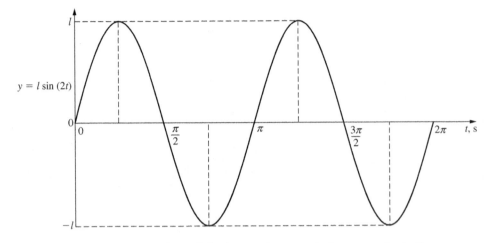

Figure 6.9 The y-component of one-link planar robot completing two cycles in 2π s or $\omega = 2$ rad/s.

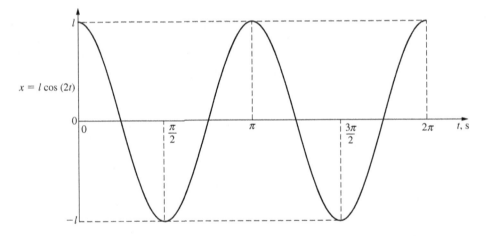

Figure 6.10 The x-component of one-link planar robot completing one cycle in 2π s or $\omega = 2$ rad/s.

6.3   PHASE ANGLE, PHASE SHIFT, AND TIME SHIFT

Suppose now that a robot of length $l = 10$ in. starts rotating from an initial position $\theta = \pi/8$ rad and takes $T = 1$ s to complete one revolution, as shown in Fig. 6.11. At any time t, the x- and y-components are given by

$$x(t) = l\cos\theta$$
$$= l\cos(\omega t + \phi)$$
$$= l\cos\left(\omega t + \frac{\pi}{8}\right)$$

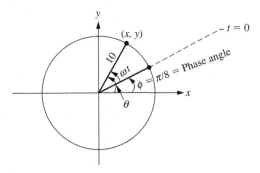

Figure 6.11 One-link planar robot starting rotation from an angle of $\pi/8$ rad.

and

$$y(t) = l \sin \theta$$
$$= l \sin (\omega t + \phi)$$
$$= l \sin \left(\omega t + \frac{\pi}{8} \right).$$

Since $l = 10$, $\omega = 2\pi/T = 2\pi/1 = 2\pi$ rad/s, and $\phi = \pi/8$ rad, the x- and y-components of the one-link robot are given by

$$x(t) = 10 \cos \left(2\pi t + \frac{\pi}{8} \right)$$

and

$$y(t) = 10 \sin \left(2\pi t + \frac{\pi}{8} \right), \tag{6.1}$$

where $\phi = \pi/8$ is called the *phase angle*. Since ϕ represents a shift from the zero phase to a phase of $\pi/8$ rad, it is sometimes called a phase shift. Therefore, the phase shift is a shift in radians or degrees. If the phase angle is positive, the sinusoid shifts to the left as shown in Fig. 6.12, but if the phase angle is negative, the sinusoid will shift to the right. For example, the value of the sinusoid given by equation (6.1) is not zero at time $t = 0$. Since the phase angle is positive, the sinusoid given by equation (6.1) is shifted to the left.

The *time shift* is the time it takes the robot moving at a speed ω to pass through the phase shift ϕ. Setting $\theta = \omega t$ gives

$$\text{time shift} = \frac{\text{Phase angle}}{\text{Angular frequency}} = \frac{\phi}{\omega}$$
$$= \frac{\frac{\pi}{8}}{2\pi}$$
$$= \frac{1}{16} \text{ s.}$$

Note that the phase angle used to calculate the time shift must be in rad.

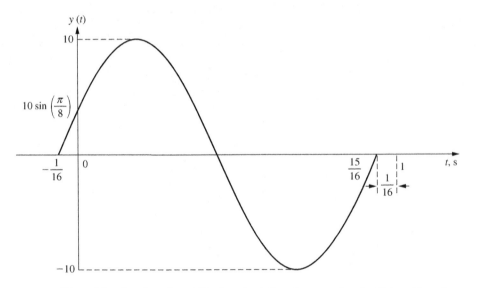

Figure 6.12 Plot of the sine function shifted to the left a phase angle of $\pi/8$ (positive phase angle).

6.4 GENERAL FORM OF A SINUSOID

The general expression of a sinusoid is

$$x(t) = A \sin(\omega t + \phi) \tag{6.2}$$

where A is the amplitude, ω is the angular frequency, and ϕ is the phase angle.

Example 6-1 Consider a cart of mass m moving on frictionless rollers as shown in Fig. 6.13. The mass is attached to the end of a spring of stiffness k.

Figure 6.13 Harmonic motion of a spring-mass system.

Suppose that the position of the mass $x(t)$ is given by

$$x(t) = 2 \sin\left(6\pi t + \frac{\pi}{2}\right) \text{ m.} \tag{6.3}$$

(a) Determine the amplitude, linear and angular frequencies, period, phase angle, and time shift.

(b) Find $x(t)$ at $t = 2.0$ s.

(c) Find the time required for the system to reach its maximum negative displacement.

(d) Plot the displacement $x(t)$ for $0 \le t \le 3$ s.

Solution   (a) Comparing the position of the mass $x(t) = 2 \sin\left(6\pi t + \dfrac{\pi}{2}\right)$ to the general expression of equation (6.2) gives

$$\text{Amplitude } A = 2 \text{ m}$$

$$\text{Angular frequency } \omega = 6\pi \text{ rad/s}$$

$$\text{Phase angle } \phi = \dfrac{\pi}{2} \text{ rad.}$$

Since the angular frequency is $\omega = 2\pi f$, the linear frequency f is given by

$$f = \frac{\omega}{2\pi} = \frac{6\pi}{2\pi} = 3 \text{ Hz},$$

and the period of the harmonic motion is given by

$$T = \frac{1}{f} = \frac{1}{3} \text{ s.}$$

The time shift can be determined from the phase angle and the angular frequency as

$$\text{Time shift} = \frac{\phi}{\omega}$$

$$= \left(\frac{\pi}{2}\right)\left(\frac{1}{6\pi}\right)$$

$$= \frac{1}{12} \text{ s.}$$

(b) To find $x(t)$ at $t = 2.0$ s, substitute $t = 2.0$ in equation (6.3), which gives

$$x(2) = 2 \sin\left(6\pi(2) + \frac{\pi}{2}\right)$$

$$= 2 \sin\left(12\pi + \frac{\pi}{2}\right)$$

$$= 2 \sin(12.5\,\pi)$$

$$= 2 \sin(12.5\,\pi - 12\,\pi)$$

$$= 2 \sin\left(\frac{\pi}{2}\right)$$

$$= 2.0 \text{ m.}$$

In obtaining $x(2)$, 12π was subtracted from the angle 12.5π to find the value of $\sin(12.5\pi)$. It should be noted that an integer multiple of 2π can always be added or subtracted from the argument of sine and cosine functions. This is done because the sine and cosine functions are periodic with a period of 2π.

(c) The displacement reaches the first maximum negative displacement of -2 m when $\sin(\theta) = \sin\left(6\pi t + \dfrac{\pi}{2}\right) = -1$ or $\theta = 6\pi t + \dfrac{\pi}{2} = \dfrac{3\pi}{2}$. Solving for t gives

$$6\pi t + \frac{\pi}{2} = \frac{3\pi}{2}$$

$$\Rightarrow \quad 6\pi t = \frac{3\pi}{2} - \frac{\pi}{2}$$

$$= \pi$$

$$\Rightarrow \quad t = \frac{\pi}{6\pi}$$

or

$$t = \frac{1}{6} \text{ s}.$$

(d) The plot of the displacement $x(t)$ for $0 \leq t \leq 3$ s is shown in Fig. 6.14.

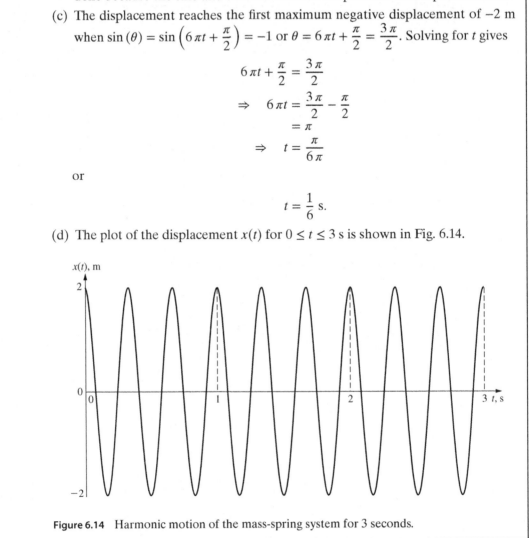

Figure 6.14   Harmonic motion of the mass-spring system for 3 seconds.

6.5   ADDITION OF SINUSOIDS OF THE SAME FREQUENCY

Adding two sinusoids of the same frequency but different amplitudes and phases results in another sinusoid (sin or cos) of the same frequency. The resulting amplitude and phase are different from the amplitude and phase of the two original sinusoids, as illustrated with the following example.

Example
6-2
Consider an electrical circuit with two elements R and L connected in series as shown in Fig. 6.15.

Figure 6.15 Addition of sinusoids in an RL circuit.

In Fig. 6.15, the current $i(t) = 6\sin(2t)$ A flowing through the circuit produces two voltages: $v_R = 6\sin(2t)$ V across the resistor and $v_L = 8\cos(2t)$ V across the inductor. The total voltage $v(t)$ across the current source can be obtained using KVL as

$$v(t) = v_R(t) + v_L(t),$$

or

$$v(t) = 6\sin(2t) + 8\cos(2t) \text{ V}. \tag{6.4}$$

The total voltage given by equation (6.4) can be written as one sinusoid (sine or cosine) of frequency 2 rad/s. The objective is to find the amplitude and the phase angle of the resulting sinusoid. In terms of a sine function,

$$v(t) = 6\sin(2t) + 8\cos(2t) = M\sin(2t + \phi), \tag{6.5}$$

where the objective is to find M and ϕ. Using the trigonometric identity $\sin(A + B) = \sin(A)\cos(B) + \cos(A)\sin(B)$ on the right side, equation (6.5) can be written as

$$6\sin(2t) + 8\cos(2t) = (M\cos\phi)\sin(2t) + (M\sin\phi)\cos(2t). \tag{6.6}$$

Equating the coefficients of $\sin(2t)$ and $\cos(2t)$ on both sides of equation (6.6) gives

$$\text{sines}: \qquad M\cos(\phi) = 6 \tag{6.7}$$
$$\text{cosines}: \qquad M\sin(\phi) = 8. \tag{6.8}$$

To determine the magnitude M and phase ϕ, equations (6.7) and (6.8) are converted to polar form as shown in Fig. 6.16. Therefore,

$$M = \sqrt{6^2 + 8^2}$$
$$= 10$$
$$\phi = \text{atan2}(8, 6)$$
$$= 53.13°.$$

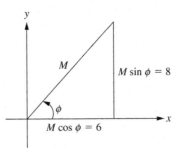

Figure 6.16   Determination of magnitude and phase of the resulting sinusoid in an RL circuit.

Therefore, $v(t) = 6\sin(2t) + 8\cos(2t) = 10\sin(2t + 53.13°)$ V. The amplitude of the voltage sinusoid is 10 V, the angular frequency is $\omega = 2$ rad/s, the linear frequency is $f = \omega/2\pi = 2/2\pi = 1/\pi$ Hz, the period is $T = \pi = 3.142$ s, the phase angle $=$ $53.13° = (53.13°)(\pi\text{rad}/180°) = 0.927$ rad, and the time shift can be calculated as

$$t = \frac{\phi}{\omega}$$
$$= \frac{0.927}{2}$$
$$= 0.464 \text{ s}.$$

The plot of the voltage and current waveforms is shown in Fig. 6.17. It can be seen from Fig. 6.17 that the voltage waveform is shifted to the left by 0.464 s (time shift).

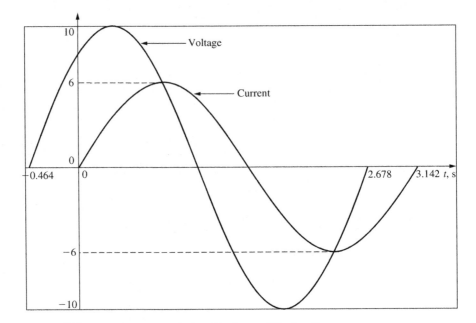

Figure 6.17   Voltage and current relationship for an RL circuit.

In other words, the voltage in the RL circuit *leads* the current by 53.3°. It will be shown later that it is opposite in an RC circuit, where the voltage waveform *lags* the current.

Note: The voltage $v(t)$ can also be represented as one sinusoid using the cosine function as

$$v(t) = 6\sin(2t) + 8\cos(2t) = M\cos(2t + \phi_1). \tag{6.9}$$

The amplitude M and the phase angle ϕ_1 can be determined using a procedure similar to that outlined above. Using the trigonometric identity $\cos(A + B) = \cos(A)\cos(B) - \sin(A)\sin(B)$ on the right side of equation (6.9) gives

$$6\sin(2t) + 8\cos(2t) = (-M\sin\phi_1)\sin(2t) + (M\cos\phi_1)\cos(2t). \tag{6.10}$$

Equating the coefficients of $\sin(2t)$ and $\cos(2t)$ on both sides of equation (6.10) gives

$$\text{sines}: \qquad M\sin(\phi_1) = -6 \tag{6.11}$$
$$\text{cosines}: \qquad M\cos(\phi_1) = 8. \tag{6.12}$$

To determine the magnitude M and phase ϕ_1, equations (6.11) and (6.12) are converted to polar form as shown in Fig. 6.18.

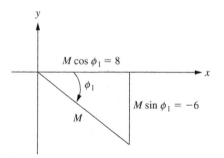

Figure 6.18   Determination of magnitude and phase for a cosine function.

Therefore,

$$M = \sqrt{6^2 + 8^2}$$
$$= 10$$
$$\phi_1 = \text{atan2}(-6, 8)$$
$$= -36.87°$$

Therefore, $v(t) = 6\sin(2t) + 8\cos(2t) = 10\cos(2t - 36.87°)$ V.

This expression can also be obtained directly from the sine function using the trig identity $\sin\theta = \cos(\theta - 90°)$. Therefore,

$$10\sin(2t + 53.13°) = 10\cos((2t + 53.13°) - 90°)$$
$$= 10\cos(2t - 36.87°).$$

In general, the results of this example can be expressed as follows:

$$A \cos \omega t + B \sin \omega t = \sqrt{A^2 + B^2} \cos(\omega t - \operatorname{atan2}(B, A))$$
$$A \cos \omega t + B \sin \omega t = \sqrt{A^2 + B^2} \sin(\omega t + \operatorname{atan2}(A, B)).$$

Example 6-3

Consider the RC circuit shown in Fig. 6.19.

Figure 6.19   Addition of sinusoids in an RC circuit.

In Fig. 6.19, the current $i(t) = \cos(120\pi t)$ mA flowing through the circuit produces two voltages: $v_R = 10 \cos(120\pi t)$ across the resistor and $v_C = 5 \sin(120\pi t)$ across the capacitor. The total voltage $v(t)$ can be obtained using KVL as

$$v(t) = v_R(t) + v_C(t)$$
$$= 10 \cos(120\pi t) + 5 \sin(120\pi t). \tag{6.13}$$

The total voltage given by equation (6.13) can be written as a single sinusoid of frequency 120π rad/s. The objective is to find the amplitude and the phase angle of the sinusoid. The total voltage can be written as a cosine function as

$$10 \cos(120\pi t) + 5 \sin(120\pi t) = M \cos(120\pi t + \phi_2)$$
$$= (M\cos\phi_2)\cos(120\pi t) + (-M\sin\phi_2)\sin(120\pi t), \tag{6.14}$$

where the trigonometric identity $\cos(A + B) = \cos A \cos B - \sin A \sin B$ is employed on the right-hand side of the equation. Equating the coefficients of $\sin(120\pi t)$ and $\cos(120\pi t)$ on both sides of equation (6.14) gives

$$\text{cosines}: \qquad M \cos(\phi_2) = 10 \tag{6.15}$$
$$\text{sines}: \qquad M \sin(\phi_2) = -5. \tag{6.16}$$

To determine the magnitude M and phase angle ϕ_2, equations (6.15) and (6.16) are converted to the polar form as shown in Fig. 6.20. Therefore,

$$M = \sqrt{10^2 + 5^2}$$
$$= 11.18$$
$$\phi_2 = \operatorname{atan2}(-5, 10)$$
$$= -26.57°,$$

which gives $v(t) = 11.18 \cos(120\pi t - 26.57°)$ V.

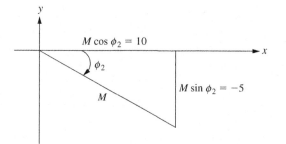

Figure 6.20 Determination of magnitude and phase of the resulting sinusoid in an RC circuit.

The amplitude of the voltage sinusoid is 11.8 V, the angular frequency is $\omega = 120\,\pi$ rad/s, the linear frequency is $f = \omega/2\pi = 60$ Hz, the period is $T = 16.7$ ms, the phase angle is $-26.57°$, and the time shift is 1.23 ms.

The plot of the voltage and current waveforms is shown in Fig. 6.21. It can be seen from the figure that the voltage waveform is shifted to the right by 1.23 ms (time shift). In other words, the voltage in this RC circuit *lags* the current by 26.57°.

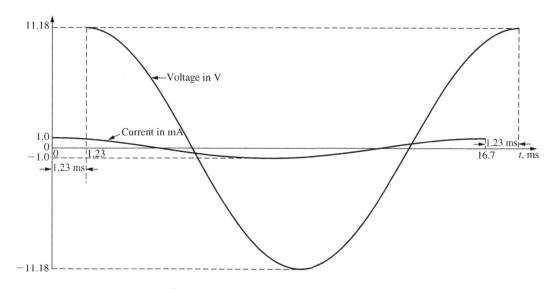

Figure 6.21 Voltage and current relationship for an RC circuit.

<table>
<tr><td>Example
6-4</td><td>A hip implant is subjected to a cyclic load during a fatigue test, as shown in Fig. 6.22. The load applied to the hip implant is given by</td></tr>
</table>

$$F(t) = 250 \sin(6\,\pi t) + 1250 \text{ N}$$

(a) Write down the amplitude, frequency (in Hertz), period (in seconds), phase angle (in degrees), time shift (in seconds), and vertical shift (in Newtons) of the load profile.

(b) Plot one cycle of $F(t)$ and indicate the earliest time when the force reaches its maximum value.

Figure 6.22   Hip implant subjected to a cyclic load.

Solution   (a) Comparing the force $F(t) = 250 \sin(\pi t) + 1250$ to the general form of equation (6.2) gives

$$\text{amplitude} = 250 \text{ N.}$$

$$\text{Frequency}: \omega = 2\pi f$$

$$= 6\pi \qquad \Rightarrow f = 3 \text{ Hz.}$$

$$\text{Period}: \quad T = \frac{1}{f}$$

$$= \frac{1}{3} \text{ s.}$$

$$\text{Phase angle}: \quad \phi = 0°.$$

$$\text{Time shift}: \quad 6\pi t = 0 \qquad \Rightarrow \quad t = 0 \text{ s (There is no time shift.)}$$

$$\text{Vertical shift} = 1250 \text{ N.}$$

(b) A plot of one cycle of the cyclic force $F(t)$ is shown in Fig. 6.23. It can be seen from this figure that the earliest time the force has a maximum value of 1500 N is at time $t_{max} = 1/12$ s. The times when the force $F(t)$ is maximum can also be found analytically as

$$250 \sin(6\pi t_{max}) + 1250 = 1500 \quad \Rightarrow \quad 250 \sin(6\pi t_{max})$$
$$= 250 \quad \Rightarrow \quad \sin(6\pi t_{max}) = 1.$$

Therefore,

$$6\pi t_{max} = \frac{\pi}{2}, \frac{3\pi}{2}, \cdots,$$

which gives

$$t_{max} = \frac{1}{12}, \frac{1}{4} \cdots \text{ s}.$$

Therefore, the earliest time the maximum force occurs is $t = 1/12$s.

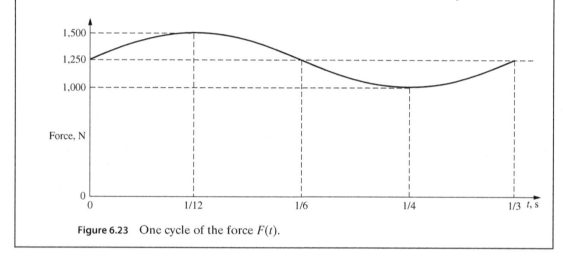

Figure 6.23   One cycle of the force $F(t)$.

PROBLEMS

6-1. The tip of a one-link robot is located at $\theta = 0$ at time $t = 0$ s as shown in Fig. P6.1. It takes 1 s for the robot to move from $\theta = 0$ to $\theta = 2\pi$ rad. If $l = 5$ in., plot the x- and y-components as a function of time. Also find the amplitude, frequency, period, phase angle, and time shift.

6-2. The tip of a one-link robot is located at $\theta = \pi/6$ rad at time $t = 0$ s, as shown in Fig. P6.2. It takes 2 s for the robot to move from $\theta = \pi/6$ rad to $\theta = \pi/6 + 2\pi$ rad. If $l = 10$ in., plot the x- and y-components as a function of time. Also find the amplitude, frequency, period, phase angle, and time shift.

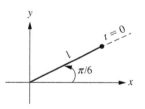

Figure P6.1   Rotating one-link robot starting at $\theta = 0°$.

Figure P6.2   Rotating one-link robot starting at $\theta = 30°$.

6-3. The tip of a one-link robot is located at $\theta = -\pi/4$ rad at time $t = 0$ s as shown in Fig. P6.3. The robot is rotating at an angular frequency of 2π rad/s. If $l = 20$ cm, plot the x- and y-components as a function of time. Also find the amplitude, frequency, period, phase angle, and time shift.

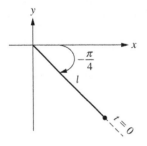

Figure P6.3 Rotating one-link robot starting at $\theta = -45°$.

6-4. The tip of a one-link robot is located at $\theta = \pi/2$ rad at $t = 0$ s as shown in Fig. P6.4. It takes 4 s for the robot to move from $\theta = \pi/2$ rad to $\theta = \pi/2 + 2\pi$ rad. If $l = 10$ cm, plot the x- and y-components as a function of time. Also find the amplitude, frequency, period, phase angle, and time shift.

Figure P6.4 Rotating one-link robot starting at $\theta = 90°$.

6-5. The tip of a one-link robot is located at $\theta = 3\pi/4$ rad at time $t = 0$ s as shown in Fig. P6.5. It takes 2 s for the robot to move from $\theta = 3\pi/4$ rad to $\theta = 3\pi/4 + 2\pi$ rad. If $l = 15$ cm, plot the x- and y-components as a function of time.

Also find the amplitude, frequency, period, phase angle, and time shift.

Figure P6.5 Rotating one-link robot starting at $\theta = 135°$.

6-6. The tip of a one-link robot is located at $\theta = \pi$ rad at time $t = 0$ s as shown in Fig. P6.6. It takes 3 s for the robot to move from $\theta = \pi$ rad to $\theta = 3\pi$ rad. If $l = 5$ cm, plot the x- and y-components as a function of time. Also find the amplitude, frequency, period, phase angle, and time shift.

Figure P6.6 Rotating one-link robot starting at $\theta = 180°$.

6-7. A spring-mass system moving in the y-direction has a sinusoidal motion as shown in Fig. P6.7. Determine the amplitude, period, frequency, and phase angle of the motion. Also, write the expression for $y(t)$.

6-8. A spring-mass system moving in the x-direction has a sinusoidal motion as shown in Fig. P6.8. Determine the amplitude, period, frequency, and phase angle of the motion. Also, write the expression for $x(t)$.

6-9. Repeat problem P6-8 for the sinusoidal motion shown in Fig. P6.9.

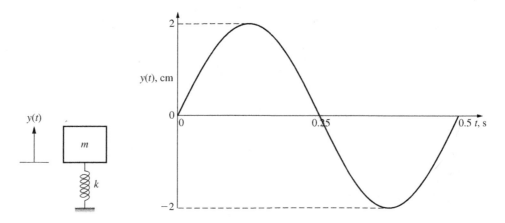

Figure P6.7   Sinusoidal motion of a spring-mass system in the y-direction for problem P6-7.

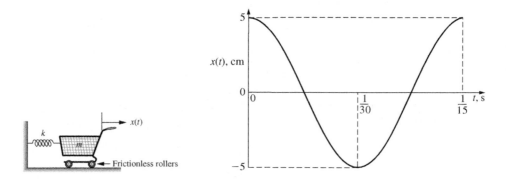

Figure P6.8   Sinusoidal motion of a spring-mass system in the x-direction for problem P6-8.

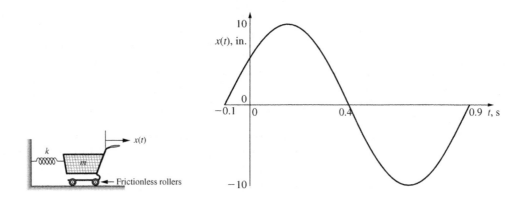

Figure P6.9   Motion of a spring-mass system in the x-direction for problem P6-9.

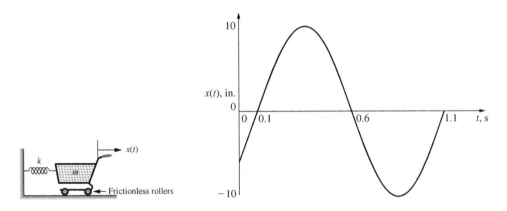

Figure P6.10 Motion of a spring-mass system in the x-direction for problem P6-10.

6-10. Repeat problem P6-8 for the sinusoidal motion shown in Fig. P6.10.

6-11. A spring-mass system is displaced $x = 10$ cm and let go. The system then vibrates under a simple harmonic motion in the horizontal direction; in other words, it travels back and forth from 10 cm to -10 cm. If it takes the system $\pi/2$ s to complete one cycle of the harmonic motion, determine

(a) The amplitude, frequency, and period of the motion.

(b) The time required for the system to reach -10 cm.

(c) Plot one complete cycle of $x(t)$, and indicate the amplitude, period, and time shift on the graph.

6-12. Suppose the spring-mass system of problem P6-11 is displaced -10 cm and takes $\pi/4$ s to complete a cycle.

(a) Find the amplitude, frequency, and period of the motion.

(b) Find the time required for the system to reach the equilibrium point (i.e, $x(t) = 0$).

(c) Plot one complete cycle of $x(t)$, and indicate the amplitude and period on the graph.

Figure P6.13 A spring-mass system for problem P6-13.

6-13. The position of a spring-mass system shown in Fig. P6.13 is given by $x(t) = 8 \sin \left(2 \pi t + \frac{\pi}{4} \right)$ cm.

(a) Find the amplitude, frequency, period, and time shift of the position of the mass.

(b) Find the time required for the system to reach the first maximum displacement.

(c) Plot one complete cycle of $x(t)$, and indicate the amplitude and the time shift on the graph.

6-14. The position of a spring-mass system shown in Fig. P6.13 is given by $x(t) = 10 \sin(4 \pi t - \frac{\pi}{2})$ cm.

(a) Find the amplitude, frequency, period, and time shift of the position $x(t)$.

(b) Find the time required for the system to reach $x(t) = 0$ cm and $x(t) = 10$ cm for the first time (after $t = 0$).

(c) Plot one complete cycle of $x(t)$, and indicate the amplitude and the time shift on the graph.

6-15. The position of a spring-mass system shown in Fig. P6.13 is given by $x(t) = 5\cos(\pi t)$ cm.

   (a) Find the amplitude, frequency, period, and time shift of the position of the mass.

   (b) Find the time required for the system to reach its first maximum negative displacement (i.e., $x(t) = -5$ cm).

   (c) Plot one complete cycle of $x(t)$, and indicate the amplitude and the time shift on the graph.

6-16. A simple pendulum of length $L = 100$ cm is shown in Fig. P6.16. The angular displacement $\theta(t)$ in radians is given by

$$\theta(t) = 0.5\cos\left(\sqrt{\frac{g}{L}}\, t\right).$$

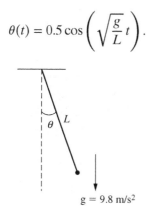

$g = 9.8 \text{ m/s}^2$

Figure P6.16   A simple pendulum.

   (a) Find the amplitude, frequency, and period of oscillation of $\theta(t)$.

   (b) Find the time required for the simple pendulum to reach its first zero angular displacement (i.e., $\theta(t) = 0$).

   (c) Plot one complete cycle of $\theta(t)$, and indicate the amplitude and period on the graph.

6-17. Repeat problem P6-16 if $L = 10$ in. and $\theta(t) = 5\sin\left(\sqrt{\frac{g}{L}}\, t + \frac{\pi}{2}\right)$. Note $g = 32.2$ ft/s^2.

6-18. A sinusoidal current $i(t) = 0.1 \sin(100\,t)$ amps is applied to the RC circuit shown in Fig. P6.18. The voltage across the resistor and capacitor are given by

$$v_R(t) = 20\sin(100\,t) \text{ V}$$
$$v_C(t) = -20\cos(100\,t) \text{ V}$$

where t is in seconds.

Figure P6.18   RC circuit for problem P6-18.

   (a) The voltage applied to the circuit is given by $v(t) = v_R(t) + v_C(t)$. Write $v(t)$ in the form $v(t) = M\sin(100\,t + \theta)$; in other words, find M and θ.

   (b) Suppose now that $v(t) = 28.28 \sin\left(100\,t - \frac{\pi}{4}\right)$ volts. Write down the amplitude, frequency (in Hz), period (in seconds), phase angle (in degrees), and time shift (in seconds) of the voltage $v(t)$.

   (c) Plot one cycle of the voltage $v(t) = 28.28 \sin\left(100\,t - \frac{\pi}{4}\right)$, and indicate the earliest time (after $t = 0$) when voltage is 28.28 V.

6-19. A sinusoidal current $i(t) = 200\cos (120\,\pi t + 14.86°)$ mA is applied to the RC circuit shown in Fig. P6.18. The voltage across the resistor and capacitor are given by

$$v_R(t) = 40\cos(120\,\pi t + 14.86°) \text{ V}$$
$$v_C(t) = 10.61\cos(120\,\pi t - 75.14°) \text{ V}$$

where t is in seconds.

(a) The voltage applied to the circuit is given by $v(t) = v_R(t) + v_C(t)$. Write $v(t)$ in the form $v(t) = M\cos(120\,\pi t + \theta)$; in other words, find M and θ.

(b) Suppose now that $v(t) = 41.38 \cos(120\,\pi t)$ volts. Write down the amplitude, frequency (in Hz), period (in seconds), phase angle (in degrees), and time shift (in seconds) of the voltage $v(t)$.

(c) Plot one cycle of the voltage $v(t) = 41.38\cos(120\,\pi t)$, and indicate the earliest time (after $t = 0$) when voltage is -41.38 V.

6-20. A series RL circuit is subjected to a sinusoidal voltage of frequency $120\,\pi$rad/s, as shown in Fig. P6.20. The current $i(t) = 10\cos(120\,\pi t)$ A is flowing through the circuit. The voltage across the resistor and inductor are given by $v_R(t) = 10\cos(120\,\pi t)$ and $v_L(t) = 12\cos\left(120\,\pi t + \frac{\pi}{2}\right)$ volts, where t is in seconds.

$$R = 1\,\Omega, L = \frac{10}{\pi}\,\text{mH}$$

Figure P6.20   A series RL circuit for problem P6-20.

(a) Write down the amplitude, frequency (in Hz), period (in seconds), phase shift (in degrees), and time shift (in seconds) of the voltage $v_L(t)$.

(b) Plot one cycle of the voltage $v_L(t)$, and indicate the earliest time after $t = 0$ when the voltage is maximum.

(c) The total voltage across the circuit is given by $v(t) = v_R(t) + v_L(t)$. Write $v(t)$ in the form $v(t) = M\cos(120\pi t + \theta)$, in other words, find M and θ.

6-21. A series RL circuit is subjected to a sinusoidal voltage of frequency $20\,\pi$rad/s, as shown in Fig. P6.20. The current $i(t) = 500\sin(20\,\pi t)$ mA is flowing through the circuit. The voltage across the resistor and inductor are given by $v_R(t) = 0.5\sin(20\,\pi t)$ and $v_L(t) = 0.1\sin\left(20\,\pi t + \frac{\pi}{2}\right)$ volts, where t is in seconds.

(a) Write down the amplitude, frequency (in Hz), period (in seconds), phase shift (in degrees), and time shift (in seconds) of the voltage $v_L(t)$.

(b) Plot one cycle of the voltage $v_L(t)$, and indicate the earliest time after $t = 0$ when the voltage is minimum.

(c) The total voltage across the circuit is given by $v(t) = v_R(t) + v_L(t)$. Write $v(t)$ in the form $v(t) = M\cos(20\pi t + \theta)$, in other words, find M and θ.

6-22. A sinusoidal voltage $v(t) = 10\sin(1000\,t)$ V is applied to the RLC circuit shown in Fig. P6.22. The current $i(t) = 0.707\sin(1000\,t + 45°)$ flowing through the circuit produces voltages across R, L, and C of

$$v_R(t) = 7.07\sin(1000\,t + 45°)\ \text{V}$$
$$v_L(t) = 7.07\sin(1000\,t + 135°)\ \text{V}$$
$$v_C(t) = 14.14\sin(1000\,t - 45°)\ \text{V}.$$

(a) Write down the amplitude, frequency (in Hz), period (in seconds), phase shift (in radians), and time shift (in ms) of the current $i(t) = 0.707\sin(1000\,t + 45°)$ A.

$i(t) = 0.707 \sin(1000t + 45°)$ A

$v(t) = 10 \sin(1000t)$ V

v_R $R = 10\,\Omega$

v_L $L = 10$ mH

v_C $C = 50\,\mu$F

Figure P6.22   RLC circuit for problem P6-22.

(b) Plot one cycle of the current $i(t) = 0.707 \sin(1000t + 45°)$ A and indicate the earliest time (after $t = 0$) when the current is 0.707 A.

(c) Using trigonometric identities, show that $v_1(t) = v_R(t) + v_C(t) = 15 \sin 1000t - 5 \cos 1000t$.

(d) Write $v_1(t)$ obtained in part (c) in the form $v_1(t) = M \cos(1000t + \theta)$; in other words, find M and θ.

6-23. A parallel RL circuit is subjected to a sinusoidal voltage of frequency $120\,\pi$rad/s, as shown in Fig. P6.23. The currents $i_1(t)$ and $i_2(t)$ are given by

$$i_1(t) = 10 \sin(120\,\pi t) \text{ A}$$

$$i_2(t) = 10 \sin\left(120\,\pi t - \frac{\pi}{2}\right) \text{ A}.$$

(a) Given that $i(t) = i_1(t) + i_2(t)$, write $i(t)$ in the form $i(t) = M \sin(120\,\pi t + \theta)$; in other words, find M and θ.

$i(t)$ $i_1(t)$

$i_2(t)$

$v(t) = 120 \sin(120\,\pi t)$ V

L $R = 12\,\Omega$

$L = \dfrac{100}{\pi}$ mH

Figure P6.23   A parallel RL circuit for problem P6-23.

(b) Suppose $i(t) = 14.14 \sin\left(120\pi t - \dfrac{\pi}{4}\right)$
A. Determine the amplitude, frequency (in Hz), period (in seconds), phase shift (in degrees), and time shift (in seconds).

(c) Given your results of part (b), plot one cycle of the current $i(t)$, and clearly indicate the earliest time after $t = 0$ at which it reaches its maximum value.

6-24. A parallel RL circuit is subjected to a sinusoidal voltage of frequency $10\,\pi$rad/s, as shown in Fig. P6.24. The currents $i_1(t)$ and $i_2(t)$ are given by

$$i_1(t) = 100 \cos(10\,\pi t) \text{ mA}$$

$$i_2(t) = 100 \sin(10\,\pi t) \text{ mA}.$$

$i(t)$ $i_1(t)$

$i_2(t)$

$v(t) = 10 \sin(10\,\pi t)$ V

L $R = 100\,\Omega$

$L = \dfrac{1}{\pi}$ mH

Figure P6.24   A parallel RL circuit for problem P6-24.

(a) Given that $i(t) = i_1(t) + i_2(t)$, write $i(t)$ in the form $i(t) = M \sin(10\,\pi t + \theta)$; in other words, find M and θ.

(b) Suppose $i(t) = 0.1414 \sin\left(10\pi t + \dfrac{\pi}{4}\right)$
mA. Determine the amplitude, frequency (in Hz), period (in seconds), phase shift (in degrees), and time shift (in seconds).

(c) Given your results of part (b), plot one cycle of the current $i(t)$, and clearly indicate the earliest time after $t = 0$ at which it reaches its maximum value.

6-25. Consider the RC circuit shown in Fig. P6.25, where the currents are $i_1(t) = 2\sin(120\pi t)$ A   and $i_2(t) = 2\sqrt{3}\sin(120\pi t + \frac{\pi}{2})$ A.

Figure P6.25   A parallel RC circuit for problem P6-25.

(a) Given that $i(t) = i_1(t) + i_2(t)$, write $i(t)$ in the form $i(t) = M\sin(120\pi t + \phi)$ A; in other words, find M and ϕ.

(b) Suppose now that $i(t) = 4\sin\left(120\pi t + \frac{\pi}{3}\right)$ A. Determine the amplitude, frequency (in Hz), period (in seconds), phase shift (in degrees), and time shift (in seconds).

(c) Given your results of part (b), plot one cycle of the current $i(t)$, and clearly indicate the earliest time after $t = 0$ at which it reaches its maximum value.

6-26. Consider the RC circuit shown in Fig. P6.25, where the currents are $i_1(t) = 100\cos(100\pi t + \frac{\pi}{4})$ mA   and $i_2(t) = 500\cos(100t + \frac{3\pi}{4})$ mA.

(a) Given that $i(t) = i_1(t) + i_2(t)$, write $i(t)$ in the form $i(t) = M\sin(100t + \phi)$ A; in other words, find M and ϕ.

(b) Suppose now that $i(t) = 0.51\sin(100t + 146.31°)$A. Determine the amplitude, frequency (in Hz), period (in seconds), phase shift

(in degrees), and time shift (in seconds).

(c) Given your results of part (b), plot one cycle of the current $i(t)$, and clearly indicate the earliest time after $t = 0$ at which it reaches its maximum value.

6-27. Two voltages $v_1(t) = 10\sin(100\pi t - 45°)$ V and $v_2(t) = 10\sin(100\pi t)$ V are applied to the Op–Amp circuit shown in Fig. P6.27.

Figure P6.27   An Op–Amp circuit for problem P6-27.

(a) Write down the amplitude, frequency (in Hz), period (in seconds), phase shift (in radians), and time shift (in seconds) of the voltage $v_1(t)$.

(b) Plot one cycle of the voltage $v_1(t)$, and indicate the earliest time after $t = 0$ when the voltage is 10 V.

(c) The output voltage $v_o(t)$ is given by $v_o(t) = -\left(\sqrt{2}\,v_1(t) + v_2(t)\right)$. Write $v_o(t)$ in the form $v_o(t) = M\cos(100\pi t + \theta°)$; in other words, find M and θ).

6-28. Repeat problem P6-27 if $v_1(t) = 10\cos(100\pi t + 90°)$ V and $v_2(t) = 3\sin(100\pi t + \frac{\pi}{4})$ V.

6-29. Two voltages $v_1(t) = 10\sqrt{2}\sin\left(500\pi t - \frac{3\pi}{4}\right)$ V and $v_2(t) = 5\sin(500\pi t)$ V are applied to the Op–Amp circuit shown in Fig. P6.29.

Figure P6.29 An Op–Amp circuit for problem P6-29.

(a) Write down the amplitude, frequency (in Hz), period (in seconds), phase angle or phase shift (in radians), and time shift (in seconds) of the voltage $v_1(t)$.

(b) Plot one cycle of the voltage $v_1(t)$, and indicate the earliest time after $t = 0$ when the voltage is $10\sqrt{2}$ V.

(c) The output voltage $v_o(t)$ is given by $v_o(t) = v_1(t) + v_2(t)$. Write $v_o(t)$ in the form $v_o(t) = M\sin(500\pi t + \theta)$ (i.e., find M and θ).

6-30. A pair of springs and masses vibrate under simple harmonic motion, as shown in Fig. P6.30. The positions of the masses in inches are given by $y_1(t) = 5\sqrt{2}\cos\left(2\pi t + \frac{\pi}{4}\right)$ and $y_2(t) = 10\cos(2\pi t)$, where t is in seconds.

(a) Write down the amplitude, frequency (in Hz), period (in seconds), phase shift (in degrees), and time shift (in seconds) of the position of the first mass $y_1(t)$.

(b) Plot one cycle of the position $y_1(t)$, and indicate the earliest time after $t = 0$ when the position is zero.

(c) The vertical distance between the two masses is given by $\delta(t) = y_1(t) - y_2(t)$. Write $\delta(t)$ in the form $\delta(t) = M\sin(2\pi t + \theta°)$; in other words, find M and θ.

6-31. Suppose the positions of the masses in problem P6-30 are given by

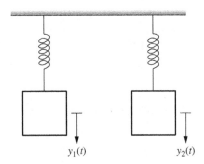

Figure P6.30 A pair of springs and masses for problem P6-30.

$$y_1(t) = 8\sin\left(4\pi t + \frac{\pi}{3}\right) \quad \text{and} \quad y_2(t) = 6\cos(4\pi t),$$ where t is in seconds.

(a) Write down the amplitude, frequency (in Hz), period (in seconds), phase shift (in degrees), and time shift (in seconds) of the position of the first mass $y_1(t)$.

(b) Plot one cycle of the position $y_1(t)$, and indicate the earliest time after $t = 0$ when the position is zero.

(c) Write $\delta(t) = y_1(t) - y_2(t)$ in the form $\delta(t) = M\cos(4\pi t + \theta°)$; in other words, find M and θ.

6-32. Two oscillating masses are connected by a spring as shown in Fig. P6.32. The positions of the masses in inches are given by $x_1(t) = 5\sqrt{2}\cos\left(2\pi t + \frac{\pi}{4}\right)$ and $x_2(t) = 10\cos(2\pi t)$, where t is in seconds.

(a) Write down the amplitude, frequency (in Hz), period (in seconds),

Figure P6.32 Two oscillating masses for problem P6-32.

phase shift (in degrees), and time shift (in seconds) of the position of the first mass $x_1(t)$.

(b) Plot one cycle of the position $x_1(t)$, and indicate the earliest time after $t = 0$ when the position is zero.

(c) The elongation of the spring is given by $\delta(t) = x_2(t) - x_1(t)$. Write $\delta(t)$ in the form $\delta(t) = M \sin(2\pi t + \phi)$; in other words, find M and ϕ.

6-33. Now assume that the positions of two masses in problem P6-31 are given by $x_1(t) = 8\sin\left(4\pi t + \dfrac{\pi}{4}\right)$ and $x_2(t) = 16\cos(4\pi t)$, where t is in seconds.

(a) Write down the amplitude, frequency (in Hz), period (in seconds), phase shift (in degrees), and time shift (in seconds) of the position of the first mass $x_1(t)$.

(b) Plot one cycle of the position $x_1(t)$, and indicate the earliest time after $t = 0$ when the position is zero.

(c) Write $\delta(t) = x_2(t) - x_1(t)$ in the form $\delta(t) = M\cos(4\pi t + \phi)$; in other words, find M and ϕ.

6-34. A manufacturing plant employs a heater and a conveyer belt motor on the same 220 V service line as shown in Fig. P6.34. The voltages across

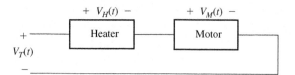

Figure P6.34 Conveyer motor and heater connected across the 220 V service line.

the heater and motor are given by $V_H(t) = 66\cos(120\pi t)$ V and $V_M(t) = 180\cos(120\pi t + \frac{\pi}{3})$ V, where t is in seconds.

(a) Write down the amplitude, frequency (in Hz), period, phase angle (in degrees), and phase shift (in seconds) of the voltage across the motor, $V_M(t)$.

(b) Plot one cycle of the motor voltage $V_M(t)$, and indicate the earliest time when the voltage is maximum.

(c) Write the voltage $V_M(t)$ in the form $V_M(t) = A\sin(120\pi t) + B\cos(120\pi t)$ (i.e., find A and B).

(d) The total voltage is $V_H(t) + V_M(t)$. Write $V_T(t)$ in the form $V_T(t) = V\cos(120\pi t + \theta)$ (i.e., find V and θ).

6-35. Repeat problem P6-34 if $V_H(t) = 100\sin(120\pi t)$ V and $V_M(t) = 164\sin(120\pi t + 70°)$ V.

6-36. In the three-phase circuit shown in Fig. P6.36, $v_{ab}(t) = v_{an}(t) - v_{bn}(t)$, where $v_{an}(t) = 120\sin(120\pi t)$ V and $v_{bn}(t) = 120\sin(120\pi t - 120°)$ V are the line-to-neutral voltages and v_{ab} is the line-to-line voltage of the three-phase system.

(a) Write down the amplitude, frequency (in Hz), period, phase angle (in degrees), and phase shift (in seconds) of the voltage, $v_{bn}(t)$.

(b) Plot one cycle of the line-to-neutral voltage $v_{bn}(t)$, and indicate the earliest time when the voltage is maximum.

(c) Write the line-to-line voltage $v_{ab}(t)$ in the form $v_{ab}(t) = A\sin(120\pi t) + B\cos(120\pi t)$ (i.e., find A and B).

(d) Write $v_{ab}(t)$ in the form $v_{ab}(t) = V\cos(120\pi t + \theta)$ (i.e., find V and θ).

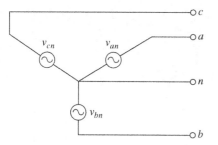

Figure P6.36 A balanced three-phase circuit.

6-37. In the three-phase circuit shown in Fig. P6.36, $v_{bc}(t) = v_{bn}(t) - v_{cn}(t)$, where $v_{bn}(t) = 120 \sin(120 \pi t - 120°)$ V and $v_{cn}(t) = 120 \sin(120 \pi t + 120°)$ V are the line-to-neutral voltages and v_{bc} is the line-to-line voltage of the three-phase system.

  (a) Write down the amplitude, frequency (in Hz), period, phase angle (in degrees), and phase shift (in seconds) of the voltage, $v_{cn}(t)$.

  (b) Plot one cycle of the line-to-neutral voltage $v_{cn}(t)$, and indicate the earliest time when the voltage is minimum.

  (c) Write the line-to-line voltage $v_{bc}(t)$ in the form $v_{bc}(t) = A \sin(120 \pi t) + B \cos(120 \pi t)$ (i.e., find A and B).

  (d) Write $v_{bc}(t)$ in the form $v_{bc}(t) = V \cos(120 \pi t + \theta)$ (i.e., find V and θ).

6-38. In the three-phase circuit shown in Fig. P6.36, $v_{ca}(t) = v_{cn}(t) - v_{an}(t)$, where $v_{cn}(t) = 120 \sin(120 \pi t + 120°)$ V and $v_{an}(t) = 120 \cos(120 \pi t - 90°)$ V are the line-to-neutral voltages and v_{ca} is the line-to-line voltage of the three-phase system.

  (a) Write down the amplitude, frequency (in Hz), period, phase angle (in degrees), and phase shift (in seconds) of the voltage, $v_{an}(t)$.

  (b) Plot one cycle of the line-to-neutral voltage $v_{an}(t)$, and indicate the earliest time when the voltage reaches 60 V.

  (c) Write the line-to-line voltage $v_{ca}(t)$ in the form $v_{ca}(t) = A \sin(120 \pi t) + B \cos(120 \pi t)$ (i.e., find A and B).

  (d) Write $v_{ca}(t)$ in the form $v_{ca}(t) = V \cos(120 \pi t + \theta)$ (i.e., find V and θ).

6-39. The hip implant shown in Fig. P6.39 is subjected to a cyclic load during a fatigue test. The load applied to the hip implant is given by

$$F(t) = 200 \sin(5 \pi t) + 1000 \, \text{N}$$

Figure P6.39 Hip implant subjected to a cyclic load.

  (a) Write down the amplitude, frequency (in Hz), period (in seconds), phase angle (in degrees), time shift (in seconds), and vertical shift (in N) of the load profile.

  (b) Plot one cycle of $F(t)$ and indicate the earliest time when the force reaches its maximum value.

6-40. Repeat problem P6-39 if $F(t) = 15 \sin(10 \pi t) + 75 \, \text{N}$.

CHAPTER 7

Systems of Equations in Engineering

7.1 INTRODUCTION

The solution of a system of linear equations is an important topic for all engineering disciplines. In this chapter, the solution of 2×2 systems of equations will be carried out using four different methods: substitution method, graphical method, matrix algebra method, and Cramer's rule. It is assumed that the students are already familiar with the substitution and graphical methods from their high school algebra course, while the matrix algebra method and Cramer's rule are explained in detail. The objective of this chapter is to be able to solve the systems of equations encountered in beginning engineering courses such as physics, statics, dynamics, and DC circuit analysis. While the examples given are limited to 2×2 systems of equations, the matrix algebra approach is applicable to linear systems having any number of unknowns and is suitable for immediate implementation in MATLAB.

7.2 SOLUTION OF A TWO-LOOP CIRCUIT

Consider a two-loop resistive circuit with unknown currents I_1 and I_2 as shown in Fig. 7.1. Using a combination of Kirchhoff's voltage law (KVL) and Ohm's law, a system of two equations with two unknowns I_1 and I_2 can be obtained as

$$10\,I_1 + 4\,I_2 = 6 \tag{7.1}$$
$$12\,I_2 + 4\,I_1 = 9. \tag{7.2}$$

Figure 7.1 A two-loop resistive circuit.

Equations (7.1) and (7.2) represent a system of equations for I_1 and I_2 that can be solved using the four different methods outlined as follows:

1. **Substitution Method:** Solving equation (7.1) for the first variable I_1 gives

$$10\,I_1 = 6 - 4\,I_2$$
$$I_1 = \frac{6 - 4\,I_2}{10}. \tag{7.3}$$

The current I_2 can now be solved by substituting I_1 from equation (7.3) into equation (7.2), which gives

$$12\,I_2 + 4\left(\frac{6 - 4\,I_2}{10}\right) = 9$$
$$12\,I_2 + 2.4 - 1.6\,I_2 = 9$$
$$10.4\,I_2 = 6.6$$
$$I_2 = \frac{6.6}{10.4}$$
$$I_2 = 0.6346 \text{ A}. \tag{7.4}$$

The current I_1 can now be obtained by substituting the value of the second variable I_2 from equation (7.4) into equation (7.3) as

$$I_1 = \frac{6 - 4(0.6346)}{10}$$
$$I_1 = 0.3462 \text{ A}.$$

Therefore, the solution of the system of equations (7.1) and (7.2) is given by

$$(I_1, I_2) = (0.3462 \text{ A}, 0.6346 \text{ A}).$$

2. **Graphical Method:** Begin by assuming I_1 as the independent variable and I_2 as the dependent variable. Solving equation (7.1) for the dependent variable I_2 gives

$$10\,I_1 + 4\,I_2 = 6$$
$$4\,I_2 = -10\,I_1 + 6$$
$$I_2 = -\frac{5}{2}\,I_1 + \frac{3}{2}. \tag{7.5}$$

Similarly, solving equation (7.2) for I_2 gives

$$4\,I_1 + 12\,I_2 = 9$$
$$12\,I_2 = -4\,I_1 + 9$$
$$I_2 = -\frac{1}{3}\,I_1 + \frac{3}{4}. \tag{7.6}$$

Equations (7.5) and (7.6) are linear equations of the form $y = mx + b$. The simultaneous solution of equations (7.5) and (7.6) is the intersection point of the two lines. The plot of the two straight lines along with their intersection point is shown in Fig. 7.2. The intersection point, $(I_1, I_2) \approx (0.35 \text{ A}, 0.63 \text{ A})$, is the solution of the 2×2 system of equations (7.1) and (7.2).

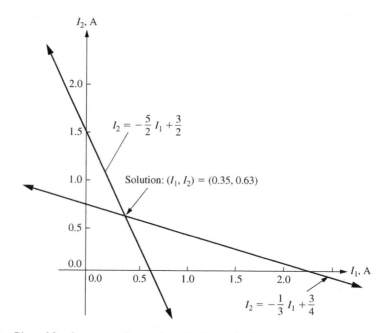

Figure 7.2 Plot of 2×2 system of equations (7.1) and (7.2).

Note that the graphical method gives only approximate results; therefore, this method is generally not used when an accurate result is needed. Also, if the two lines do not intersect, then one of the following two possibilities exist:

(i) The two lines are parallel lines (same slope but different y-intercepts) and the system of equations has no solution.

(ii) The two lines are parallel lines with same slope and y-intercept (the two lines lie on top of each other; they are the same line) and the system of equations has infinitely many solutions. In this case, the two equations are dependent (i.e., one equation can be obtained by performing linear operations on the other equation).

3. **Matrix Algebra Method:** The matrix algebra method can also be used to solve the system of equations given by equations (7.1) and (7.2). Rewriting the system of equations (7.1) and (7.2) in the form so that the two variables line up gives

$$10\,I_1 + \ 4\,I_2 = 6 \tag{7.7}$$
$$4\,I_1 + 12\,I_2 = 9. \tag{7.8}$$

Now, writing equations (7.7) and (7.8) in matrix form yields

$$\begin{bmatrix} 10 & 4 \\ 4 & 12 \end{bmatrix} \begin{bmatrix} I_1 \\ I_2 \end{bmatrix} = \begin{bmatrix} 6 \\ 9 \end{bmatrix}. \tag{7.9}$$

Equation (7.9) is of the form $\mathbf{Ax} = \mathbf{b}$, where

$$\mathbf{A} = \begin{bmatrix} 10 & 4 \\ 4 & 12 \end{bmatrix} \tag{7.10}$$

is a 2×2 coefficient matrix,

$$\mathbf{x} = \begin{bmatrix} I_1 \\ I_2 \end{bmatrix} \tag{7.11}$$

is a 2×1 matrix (column vector) of unknowns, and

$$\mathbf{b} = \begin{bmatrix} 6 \\ 9 \end{bmatrix} \tag{7.12}$$

is a 2×1 matrix on the right-hand side (RHS) of equation (7.9). For any system of the form $\mathbf{Ax} = \mathbf{b}$, the solution is given by

$$\mathbf{x} = \mathbf{A}^{-1}\mathbf{b}$$

where $\mathbf{A}^{-1}$ is the inverse of the matrix $\mathbf{A}$. For a 2×2 system of equations where

$$\mathbf{A} = \begin{bmatrix} a & b \\ c & d \end{bmatrix},$$

the inverse of the matrix $\mathbf{A}$ is given by

$$A^{-1} = \frac{1}{\Delta} \begin{bmatrix} d & -b \\ -c & a \end{bmatrix}$$

where $\Delta = |\mathbf{A}|$ is the determinant of matrix $\mathbf{A}$ and is given by

$$\Delta = \begin{vmatrix} a & b \\ c & d \end{vmatrix}$$
$$= ad - bc.$$

Note that if $\Delta = |\mathbf{A}| = 0$, $\mathbf{A}^{-1}$ does not exist. In other words, the system of equations $\mathbf{Ax} = \mathbf{b}$ has no solution. Now, for the two-loop circuit problem,

$$\mathbf{A} = \begin{bmatrix} 10 & 4 \\ 4 & 12 \end{bmatrix}$$
$$= \begin{bmatrix} a & b \\ c & d \end{bmatrix}.$$

The inverse of matrix $\mathbf{A}$ is given by

$$\mathbf{A}^{-1} = \frac{1}{\Delta} \begin{bmatrix} d & -b \\ -c & a \end{bmatrix}$$
$$= \frac{1}{\Delta} \begin{bmatrix} 12 & -4 \\ -4 & 10 \end{bmatrix}$$

where $\Delta = |\mathbf{A}| = a\,d - c\,b = (10)\,(12) - (4)\,(4) = 104$. Therefore, the inverse of matrix $\mathbf{A}$ can be calculated as

$$\mathbf{A}^{-1} = \frac{1}{104} \begin{bmatrix} 12 & -4 \\ -4 & 10 \end{bmatrix}$$

$$= \begin{bmatrix} \dfrac{3}{26} & -\dfrac{1}{26} \\ -\dfrac{1}{26} & \dfrac{5}{52} \end{bmatrix}. \tag{7.13}$$

The solution of the system of equations $\mathbf{x} = \begin{bmatrix} I_1 \\ I_2 \end{bmatrix}$ can now be found by multiplying A^{-1} (given by equation (7.13)) with the column matrix b (given by equation (7.12)) as

$$\mathbf{x} = \mathbf{A}^{-1}\,\mathbf{b}$$

$$\begin{bmatrix} I_1 \\ I_2 \end{bmatrix} = \begin{bmatrix} \dfrac{3}{26} & -\dfrac{1}{26} \\ -\dfrac{1}{26} & \dfrac{5}{52} \end{bmatrix} \begin{bmatrix} 6 \\ 9 \end{bmatrix}$$

$$= \begin{bmatrix} \dfrac{3}{26}\,(6) + \left(-\dfrac{1}{26}\right)\,(9) \\ \left(-\dfrac{1}{26}\right)\,(6) + \left(\dfrac{5}{52}\right)\,(9) \end{bmatrix}$$

$$= \begin{bmatrix} \dfrac{18-9}{26} \\ \dfrac{-12+45}{52} \end{bmatrix}$$

$$= \begin{bmatrix} 0.3462 \\ 0.6346 \end{bmatrix}.$$

The solution of the system of equations (7.1) and (7.2) is therefore given by

$$(I_1, I_2) = (0.3462\text{A}, 0.6346\text{A}).$$

4. **Cramer's Rule:** For any system $\mathbf{Ax} = \mathbf{b}$, the solution of the system of equations is given by

$$x_1 = \frac{|A_1|}{|A|}, \quad x_2 = \frac{|A_2|}{|A|}, \ldots x_i = \frac{|A_i|}{|A|}$$

where $|\mathbf{A}_i|$ is obtained by replacing the ith column of the matrix $\mathbf{A}$ with the column vector $\mathbf{b}$. Writing the 2×2 system of equations

$$a_{11}\,x_1 + a_{12}\,x_2 = b_1$$
$$a_{21}\,x_1 + a_{22}\,x_2 = b_2$$

in matrix form as

$$\begin{bmatrix} a_{11} & a_{12} \\ a_{21} & a_{22} \end{bmatrix} \begin{bmatrix} x_1 \\ x_2 \end{bmatrix} = \begin{bmatrix} b_1 \\ b_2 \end{bmatrix},$$

Cramer's rule gives the solution of the system of equations as

$$x_1 = \frac{\begin{vmatrix} b_1 & a_{12} \\ b_2 & a_{22} \end{vmatrix}}{\begin{vmatrix} a_{11} & a_{12} \\ a_{21} & a_{22} \end{vmatrix}}$$

$$= \frac{a_{22} b_1 - a_{12} b_2}{a_{11} a_{22} - a_{12} a_{21}},$$

$$x_2 = \frac{\begin{vmatrix} a_{11} & b_1 \\ a_{21} & b_2 \end{vmatrix}}{\begin{vmatrix} a_{11} & a_{12} \\ a_{21} & a_{22} \end{vmatrix}}$$

$$= \frac{a_{11} b_2 - a_{21} b_1}{a_{11} a_{22} - a_{12} a_{21}}.$$

For the two-loop circuit, the 2×2 system of equations is

$$\begin{bmatrix} 10 & 4 \\ 4 & 12 \end{bmatrix} \begin{bmatrix} I_1 \\ I_2 \end{bmatrix} = \begin{bmatrix} 6 \\ 9 \end{bmatrix}.$$

Using Cramer's rule, the currents I_1 and I_2 can be determined as

$$I_1 = \frac{\begin{vmatrix} 6 & 4 \\ 9 & 12 \end{vmatrix}}{\begin{vmatrix} 10 & 4 \\ 4 & 12 \end{vmatrix}}$$

$$= \frac{6(12) - 9(4)}{10(12) - 4(4)}$$

$$= \frac{36}{104}$$

$$= 0.3462 \text{ A},$$

$$I_2 = \frac{\begin{vmatrix} 10 & 6 \\ 4 & 9 \end{vmatrix}}{\begin{vmatrix} 10 & 4 \\ 4 & 12 \end{vmatrix}}$$

$$= \frac{10(9) - 4(6)}{10(12) - 4(4)}$$

$$= \frac{66}{104}$$

$$= 0.6346 \text{ A}.$$

Therefore, $I_1 = 0.3462$ A and $I_2 = 0.6346$ A. Note that Cramer's rule is probably fastest for solving 2×2 systems but not faster than MATLAB.

An object weighing 95 N is hanging from a roof with two cables as shown in Fig. 7.3. Determine the tension in each cable using the substitution, matrix algebra, and Cramer's rule methods.

Since the system shown in Fig. 7.3 is in equilibrium, the sum of all the forces shown in the free-body diagram must be equal to zero. This implies that all the forces in the x- and y-directions are equal to zero (see Chapter 4). The components of the tension $\vec{T}_1$ in the x- and y-directions are given by $-T_1 \cos(45°)$ N and $T_1 \sin(45°)$ N, respectively. Similarly, the components of the tension $\vec{T}_2$ in the x- and y-directions are given by $T_2 \cos(30°)$ N and $T_2 \sin(30°)$ N, respectively. The components of the object weight is 0 N in the x-direction and -95 N in the y-direction. Summing all the forces in the x-direction gives

$$-T_1 \cos(45°) + T_2 \cos(30°) = 0$$
$$-0.7071\, T_1 + 0.8660\, T_2 = 0. \tag{7.14}$$

Figure 7.3   A 95 N object hanging from two cables.

Similarly, summing the forces in the y-direction yields

$$T_1 \sin(45°) + T_2 \sin(30°) = 95$$
$$0.7071\, T_1 + 0.5\, T_2 = 95. \tag{7.15}$$

Equations (7.14) and (7.15) make a 2×2 system of equations with two unknowns T_1 and T_2 that can be written in matrix form as

$$\begin{bmatrix} -0.7071 & 0.8660 \\ 0.7071 & 0.5 \end{bmatrix} \begin{bmatrix} T_1 \\ T_2 \end{bmatrix} = \begin{bmatrix} 0 \\ 95 \end{bmatrix}. \tag{7.16}$$

The solution of the system of equations (T_1 and T_2) will now be obtained using three methods: the substitution method, the matrix algebra method, and Cramer's rule.

1. Substitution Method: Using equation (7.14), the second variable T_2 is found in terms of the first variable T_1 as

$$0.8660\, T_2 = 0.7071\, T_1$$
$$T_2 = 0.8165\, T_1. \tag{7.17}$$

Substituting T_2 from equation (7.17) into equation (7.15) gives

$$0.7071\,T_1 + 0.5\,(0.8165\,T_1) = 95$$
$$1.115\,T_1 = 95$$
$$T_1 = 85.17 \text{ N}. \tag{7.18}$$

Now, substituting T_1 obtained in equation (7.18) into equation (7.17) yields

$$T_2 = 0.8165\,(85.17)$$
$$= 69.55 \text{ N}.$$

Therefore, $T_1 = 85.2$ N and $T_2 = 69.6$ N.

2. **Matrix Algebra Method:** The two unknowns (T_1 and T_2) in the 2×2 system of equations (7.14) and (7.15) are now determined using the matrix algebra method. Write equations (7.14) and (7.15) in the matrix form as

$$\mathbf{A}\,\mathbf{x} = \mathbf{b} \tag{7.19}$$

where matrices $\mathbf{A}$, $\mathbf{x}$, and $\mathbf{b}$ are given by

$$\mathbf{A} = \begin{bmatrix} -0.7071 & 0.8660 \\ 0.7071 & 0.5 \end{bmatrix}$$

$$\mathbf{x} = \begin{bmatrix} T_1 \\ T_2 \end{bmatrix}$$

$$\mathbf{b} = \begin{bmatrix} 0 \\ 95 \end{bmatrix}. \tag{7.20}$$

Therefore, the solution of the 2×2 system of equations $\mathbf{x} = \begin{bmatrix} T_1 \\ T_2 \end{bmatrix}$ can be found by solving equation (7.19) as

$$\mathbf{x} = \mathbf{A}^{-1}\mathbf{b} \tag{7.21}$$

where $\mathbf{A}^{-1}$ is the inverse of matrix $\mathbf{A}$. If $\mathbf{A} = \begin{bmatrix} a & b \\ c & d \end{bmatrix}$, then

$$\mathbf{A}^{-1} = \frac{1}{\Delta} \begin{bmatrix} d & -b \\ -c & a \end{bmatrix}$$

where $\Delta = ad - bc$. Since, for this example, $a = -0.7071$, $b = 0.8660$, $c = 0.7071$, and $d = 0.5$, therefore,

$$\Delta = (-0.7071)\,(0.5) - (0.7071)\,(0.8660)$$
$$= -0.9659$$

$$\mathbf{A}^{-1} = \frac{1}{-0.9659} \begin{bmatrix} 0.5 & -0.8660 \\ -0.7071 & -0.7071 \end{bmatrix}$$

$$= \begin{bmatrix} -0.5177 & 0.8966 \\ 0.7321 & 0.7321 \end{bmatrix}. \tag{7.22}$$

Substituting matrices $\mathbf{A}^{-1}$ from equation (7.22) and $\mathbf{b}$ from equation (7.20) into equation (7.21) gives

$$\mathbf{x} = \begin{bmatrix} -0.5177 & 0.8966 \\ 0.7321 & 0.7321 \end{bmatrix} \begin{bmatrix} 0 \\ 95 \end{bmatrix}$$

$$\begin{bmatrix} T_1 \\ T_2 \end{bmatrix} = \begin{bmatrix} 0 + 0.8966(95) \\ 0 + 0.7321(95) \end{bmatrix}$$

$$= \begin{bmatrix} 85.2 \\ 69.6 \end{bmatrix}.$$

Therefore, $T_1 = 85.2$ N and $T_2 = 69.6$ N.

3. **Cramer's Rule:** The two unknowns (T_1 and T_2) in the 2×2 system of equations (7.14) and (7.15) are now determined using Cramer's rule. Using matrix equation (7.19), the tensions T_1 and T_2 can be found as

$$T_1 = \frac{\begin{vmatrix} 0 & 0.8660 \\ 95 & 0.5 \end{vmatrix}}{\begin{vmatrix} -0.7071 & 0.866 \\ 0.7071 & 0.5 \end{vmatrix}}$$

$$= \frac{0 - 95(0.8660)}{-0.7071(0.5) - 0.7071(0.8660)}$$

$$= \frac{-82.27}{-0.9659}$$

$$= 85.2 \text{ N}$$

$$T_2 = \frac{\begin{vmatrix} -0.7071 & 0 \\ 0.7071 & 95 \end{vmatrix}}{-0.9659}$$

$$= \frac{-0.7071(95) - 0}{-0.9659}$$

$$= \frac{-67.16}{-0.9659}$$

$$= 69.6 \text{ N}.$$

Therefore, $T_1 = 85.2$ N and $T_2 = 69.6$ N.

7.4 | **FURTHER EXAMPLES OF SYSTEMS OF EQUATIONS IN ENGINEERING**

Example 7-1

Reaction Forces on a Vehicle: The weight of a vehicle is supported by reaction forces at its front and rear wheels as shown in Fig. 7.4. If the weight is $W = 4800$ lb, the reaction forces R_1 and R_2 satisfy the equation:

$$R_1 + R_2 - 4800 = 0 \qquad (7.23)$$

Also, suppose that

$$6R_1 - 4R_2 = 0. \qquad (7.24)$$

(a) Find R_1 and R_2 using the substitution method.

(b) Write the system of equations (7.23) and (7.24) in the matrix form $\mathbf{A}\,\mathbf{x} = \mathbf{b}$, where $\mathbf{x} = \begin{bmatrix} R_1 \\ R_2 \end{bmatrix}$.

(c) Find R_1 and R_2 using the matrix algebra method. Perform all computations by hand and show all steps.

(d) Find R_1 and R_2 using Cramer's rule.

R_1 R_2

Figure 7.4   Reaction forces acting on a vehicle.

Solution

(a) **Substitution Method:** Using equation (7.23), find R_1 in terms of R_2 as

$$R_1 = 4800 - R_2. \qquad (7.25)$$

Substituting R_1 from equation (7.25) into equation (7.24) gives

$$6(4800 - R_2) - 4R_2 = 0$$
$$28,800 - 6R_2 - 4R_2 = 0$$
$$10R_2 = 28,800$$
$$R_2 = 2880 \text{ lb.} \qquad (7.26)$$

Now, substituting R_2 from equation (7.26) into equation (7.25) yields

$$R_1 = 4800 - 2880$$
$$= 1920 \text{ lb.}$$

Therefore, $R_1 = 1920$ lb and $R_2 = 2880$ lb.

(b) Writing equations (7.23) and (7.24) in the matrix form gives

$$\begin{bmatrix} 1 & 1 \\ 6 & -4 \end{bmatrix} \begin{bmatrix} R_1 \\ R_2 \end{bmatrix} = \begin{bmatrix} 4800 \\ 0 \end{bmatrix}. \tag{7.27}$$

(c) **Matrix Algebra Method:** From the matrix equation (7.27), the matrices $\mathbf{A}$, $\mathbf{x}$, and $\mathbf{b}$ are given by

$$\mathbf{A} = \begin{bmatrix} 1 & 1 \\ 6 & -4 \end{bmatrix} \tag{7.28}$$

$$\mathbf{b} = \begin{bmatrix} 4800 \\ 0 \end{bmatrix} \tag{7.29}$$

$$\mathbf{x} = \begin{bmatrix} R_1 \\ R_2 \end{bmatrix}. \tag{7.30}$$

The reaction forces can be found by finding the inverse of matrix $\mathbf{A}$ and then multiplying this with column matrix $\mathbf{b}$ as

$$\mathbf{x} = \mathbf{A}^{-1}\mathbf{b}$$

where

$$\mathbf{A}^{-1} = \frac{1}{|\mathbf{A}|} \begin{bmatrix} -4 & -1 \\ -6 & 1 \end{bmatrix} \tag{7.31}$$

and

$$|\mathbf{A}| = \begin{vmatrix} 1 & 1 \\ 6 & -4 \end{vmatrix}$$
$$= (1)(-4) - (6)(1)$$
$$= -10. \tag{7.32}$$

Substituting equation (7.32) in equation (7.31), the inverse of matrix $\mathbf{A}$ is given by

$$\mathbf{A}^{-1} = \frac{1}{-10} \begin{bmatrix} -4 & -1 \\ -6 & 1 \end{bmatrix}$$
$$= \begin{bmatrix} 0.4 & 0.1 \\ 0.6 & -0.1 \end{bmatrix}. \tag{7.33}$$

The reaction forces can now be found by multiplying $\mathbf{A}^{-1}$ in equation (7.33) with matrix $\mathbf{b}$ given in equation (7.29) as

$$\mathbf{x} = \begin{bmatrix} 0.4 & 0.1 \\ 0.6 & -0.1 \end{bmatrix} \begin{bmatrix} 4800 \\ 0 \end{bmatrix}$$
$$\begin{bmatrix} R_1 \\ R_2 \end{bmatrix} = \begin{bmatrix} (0.4)(4800) + 0 \\ (0.6)(4800) + 0 \end{bmatrix}$$
$$= \begin{bmatrix} 1920 \\ 2880 \end{bmatrix}.$$

Therefore, $R_1 = 1920$ lb and $R_2 = 2880$ lb.

(d) **Cramer's Rule:** The reaction forces R_1 and R_2 can be found by solving the system of equations (7.23) and (7.24) using Cramer's rule as

$$R_1 = \frac{\begin{vmatrix} 4800 & 1 \\ 0 & -4 \end{vmatrix}}{-10}$$

$$= \frac{(4800)(-4) - (0)(1)}{-10}$$

$$= 1920.$$

$$R_2 = \frac{\begin{vmatrix} 1 & 4800 \\ 6 & 0 \end{vmatrix}}{-10}$$

$$= \frac{(1)(0) - (6)(4800)}{-10}$$

$$= 2880.$$

Therefore, $R_1 = 1920$ lb and $R_2 = 2880$ lb.

Example 7-2

External Forces Acting on a Truss: A two-bar truss is subjected to external forces in both the horizontal and vertical directions as shown in Fig. 7.5. The forces F_1 and F_2 satisfy the following system of equations:

$$0.8\,F_1 + 0.8\,F_2 - 200 = 0 \tag{7.34}$$
$$0.6\,F_1 - 0.6\,F_2 - 100 = 0. \tag{7.35}$$

(a) Find F_1 and F_2 using the substitution method.

(b) Write the system of equations (7.34) and (7.35) in the matrix form $\mathbf{A}\mathbf{x} = \mathbf{b}$, where $\mathbf{x} = \begin{bmatrix} F_1 \\ F_2 \end{bmatrix}$.

(c) Find F_1 and F_2 using the matrix algebra method. Perform all computations by hand and show all steps.

(d) Find F_1 and F_2 using Cramer's rule.

Figure 7.5   A truss subjected to external forces.

Solution   (a) **Substitution Method:** Using equation (7.34), find force F_1 in terms of F_2 as

$$0.8\,F_1 = 200 - 0.8\,F_2$$
$$F_1 = 250 - F_2. \tag{7.36}$$

Substituting F_1 from equation (7.36) into equation (7.35) gives

$$0.6\,(250 - F_2) - 0.6\,F_2 = 100$$
$$250 - 2\,F_2 = 166.67$$
$$F_2 = \frac{(250 - 166.67)}{2}$$
$$= 41.67 \text{ lb.} \tag{7.37}$$

Now, substituting F_2 from equation (7.37) into equation (7.36) yields

$$F_1 = 250 - 41.67$$
$$= 208.33 \text{ lb.}$$

Therefore, $F_1 = 208.33$ lb and $F_2 = 41.67$ lb.

(b) Writing equations (7.34) and (7.35) in the matrix form yields

$$\begin{bmatrix} 0.8 & 0.8 \\ 0.6 & -0.6 \end{bmatrix} \begin{bmatrix} F_1 \\ F_2 \end{bmatrix} = \begin{bmatrix} 200 \\ 100 \end{bmatrix}. \tag{7.38}$$

(c) **Matrix Algebra Method:** Writing matrix equation in (7.38) in the form $\mathbf{A}\,\mathbf{x} = \mathbf{b}$ gives

$$\mathbf{A} = \begin{bmatrix} 0.8 & 0.8 \\ 0.6 & -0.6 \end{bmatrix} \tag{7.39}$$

$$\mathbf{b} = \begin{bmatrix} 200 \\ 100 \end{bmatrix} \tag{7.40}$$

$$\mathbf{x} = \begin{bmatrix} F_1 \\ F_2 \end{bmatrix}. \tag{7.41}$$

The forces F_1 and F_2 can be found by finding the inverse of matrix $\mathbf{A}$ and then multiplying this with matrix $\mathbf{b}$ as

$$\mathbf{x} = \mathbf{A}^{-1}\mathbf{b}$$

where

$$\mathbf{A}^{-1} = \frac{1}{|A|} \begin{bmatrix} -0.6 & -0.8 \\ -0.6 & 0.8 \end{bmatrix} \tag{7.42}$$

and

$$|\mathbf{A}| = \begin{vmatrix} 0.8 & 0.8 \\ 0.6 & -0.6 \end{vmatrix}$$
$$= (0.8)(-0.6) - (0.6)(0.8)$$
$$= -0.96. \tag{7.43}$$

Substituting equation (7.43) in equation (7.42), the inverse of matrix **A** is given by

$$\mathbf{A}^{-1} = \frac{1}{-0.96} \begin{bmatrix} -0.6 & -0.8 \\ -0.6 & 0.8 \end{bmatrix}$$

$$= \begin{bmatrix} 0.625 & 0.833 \\ 0.625 & -0.833 \end{bmatrix}. \tag{7.44}$$

The forces F_1 and F_2 can now be found by multiplying $\mathbf{A}^{-1}$ in equation (7.44) by column matrix **b** given in equation (7.40) as

$$\mathbf{x} = \begin{bmatrix} 0.625 & 0.833 \\ 0.625 & -0.833 \end{bmatrix} \begin{bmatrix} 200 \\ 100 \end{bmatrix}$$

$$\begin{bmatrix} F_1 \\ F_2 \end{bmatrix} = \begin{bmatrix} (0.625)(200) + (0.833)(100) \\ (0.625)(200) + (-0.833)(100) \end{bmatrix}$$

$$= \begin{bmatrix} 208.33 \\ 41.67 \end{bmatrix} \text{ lb}$$

Therefore, $F_1 = 208.33$ lb and $F_2 = 41.67$ lb.

(d) **Cramer's Rule:** The forces F_1 and F_2 can be found by solving the system of equations (7.34) and (7.35) using Cramer's rule as

$$F_1 = \frac{\begin{vmatrix} 200 & 0.8 \\ 100 & -0.6 \end{vmatrix}}{-0.96}$$

$$= \frac{(200)(-0.6) - (100)(0.8)}{-0.96}$$

$$= 208.33 \text{ lb}$$

$$F_2 = \frac{\begin{vmatrix} 0.8 & 200 \\ 0.6 & 100 \end{vmatrix}}{-0.96}$$

$$= \frac{(0.8)(100) - (0.6)(200)}{-0.96}$$

$$= 41.67 \text{ lb}$$

Therefore, $F_1 = 208.33$ lb and $R_2 = 41.67$ lb.

Example 7-3

Summing Op-Amp Circuit: A summing Op-Amp circuit is shown in Fig. 7.6. An analysis of the Op-Amp circuit shows that the conductances G_1 and G_2 in mho (℧) satisfy the following system of equations:

$$10\,G_1 + 5\,G_2 = 125 \tag{7.45}$$

$$9\,G_1 - 19 = 4\,G_2. \tag{7.46}$$

(a) Find G_1 and G_2 using the substitution method.

(b) Write the system of equations (7.45) and (7.46) in the matrix form $\mathbf{A}\mathbf{x} = \mathbf{b}$, where $\mathbf{x} = \begin{bmatrix} G_1 \\ G_2 \end{bmatrix}$.

(c) Find G_1 and G_2 using the matrix algebra method. Perform all computations by hand and show all steps.

(d) Find G_1 and G_2 using Cramer's rule.

Figure 7.6   A summing Op-Amp circuit.

Solution

(a) **Substitution Method:** Using equation (7.45), find the admittance G_1 in terms of G_2 as

$$10\,G_1 = 125 - 5\,G_2$$
$$G_1 = 12.5 - 0.5\,G_2. \tag{7.47}$$

Substituting G_1 from equation (7.47) into equation (7.46) gives

$$9\,(12.5 - 0.5\,G_2) - 19 = 4\,G_2$$
$$93.5 - 4.5\,G_2 = 4\,G_2$$
$$93.5 = 8.5\,G_2$$
$$G_2 = 11\ \mho. \tag{7.48}$$

Now, substituting G_2 from equation (7.48) into equation (7.47) yields

$$G_1 = 12.5 - 0.5(11)$$
$$= 7.0\ \mho.$$

Therefore, $G_1 = 7\ \mho$ and $G_2 = 11\ \mho$.

(b) Rewrite equations (7.45) and (7.46) in the form

$$10\,G_1 + 5\,G_2 = 125 \tag{7.49}$$
$$9\,G_1 - 4\,G_2 = 19. \tag{7.50}$$

Now, write equations (7.49) and (7.50) in matrix form as

$$\begin{bmatrix} 10 & 5 \\ 9 & -4 \end{bmatrix} \begin{bmatrix} G_1 \\ G_2 \end{bmatrix} = \begin{bmatrix} 125 \\ 19 \end{bmatrix}. \tag{7.51}$$

(c) **Matrix Algebra Method:** Writing the matrix equation in (7.64) in the form $\mathbf{A}\mathbf{x} = \mathbf{b}$ gives

$$\mathbf{A} = \begin{bmatrix} 10 & 5 \\ 9 & -4 \end{bmatrix} \tag{7.52}$$

$$\mathbf{b} = \begin{bmatrix} 125 \\ 19 \end{bmatrix} \tag{7.53}$$

$$\mathbf{x} = \begin{bmatrix} G_1 \\ G_2 \end{bmatrix}. \tag{7.54}$$

The admittance G_1 and G_2 can be found by finding the inverse of matrix $\mathbf{A}$ and then multiplying this with matrix $\mathbf{b}$ as

$$\mathbf{x} = \mathbf{A}^{-1}\mathbf{b}$$

where

$$\mathbf{A}^{-1} = \frac{1}{|\mathbf{A}|} \begin{bmatrix} -4 & -5 \\ -9 & 10 \end{bmatrix} \tag{7.55}$$

and

$$\begin{aligned} |\mathbf{A}| &= \begin{vmatrix} 10 & 5 \\ 9 & -4 \end{vmatrix} \\ &= (10)(-4) - (5)(9) \\ &= -85. \end{aligned} \tag{7.56}$$

Substituting equation (7.56) in equation (7.55), the inverse of matrix $\mathbf{A}$ is given as

$$\begin{aligned} \mathbf{A}^{-1} &= \frac{1}{-85} \begin{bmatrix} -4 & -5 \\ -9 & 10 \end{bmatrix} \\ &= \begin{bmatrix} \dfrac{4}{85} & \dfrac{5}{85} \\ \dfrac{9}{85} & -\dfrac{10}{85} \end{bmatrix}. \end{aligned} \tag{7.57}$$

The admittance G_1 and G_2 can now be found by multiplying $\mathbf{A}^{-1}$ in equation (7.57) and the column matrix $\mathbf{b}$ given in equation (7.53) as

$$\mathbf{x} = \begin{bmatrix} \dfrac{4}{85} & \dfrac{5}{85} \\ \dfrac{9}{85} & -\dfrac{10}{85} \end{bmatrix} \begin{bmatrix} 125 \\ 19 \end{bmatrix}$$

$$\begin{bmatrix} G_1 \\ G_2 \end{bmatrix} = \begin{bmatrix} \left(\dfrac{4}{85} \right)(125) + \left(\dfrac{5}{85} \right)(19) \\ \left(\dfrac{9}{85} \right)(125) + \left(-\dfrac{10}{85} \right)(19) \end{bmatrix}$$

$$= \begin{bmatrix} \dfrac{(100 + 19)}{17} \\ \dfrac{(225 - 38)}{17} \end{bmatrix} = \begin{bmatrix} 7 \\ 11 \end{bmatrix} \text{℧}.$$

Therefore, $G_1 = 7$ ℧ and $G_2 = 11$ ℧.

(d) **Cramer's Rule:** The admittance G_1 and G_2 can be found by solving the system of equations (7.45) and (7.46) using Cramer's rule as

$$G_1 = \frac{\begin{vmatrix} 125 & 5 \\ 19 & -4 \end{vmatrix}}{-85}$$

$$= \frac{(125)(-4) - (19)(5)}{-85}$$

$$= 7 \, \mho$$

$$G_2 = \frac{\begin{vmatrix} 10 & 125 \\ 9 & 19 \end{vmatrix}}{-85}$$

$$= \frac{(10)(19) - (9)(125)}{-85}$$

$$= 11 \, \mho.$$

Therefore, $G_1 = 7 \, \mho$ and $G_2 = 11 \, \mho$.

Example 7-4

Force on the Gastrocnemius Muscle: A driver applies a steady force of $F_P = 30$ N against a gas pedal, as shown in Fig. 7.7. The free-body diagram of the driver's foot is also shown. Based on the x-y coordinate system shown, the force of the gastrocnemius muscle F_m and the weight of the foot W_F satisfy the following system of equations:

$$F_m \cos 60° - W_F \cos 30° = R_x \tag{7.58}$$

$$F_m \sin 60° - W_F \sin 30° = R_y - F_P. \tag{7.59}$$

where R_x and R_y are the reactions at the ankle.

(a) Suppose $R_x = \dfrac{70}{\sqrt{3}}$ N and $R_y = 120$ N. Write the system of equations in terms of F_m and W_F.

(b) Find F_m and W_F using the substitution method.

(c) Write the system of equations (7.58) and (7.59) in the matrix form $\mathbf{A}\mathbf{x} = \mathbf{b}$, where $\mathbf{x} = \begin{bmatrix} F_m \\ W_F \end{bmatrix}$.

(d) Find F_m and W_F using the matrix algebra method. Perform all computations by hand and show all steps.

(e) Find F_m and W_F using Cramer's rule.

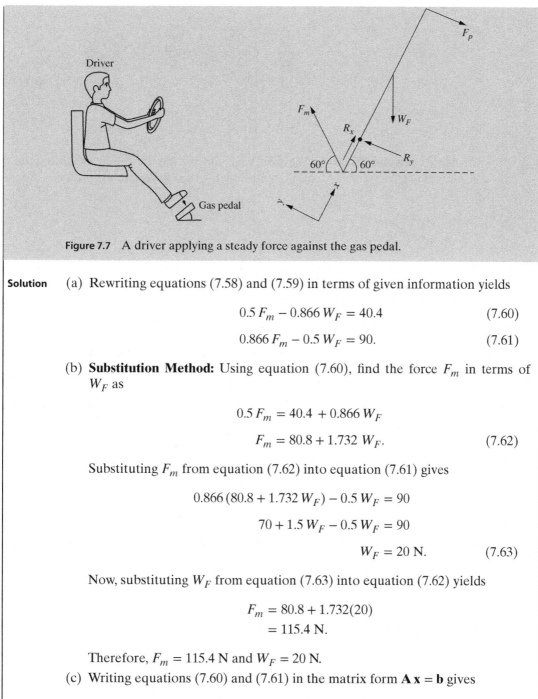

Figure 7.7 A driver applying a steady force against the gas pedal.

Solution (a) Rewriting equations (7.58) and (7.59) in terms of given information yields

$$0.5\,F_m - 0.866\,W_F = 40.4 \qquad (7.60)$$

$$0.866\,F_m - 0.5\,W_F = 90. \qquad (7.61)$$

(b) **Substitution Method:** Using equation (7.60), find the force F_m in terms of W_F as

$$0.5\,F_m = 40.4 + 0.866\,W_F$$

$$F_m = 80.8 + 1.732\,W_F. \qquad (7.62)$$

Substituting F_m from equation (7.62) into equation (7.61) gives

$$0.866\,(80.8 + 1.732\,W_F) - 0.5\,W_F = 90$$

$$70 + 1.5\,W_F - 0.5\,W_F = 90$$

$$W_F = 20\text{ N}. \qquad (7.63)$$

Now, substituting W_F from equation (7.63) into equation (7.62) yields

$$F_m = 80.8 + 1.732(20)$$

$$= 115.4\text{ N}.$$

Therefore, $F_m = 115.4$ N and $W_F = 20$ N.

(c) Writing equations (7.60) and (7.61) in the matrix form $\mathbf{A}\mathbf{x} = \mathbf{b}$ gives

$$\begin{bmatrix} 0.5 & -0.866 \\ 0.866 & -0.5 \end{bmatrix} \begin{bmatrix} F_m \\ W_F \end{bmatrix} = \begin{bmatrix} 40.4 \\ 90 \end{bmatrix} \qquad (7.64)$$

where

$$\mathbf{A} = \begin{bmatrix} 0.5 & -0.866 \\ 0.866 & -0.5 \end{bmatrix}, \tag{7.65}$$

$$\mathbf{b} = \begin{bmatrix} 40.4 \\ 90 \end{bmatrix}, \tag{7.66}$$

$$\text{and} \quad \mathbf{x} = \begin{bmatrix} F_m \\ W_F \end{bmatrix}. \tag{7.67}$$

(d) **Matrix Algebra Method:** The forces F_m and W_F can be found by finding the inverse of the matrix $\mathbf{A}$ and then multiplying this by vector $\mathbf{b}$ as

$$\mathbf{x} = \mathbf{A}^{-1}\mathbf{b},$$

where

$$\mathbf{A}^{-1} = \frac{1}{|\mathbf{A}|} \begin{bmatrix} -0.5 & 0.866 \\ -0.866 & 0.5 \end{bmatrix}, \tag{7.68}$$

$$\text{and} \quad |\mathbf{A}| = \begin{vmatrix} 0.5 & -0.866 \\ 0.866 & -0.5 \end{vmatrix}$$

$$= (0.5)(-0.5) - (0.866)(-0.866)$$

$$= 0.5. \tag{7.69}$$

Substituting equation (7.69) in equation (7.68), the inverse of matrix $\mathbf{A}$ is given as

$$\mathbf{A}^{-1} = \frac{1}{0.5} \begin{bmatrix} -0.5 & 0.866 \\ -0.866 & 0.5 \end{bmatrix}$$

$$= \begin{bmatrix} -1.0 & 1.732 \\ -1.732 & 1.0 \end{bmatrix}. \tag{7.70}$$

The forces F_m and W_F can now be found by multiplying $\mathbf{A}^{-1}$ in equation (7.70) and the column matrix $\mathbf{b}$ given in equation (7.66) as

$$\mathbf{x} = \begin{bmatrix} -1.0 & 1.732 \\ -1.732 & 1.0 \end{bmatrix} \begin{bmatrix} 40.4 \\ 90 \end{bmatrix}$$

$$\begin{bmatrix} F_m \\ W_F \end{bmatrix} = \begin{bmatrix} -1.0\,(40.4) + 1.732\,(90) \\ -1.732\,(40.4) + 1.0\,(90) \end{bmatrix}$$

$$= \begin{bmatrix} 115.4 \\ 20.0 \end{bmatrix} \text{N}.$$

Therefore, $F_m = 115.4$ N and $W_F = 20$ N.

(e) **Cramer's Rule:** The forces F_m and W_F can be found by solving the system of equations (7.60) and (7.61) using Cramer's rule as

$$F_m = \frac{\begin{vmatrix} 40.4 & -0.866 \\ 90 & -0.5 \end{vmatrix}}{0.5}$$

$$= \frac{(40.4)(-0.5) - (90)(-0.866)}{0.5}$$

$$= 115.4 \text{ N}$$

$$W_F = \frac{\begin{vmatrix} 0.5 & 40.4 \\ 0.866 & 90 \end{vmatrix}}{0.5}$$

$$= \frac{(0.5)(90) - (0.866)(40.4)}{0.5}$$

$$= 20.0 \text{ N}.$$

Therefore, $F_m = 115.4$ N and $W_F = 20$ N.

Example 7-5

Two-Component Blending of Liquids: An environmental engineer wishes to blend a single mixture of insecticide spray solution of volume $V = 1000$ L and concentration $C = 0.15$ from two spray solutions of concentrations $c_1 = 0.12$ and $c_2 = 0.17$. The required volumes of the two spray solutions v_1 and v_2 can be determined from a system of equations describing conditions for volume and concentration, respectively, as

$$v_1 + v_2 = V \tag{7.71}$$
$$c_1 v_1 + c_2 v_2 = C V \tag{7.72}$$

(a) Knowing that $V = 1000$ L, $c_1 = 0.12$, $c_2 = 0.17$, and $C = 0.15$, rewrite the system of equations in terms of v_1 and v_2.

(b) Find v_1 and v_2 using the substitution method.

(c) Write the system of equations (7.71) and (7.72) in the matrix form $\mathbf{A}\mathbf{x} = \mathbf{b}$, where $\mathbf{x} = \begin{bmatrix} v_1 \\ v_2 \end{bmatrix}$.

(d) Find v_1 and v_2 using the matrix algebra method. Perform all computations by hand and show all steps.

(e) Find v_1 and v_2 using Cramer's rule.

Solution

(a) Rewriting equations (7.71) and (7.72) in terms of the given information yields

$$v_1 + v_2 = 1000 \tag{7.73}$$
$$0.12 v_1 + 0.17 v_2 = 150. \tag{7.74}$$

(b) **Substitution Method:** Using equation (7.73), find the volume v_2 in terms of v_1 as

$$v_2 = 1000 - v_1. \tag{7.75}$$

Substituting v_2 from equation (7.75) into equation (7.74) gives

$$
\begin{aligned}
0.12\,v_1 + 0.17\,(1000 - v_1) &= 150 \\
0.12\,v_1 + 170 - 0.17\,v_1 &= 150 \\
-0.05\,v_1 &= -20 \\
v_1 &= 400.
\end{aligned}
\tag{7.76}
$$

Now, substituting v_1 from equation (7.76) into equation (7.75) yields

$$
\begin{aligned}
v_2 &= 1000 - 400 \\
&= 600.
\end{aligned}
$$

Therefore, $v_1 = 400$ L and $v_2 = 600$ L.

(c) Writing equations (7.73) and (7.74) in matrix form $\mathbf{A}\,\mathbf{x} = \mathbf{b}$ gives

$$
\begin{bmatrix} 1 & 1 \\ 0.12 & 0.17 \end{bmatrix}
\begin{bmatrix} v_1 \\ v_2 \end{bmatrix}
=
\begin{bmatrix} 1000 \\ 150 \end{bmatrix},
\tag{7.77}
$$

where

$$
\mathbf{A} = \begin{bmatrix} 1 & 1 \\ 0.12 & 0.17 \end{bmatrix},
\tag{7.78}
$$

$$
\mathbf{b} = \begin{bmatrix} 1000 \\ 150 \end{bmatrix},
\tag{7.79}
$$

$$
\text{and} \quad \mathbf{x} = \begin{bmatrix} v_1 \\ v_2 \end{bmatrix}.
\tag{7.80}
$$

(d) **Matrix Algebra Method:** The volumes v_1 and v_2 can be found by finding the inverse of the matrix $\mathbf{A}$ and then multiplying this by column vector $\mathbf{b}$ as

$$\mathbf{x} = \mathbf{A}^{-1}\mathbf{b},$$

where

$$
\mathbf{A}^{-1} = \frac{1}{|\mathbf{A}|} \begin{bmatrix} 0.17 & -1 \\ -0.12 & 1 \end{bmatrix}
\tag{7.81}
$$

$$
\text{and} \quad |\mathbf{A}| = \begin{vmatrix} 1 & 1 \\ 0.12 & 0.17 \end{vmatrix}
$$

$$
\begin{aligned}
&= (1)(0.17) - (0.12)(1) \\
&= 0.05.
\end{aligned}
\tag{7.82}
$$

Substituting equation (7.82) in equation (7.81), the inverse of matrix $\mathbf{A}$ is given as

$$\mathbf{A}^{-1} = \frac{1}{0.05} \begin{bmatrix} 0.17 & -1 \\ -0.12 & 1 \end{bmatrix}$$

$$= \begin{bmatrix} 3.4 & -20 \\ -2.4 & 20 \end{bmatrix}. \tag{7.83}$$

The volumes v_1 and v_2 can now be found by multiplying $\mathbf{A}^{-1}$ in equation (7.83) and the column vector $\mathbf{b}$ given in equation (7.79) as

$$\mathbf{x} = \begin{bmatrix} 3.4 & -20 \\ -2.4 & 20 \end{bmatrix} \begin{bmatrix} 1000 \\ 150 \end{bmatrix}$$

$$\begin{bmatrix} v_1 \\ v_2 \end{bmatrix} = \begin{bmatrix} 3.4\,(1000) - 20\,(150) \\ -2.4\,(1000) + 20\,(150) \end{bmatrix}$$

$$= \begin{bmatrix} 400 \\ 600 \end{bmatrix}.$$

Therefore, $v_1 = 400$ L and $v_2 = 600$ L.

(e) **Cramer's Rule:** The volumes v_1 and v_2 can be found by solving the system of equations (7.73) and (7.74) using Cramer's rule as

$$v_1 = \frac{\begin{vmatrix} 1000 & 1 \\ 150 & 0.17 \end{vmatrix}}{0.05}$$

$$= \frac{(1000)(0.17) - (150)(1)}{0.05}$$

$$= 400$$

$$v_2 = \frac{\begin{vmatrix} 1 & 1000 \\ 0.12 & 150 \end{vmatrix}}{0.05}$$

$$= \frac{(1)(150) - (0.12)(1000)}{0.05}$$

$$= 600.$$

Therefore, $v_1 = 400$ L and $v_2 = 600$ L.

PROBLEMS

7-1. Consider the two-loop circuit shown in Fig. P7.1. The currents I_1 and I_2 (in A) satisfy the following system of equations:

$$16 I_1 - 9 I_2 = 110 \qquad (7.84)$$
$$20 I_2 - 9 I_1 + 110 = 0. \qquad (7.85)$$

Figure P7.1   Two-loop circuit for problem P7-1.

(a) Find I_1 and I_2 using the substitution method.

(b) Write the system of equations (7.84) and (7.85) in the matrix form $A I = b$, where $I = \begin{bmatrix} I_1 \\ I_2 \end{bmatrix}$.

(c) Find I_1 and I_2 using the matrix algebra method. Perform all computations by hand and show all steps.

(d) Find I_1 and I_2 using Cramer's rule.

7-2. Consider the two-loop circuit shown in Fig. P7.2. The currents I_1 and I_2 (in A) satisfy the following system of equations:

$$18 I_1 - 10 I_2 - 246 = 0 \qquad (7.86)$$
$$22 I_2 - 10 I_1 = -334. \qquad (7.87)$$

Figure P7.2   Two-loop circuit for problem P7-2.

(a) Find I_1 and I_2 using the substitution method.

(b) Write the system of equations (7.86) and (7.87) in the matrix form $A I = b$, where $I = \begin{bmatrix} I_1 \\ I_2 \end{bmatrix}$.

(c) Find I_1 and I_2 using the matrix algebra method. Perform all computations by hand and show all steps.

(d) Find I_1 and I_2 using Cramer's rule.

7-3. Consider the two-loop circuit shown in Fig. P7.3. The currents I_1 and I_2 (in A) satisfy the following system of equations:

$$1100 I_1 + 1000 I_2 - 9 = 0 \qquad (7.88)$$
$$1100 I_2 + 1000 I_1 = 0. \qquad (7.89)$$

Figure P7.3   Two-loop circuit for problem P7-3.

(a) Find I_1 and I_2 using the substitution method.

(b) Write the system of equations (7.88) and (7.89) in the matrix form $AI = b$, where $I = \begin{bmatrix} I_1 \\ I_2 \end{bmatrix}$.

(c) Find I_1 and I_2 using the matrix algebra method. Perform all computations by hand and show all steps.

(d) Find I_1 and I_2 using Cramer's rule.

7-4. Consider the two-node circuit shown in Fig. P7.4. The voltages V_1 and V_2 (in V) satisfy the following system of equations:

$$4V_1 - V_2 = 20 \qquad (7.90)$$
$$-3V_1 + 8V_2 = 40. \qquad (7.91)$$

Figure P7.4   Two-node circuit for problem P7-4.

(a) Find V_1 and V_2 using the substitution method.

(b) Write the system of equations (7.90) and (7.91) in the matrix form $AV = b$, where $V = \begin{bmatrix} V_1 \\ V_2 \end{bmatrix}$.

(c) Find V_1 and V_2 using the matrix algebra method. Perform all computations by hand and show all steps.

(d) Find V_1 and V_2 using Cramer's rule.

7-5. Consider the two-node circuit shown in Fig. P7.5. The voltages V_1 and V_2 (in V) satisfy the following system of equations:

$$17V_1 = 10V_2 + 50 \quad (7.92)$$
$$11V_2 - 6V_1 - 42 = 0. \qquad (7.93)$$

Figure P7.5   Two-node circuit for problem P7-5.

(a) Find V_1 and V_2 using the substitution method.

(b) Write the system of equations (7.92) and (7.93) in the matrix form $AV = b$, where $V = \begin{bmatrix} V_1 \\ V_2 \end{bmatrix}$.

(c) Find V_1 and V_2 using the matrix algebra method. Perform all computations by hand and show all steps.

(d) Find V_1 and V_2 using Cramer's rule.

7-6. Consider the two-node circuit shown in Fig. P7.6. The voltages V_1 and V_2 (in V) satisfy the following system of equations:

$$0.2V_1 - 0.1V_2 = 4 \qquad (7.94)$$
$$0.3V_2 - 0.1V_1 + 2 = 0. \qquad (7.95)$$

Figure P7.6   Two-node circuit for problem P7-6.

(a) Find V_1 and V_2 using the substitution method.

(b) Write the system of equations (7.94) and (7.95) in the matrix form $AV = b$, where $V = \begin{bmatrix} V_1 \\ V_2 \end{bmatrix}$.

(c) Find V_1 and V_2 using the matrix algebra method. Perform all computations by hand and show all steps.

(d) Find V_1 and V_2 using Cramer's rule.

7-7. An analysis of the circuit shown in Fig. P7.7 yields the following system of equations:

$$-4V_2 + 7V_1 = 0 \qquad (7.96)$$
$$2V_1 - 7V_2 + 10 = 0. \qquad (7.97)$$

Figure P7.7 Two-node circuit for problem P7-7.

(a) Find V_1 and V_2 using the substitution method.

(b) Write the system of equations (7.96) and (7.97) in the matrix form $AV = b$, where $V = \begin{bmatrix} V_1 \\ V_2 \end{bmatrix}$.

(c) Find V_1 and V_2 using the matrix algebra method. Perform all computations by hand and show all steps.

(d) Find V_1 and V_2 using Cramer's rule.

7-8. A summing Op-Amp circuit is shown in Fig. 7.6. An analysis of the Op-Amp circuit shows that the admittances G_1 and G_2 in mho (℧) satisfy the following system of equations:

$$5G_1 - 145 = -10G_2 \qquad (7.98)$$
$$-9G_2 + 71 = -4G_1. \qquad (7.99)$$

(a) Find G_1 and G_2 using the substitution method.

(b) Write the system of equations (7.98) and (7.99) in the matrix form $AG = b$, where $G = \begin{bmatrix} G_1 \\ G_2 \end{bmatrix}$.

(c) Find G_1 and G_2 using the matrix algebra method. Perform all computations by hand and show all steps.

(d) Find G_1 and G_2 using Cramer's rule.

7-9. A summing Op-Amp circuit is shown in Fig. 7.6. An analysis of the Op-Amp circuit shows that the admittances G_1 and G_2 in mho (℧) satisfy the following system of equations:

$$10G_1 + 10G_2 = 0.2 \qquad (7.100)$$
$$5G_1 + 15G_2 = 0.125. \qquad (7.101)$$

(a) Find G_1 and G_2 using the substitution method.

(b) Write the system of equations (7.100) and (7.101) in the matrix form $AG = b$, where $G = \begin{bmatrix} G_1 \\ G_2 \end{bmatrix}$.

(c) Find G_1 and G_2 using the matrix algebra method. Perform all computations by hand and show all steps.

(d) Find G_1 and G_2 using Cramer's rule.

7-10. A 20 kg object is suspended by two cables as shown in Fig. P7.10. The tensions T_1 and T_2 satisfy the following system of equations:

$$0.5T_1 = 0.866T_2 \qquad (7.102)$$
$$0.5T_2 + 0.866T_1 = 196. \qquad (7.103)$$

Figure P7.10 A 20 kg object suspended by two cables for problem P7-10.

(a) Write the system of equations (7.100) and (7.101) in the matrix form $AT = b$, where

$$T = \begin{bmatrix} T_1 \\ T_2 \end{bmatrix}.$$

In other words, find matrices A and b. What are the dimensions of A and b?

(b) Find T_1 and T_2 using the matrix algebra method. Perform all matrix computation by hand.

(c) Find T_1 and T_2 using Cramer's rule.

7-11. A 100 lb weight is suspended by two cables as shown in Fig. P7.11. The tensions T_1 and T_2 satisfy the following system of equations:

$$0.6\,T_1 + 0.8\,T_2 = 100 \qquad (7.104)$$
$$0.8\,T_1 - 0.6\,T_2 = 0. \qquad (7.105)$$

(a) Write the system of equations (7.104) and (7.105) in the matrix form $A\,x = b$, where

$$\mathbf{x} = \begin{bmatrix} T_1 \\ T_2 \end{bmatrix}.$$

In other words, find matrices A and b.

(b) Find T_1 and T_2 using the matrix algebra method. Perform all matrix computation by hand and show all steps.

(c) Find T_1 and T_2 using Cramer's rule.

Figure P7.11 A 100 lb weight suspended by two cables for problem P7-11.

7-12. A two-bar truss supports a weight of $W = 750$ lb as shown in Fig. P7.12. The forces F_1 and F_2 satisfy the following system of equations:

$$0.866\,F_1 = F_2 \qquad (7.106)$$
$$0.5\,F_1 = 750. \qquad (7.107)$$

(a) Write the system of equations (7.106) and (7.107) in the matrix form $A\,F = b$, where

$$\mathbf{F} = \begin{bmatrix} F_1 \\ F_2 \end{bmatrix}.$$

In other words, find matrices A and b.

(b) Find F_1 and F_2 using the matrix algebra method. Perform all matrix computation by hand and show all steps.

(c) Find F_1 and F_2 using Cramer's rule.

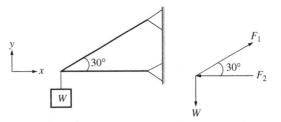

Figure P7.12 A two-bar truss supporting a weight for problem P7-12.

7-13. A 10 kg object is suspended from a two-bar truss shown in Fig. P7.13. The forces F_1 and F_2 satisfy the following system of equations:

$$0.707\,F_1 = F_2 \qquad (7.108)$$
$$0.707\,F_1 = 98.1. \qquad (7.109)$$

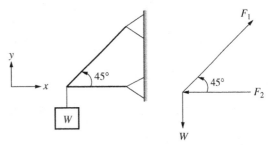

Figure P7.13 An object suspended from a two-bar truss for problem P7-13.

(a) Write the system of equations (7.108) and (7.109) in the matrix form $A\,F = b$, where

$$F = \begin{bmatrix} F_1 \\ F_2 \end{bmatrix}.$$

In other words, find matrices A and b.

(b) Find F_1 and F_2 using the matrix algebra method. Perform all matrix computation by hand and show all steps.

(c) Find F_1 and F_2 using Cramer's rule.

7-14. A $F = 100$ N force is applied to a two-bar truss as shown in Fig. P7.14. The forces F_1 and F_2 satisfy the following system of equations:

$$-0.5548\,F_1 - 0.8572\,F_2 = -100 \qquad (7.110)$$
$$0.832\,F_1 = 0.515F_2. \qquad (7.111)$$

(a) Write the system of equations (7.110) and (7.111) in the matrix form $A\,F = b$, where

$$F = \begin{bmatrix} F_1 \\ F_2 \end{bmatrix}.$$

In other words, find matrices A and b.

(b) Find F_1 and F_2 using the matrix algebra method. Perform all matrix computation by hand and show all steps.

(c) Find F_1 and F_2 using Cramer's rule.

Figure P7.14   A 100 N force applied to a two-bar truss for problem P7-14.

7-15. A force $F = 200$ lb is applied to a two-bar truss as shown in Fig. P7.15. The forces F_1 and F_2 satisfy the following system of equations:

$$0.866\,F_1 + 0.707\,F_2 = 200 \qquad (7.112)$$
$$0.5\,F_1 - 0.707\,F_2 = 0. \qquad (7.113)$$

(a) Write the system of equations (7.112) and (7.113) in the matrix form $A\,F = b$, where

$$F = \begin{bmatrix} F_1 \\ F_2 \end{bmatrix}.$$

In other words, find matrices A and b.

(b) Find F_1 and F_2 using the matrix algebra method. Perform all matrix computation by hand and show all steps.

(c) Find F_1 and F_2 using Cramer's rule.

Figure P7.15   A 200 lb force applied to a two-bar truss for problem P7-15.

7-16. The weight of a vehicle is supported by reaction forces at its front and rear wheels as shown in Fig. P7.16. If the weight of the vehicle is $W = 4800$ lb, the reaction forces R_1 and R_2 satisfy the following system of equations:

$$R_1 + R_2 - 4800 = 0 \qquad (7.114)$$
$$6\,R_1 - 4\,R_2 = 0. \qquad (7.115)$$

(a) Find R_1 and R_2 using the substitution method.

(b) Write the system of equations (7.114) and (7.115) in the matrix

form $A\,x = b$, where

$$\mathbf{x} = \begin{bmatrix} R_1 \\ R_2 \end{bmatrix}.$$

In other words, find matrices A and b.

(c) Find R_1 and R_2 using the matrix algebra method. Perform all matrix computation by hand and show all steps.

(d) Find R_1 and R_2 using Cramer's rule.

Figure P7.16   A vehicle supported by reaction forces.

7-17. Solve problem P7-16 if the system of equations are given by equations (7.116) and (7.117).

$$R_1 + R_2 - 3000 = 0 \qquad (7.116)$$
$$7\,R_1 - 3\,R_2 = 0. \qquad (7.117)$$

7-18. Solve problem P7-16 if the system of equations are given by (7.118) and (7.119).

$$R_1 + R_2 = 5000 \qquad (7.118)$$
$$4\,R_1 = 6\,R_2. \qquad (7.119)$$

7-19. A vehicle weighing $W = 10$ kN is parked on an inclined driveway ($\theta = 21.8°$) as shown in Fig. P7.19. The forces F and N satisfy the following system of equations:

$$0.9285\,F = 0.3714\,N \qquad (7.120)$$
$$0.9285\,N + 0.3714\,F - 10 = 0 \qquad (7.121)$$

where forces F and N are in kN.

(a) Write the system of equations (7.120) and (7.121). in the matrix form $A\,x = b$, where

$$\mathbf{x} = \begin{bmatrix} F \\ N \end{bmatrix}.$$

In other words, find matrices A and b.

(b) Find F and N using the matrix algebra method. Perform all matrix computations by hand and show all steps.

(c) Find F and N using Cramer's rule.

Figure P7.19   A truck parked on an inclined driveway.

7-20. A vehicle weighing $W = 2000$ lb is parked on an inclined driveway ($\theta = 35°$) as shown in Fig. P7.19. The forces F and N satisfy the following system of equations:

$$-0.8192\,F + 0.5736\,N = 0 \qquad (7.122)$$
$$0.8192\,N + 0.5736\,F = 2000. \qquad (7.123)$$

where forces F and N are in lb. Solve parts (a)–(c) of problem P7-19 for system of equations given by (7.122) and (7.123).

7-21. A 100 lb crate weighing $W = 100$ lb is pushed against an incline ($\theta = 30°$) with a force of $F_a = 30$ lb as shown in Fig. P7.21. The forces F_f and N satisfy the following system of equations:

$$0.866\,F_f - 0.5N + 30 = 0 \qquad (7.124)$$
$$0.866\,N + 0.5\,F_f - 100 = 0. \qquad (7.125)$$

(a) Write the system of equations (7.124) and (7.125) in the matrix form $A x = b$, where

$$\mathbf{x} = \begin{bmatrix} F_f \\ N \end{bmatrix}.$$

In other words, find matrices A and b.

(b) Find F_f and N using the matrix algebra method. Perform all matrix computation by hand and show all steps.

(c) Find F_f and N using Cramer's rule.

(a) Find N_f and a using the substitution method.

(b) Write the system of equations (7.128) and (7.129) in the matrix form $A x = b$, where

$$\mathbf{x} = \begin{bmatrix} N_f \\ a \end{bmatrix}.$$

In other words, find matrices A and b.

(c) Find N_f and a using the matrix algebra method.

(d) Find N_f and a using Cramer's rule.

Figure P7.21   A crate being pushed by a force.

Figure P7.23   Desk being pulled by a force F.

7-22. A crate weighing W = 500 N is pushed against an incline ($\theta = 20°$) with a force of $F_a = 100$ N as shown in Fig. P7.21. The forces F_f and N satisfy the following system of equations:

$$0.9397\, F_f + 100 = 0.342\, N \quad (7.126)$$
$$0.9397\, N + 0.342\, F_f - 500 = 0. \quad (7.127)$$

Solve parts (a)–(c) of problem P7-21 for system of equations given by (7.126) and (7.127).

7-23. A 100 kg desk rests on a horizontal plane with a coefficient of friction of $\mu = 0.4$. It is pulled with a force of $F = 800$ N at an angle of $\theta = 45°$ as shown in Fig. P7.23. The normal force N_f and acceleration a in m/s^2 satisfy the following system of equations:

$$565.6 - 0.4\, N_f = 100\, a \quad (7.128)$$
$$N_f + 565.5 = 9.81 \times 100. \quad (7.129)$$

7-24. A 200 kg desk rests on a horizontal plane with a coefficient of friction of $\mu = 0.3$. It is pulled with a force of $F = 1000$ N at an angle of $\theta = 20°$ as shown in Fig. P7.23. The normal force N_f and acceleration a in m/s^2 satisfy the following system of equations:

$$939.7 - 0.3\, N_f = 200\, a \quad (7.130)$$
$$N_f + 342 = 9.81 \times 200. \quad (7.131)$$

Solve parts (a)–(d) of problem P7-23 for system of equations (7.130) and (7.131).

7-25. The weight of a vehicle is supported by reaction forces at its front and rear wheels as shown in Fig. P7.25. The reaction forces R_1 and R_2 satisfy the following system of equations:

$$R_1 + R_2 - m\, g = 0 \quad (7.132)$$
$$l_1\, R_1 - l_2\, R_2 - m\, a\, k = 0. \quad (7.133)$$

(a) Write the system of equations (7.132) and (7.133) if $l_1 = 2$ m, $l_2 = 1.5$ m, $k = 1.5$ m, $g = 9.81$ m/s^2, $m = 1000$ kg, and $a = 5$ m/s^2.

(b) For the system of equations found in part (a), find R_1 and R_2 using the substitution method.

(c) Write the system of equations in the matrix form $A x = b$, where

$$\mathbf{x} = \begin{bmatrix} R_1 \\ R_2 \end{bmatrix}.$$

In other words, find matrices A and b.

(d) Find R_1 and R_2 using the matrix algebra method.

(e) Find R_1 and R_2 using Cramer's rule.

Figure P7.25 A vehicle supported by reaction forces.

7-26. Repeat problem P7-25, if $l_1 = 2$ m, $l_2 = 1.5$ m, $k = 1.5$ m, $g = 9.81$ m/s^2, $m = 1000$ kg, and $a = -5$ m/s^2.

7-27. Repeat problem P7-25, if $l_1 = 2$ m, $l_2 = 2$ m, $k = 1.5$ m, $g = 9.81$ m/s^2, $m = 1200$ kg, and $a = 4.5$ m/s^2.

7-28. Repeat problem P7-25, if $l_1 = 2$ m, $l_2 = 2$ m, $k = 1.5$ m, $g = 9.81$ m/s^2, $m = 1200$ kg, and $a = -4.5$ m/s^2

7-29. A driver applies a steady force of $F_P = 25$ N against a gas pedal, as shown in Fig. P7.29. The free-body diagram of the driver's foot is also shown. Based on the

x-y coordinate system shown, the force of the gastrocnemius muscle F_m and the weight of the foot W_F satisfy the following system of equations:

$$F_m \cos 45° - W_F \cos 30° = R_x \quad (7.134)$$
$$F_m \sin 45° - W_F \sin 30° = R_y - F_P. \quad (7.135)$$

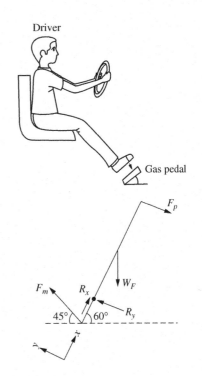

Figure P7.29 A driver applying a steady force against the gas pedal.

(a) Knowing that $R_x = 70/\sqrt{2}$ N and $R_y = 105$ N, rewrite the system of equations (7.134) and (7.135) in terms of F_m and W_F.

(b) Find F_m and W_F using the substitution method.

(c) Write the system of equations obtained in part (a) in the matrix form $\mathbf{A x = b}$, where $\mathbf{x} = \begin{bmatrix} F_m \\ W_F \end{bmatrix}$.

(d) Find F_m and W_F using the matrix algebra method. Perform all computations by hand and show all steps.

(e) Find F_m and W_F using Cramer's rule.

7-30. An environmental engineer wishes to blend a single mixture of insecticide spray solution of volume $V = 500$ L with a specified concentration $C = 0.2$ from two spray solutions of concentration $c_1 = 0.15$ and $c_2 = 0.25$. The required volumes of the two spray solutions v_1 and v_2 can be determined from a system of equations describing conditions for volume and concentration, respectively as

$$v_1 + v_2 = V \qquad (7.136)$$

$$c_1 v_1 + c_2 v_2 = C V \qquad (7.137)$$

(a) Knowing that $V = 500$ L, $c_1 = 0.15$, $c_2 = 0.25$, and $C = 0.2$, rewrite the system of equations (7.136) and (7.137) in terms of v_1 and v_2.

(b) Find v_1 and v_2 using the substitution method.

(c) Write the system of equations obtained in part (a) in the matrix form $\mathbf{A}\mathbf{x} = \mathbf{b}$, where $\mathbf{x} = \begin{bmatrix} v_1 \\ v_2 \end{bmatrix}$.

(d) Find v_1 and v_2 using the matrix algebra method. Perform all computations by hand and show all steps.

(e) Find v_1 and v_2 using Cramer's rule.

7-31. Repeat problem P7-30 if $V = 400$ L, $C = 0.25$, $c_1 = 0.2$, and $c_2 = 0.4$.

7-32. Consider the two-loop circuit shown in Fig. P7.32. The currents I_1 and I_2 (in A) satisfy the following system of

equations:

$$(0.1\,s + 1)\,I_1 - I_2 = \frac{100}{s} \qquad (7.138)$$

$$-1\,I_1 + \left(1 + \frac{2}{s}\right) I_2 = 0. \qquad (7.139)$$

Figure P7.32   Two-loop circuit for problem P7-32.

(a) Find I_1 and I_2 using the substitution method.

(b) Write the system of equations (7.138) and (7.139) in the matrix form $A\,I = b$, where $I = \begin{bmatrix} I_1 \\ I_2 \end{bmatrix}$.

(c) Find I_1 and I_2 using the matrix algebra method.

(d) Find I_1 and I_2 using Cramer's rule.

7-33. Consider the two-loop circuit shown in Fig. P7.33. The currents I_1 and I_2 (in A) satisfy the following system of equations:

$$(0.2\,s + 10)\,I_1 - 0.2\,s\,I_2 = \frac{100}{s} \qquad (7.140)$$

$$-0.2\,s\,I_1 + \left(0.2\,s + \frac{10}{s}\right) I_2 = 0. \qquad (7.141)$$

Figure P7.33   Two-loop circuit for problem P7-33.

(a) Find I_1 and I_2 using the substitution method.

(b) Write the system of equations (7.140) and (7.141) in the matrix form $A\,I = b$, where $I = \begin{bmatrix} I_1 \\ I_2 \end{bmatrix}$.

(c) Find I_1 and I_2 using the matrix algebra method.

(d) Find I_1 and I_2 using Cramer's rule.

7-34. Consider the two-node circuit shown in Fig. P7.34. The voltages V_1 and V_2 (in V) satisfy the following system of equations:

$$\left(\frac{5}{s} + 0.1\right) V_1 - 0.1\, V_2 = \frac{0.1}{s} \quad (7.142)$$

$$\left(0.1 + \frac{5}{s}\right) V_2 - 0.1\, V_1 = \frac{0.2}{s}. \quad (7.143)$$

Figure P7.34 Two-loop circuit for problem P7-34.

(a) Write the system of equations (7.142) and (7.143) in the matrix form $A\,V = b$, where $V = \begin{bmatrix} V_1 \\ V_2 \end{bmatrix}$.

(b) Find V_1 and V_2 using the matrix algebra method.

(c) Find V_1 and V_2 using Cramer's rule.

7-35. A mechanical system along with its free body diagram is shown in Fig. P7.35, where $f = 50\,\text{N}$ is the applied force. The linear displacements $X_1(s)$ and $X_2(s)$ of the two springs in the s-domain satisfy the following system of equations:

$$40\,X_1(s) - 40\,X_2(s) = \frac{50}{s} \quad (7.144)$$

$$-40\,X_1(s) + (s^2 + 65)\,X_2(s) = 0 \quad (7.145)$$

(a) Write the system of equations (7.144) and (7.145) in the matrix form $A\,X(s) = b$, where $X(s) = \begin{bmatrix} X_1(s) \\ X_2(s) \end{bmatrix}$.

(b) Find the expressions of $X_1(s)$ and $X_2(s)$ using the matrix algebra method.

(c) Find the expressions of $X_1(s)$ and $X_2(s)$ using Cramer's rule.

7-36. Fig. P7.36 shows a system with two-mass elements and two springs. The mass m_1 is pulled with a force $f = 100\,\text{N}$, and the displacements $X_1(s)$ and $X_2(s)$ of the two masses in the s-domain satisfy the

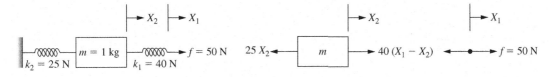

Figure P7.35 Mechanical system for problem P7-35.

Figure P7.36 Mechanical system for problem P7-36.

following system of equations:

$$(2s^2 + 50)\,X_1(s) - 50\,X_2(s) = \frac{100}{s}$$
(7.146)

$$-50\,X_1(s) + (2s^2 + 100)\,X_2(s) = 0$$
(7.147)

(a) Write the system of equations (7.146) and (7.146) in the matrix form $AX(s) = b$, where $X(s) = \begin{bmatrix} X_1(s) \\ X_2(s) \end{bmatrix}$.

(b) Find the expressions of $X_1(s)$ and $X_2(s)$ using the matrix algebra method.

(c) Find the expressions of $X_1(s)$ and $X_2(s)$ using Cramer's rule.

7-37. A co-op student at a composite materials manufacturing company is attempting to determine how much of two different composite materials to make from the available quantities of carbon fiber and resin. The student knows that Composite A requires 0.15 pounds of carbon fiber and 1.6 ounces of polymer resin while Composite B requires 3.2 pounds of carbon fiber and 2.0 ounces of polymer resin. If the available quantities of carbon fiber and resin are 15 pounds and 10 ounces, respectively, the amount of each composite to manufacture satisfies the system of equations:

$$0.15\,x_A + 3.2\,x_B = 15 \qquad (7.148)$$
$$1.6\,x_A + 2.0\,x_B = 10, \qquad (7.149)$$

where x_A and x_B are the amounts of Composite A and B (in pounds), respectively.

(a) Find x_A and x_B using the substitution method.

(b) Write the system of equations (7.148) and (7.149) in the matrix form $A\,x = b$, where $x = \begin{bmatrix} x_A \\ x_B \end{bmatrix}$.

(c) Find x_A and x_B using the matrix algebra method. Perform computation by hand and show all steps.

(d) Find x_A and x_B using Cramer's rule.

7-38. Repeat parts (a)–(d) of problem P7-37 if Composite A requires 1.6 pounds of carbon fiber and 0.5 ounces of polymer resin while Composite B requires 1.75 pounds of carbon fiber and 0.78 ounces of polymer resin. If the available quantities of carbon fiber and resin are 17 pounds and 12 ounces, respectively, the amount of each composite to manufacture satisfies the system of equations (7.150) and (7.151).

$$0.16\,x_A + 2.5\,x_B = 17 \qquad (7.150)$$
$$1.75\,x_A + 0.78\,x_B = 12. \qquad (7.151)$$

7-39. A structural engineer is performing a finite element analysis on an aluminum truss support structure subjected to a load as shown in Fig. P7.39. By the finite

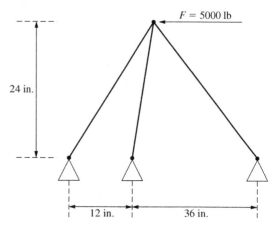

Figure P7.39 Finite element idealization of truss structure subject to loading.

element method, the displacements in the horizontal and vertical directions at the node where the load is applied can be determined from the system of equations:

$$2.57\,u + 3.33\,v = -0.05 \qquad (7.152)$$
$$3.33\,u + 6.99\,v = 0, \qquad (7.153)$$

where u and v are the the displacements in the horizontal and vertical directions, respectively.

(a) Find u and v using the substitution method.

(b) Write the system of equations (7.152) and (7.153) in the matrix form $A\,x = b$, where $x = \begin{bmatrix} u \\ v \end{bmatrix}$.

(c) Find u and v using the matrix algebra method. Perform computation by hand and show all steps.

(d) Find u and v using Cramer's rule.

7-40. A structural engineer is designing a crane that is to be used to unload barges at a dock. She is using a technique called finite element analysis to determine the displacement at the end of the crane. The finite element idealization of the crane structure picking up a 2000 lb load is shown in Fig. P7.40. The displacement in the horizontal and vertical directions at the end of the crane

can be determined from the system of equations:

$$2.47\,u + 1.08\,v = 0 \qquad (7.154)$$
$$1.08\,u + 0.49\,v = -0.02, \qquad (7.155)$$

where u and v are the the displacements in the horizontal and vertical directions, respectively.

(a) Find u and v using the substitution method.

(b) Write the system of equations (7.154) and (7.155) in the matrix form $A\,x = b$, where $x = \begin{bmatrix} u \\ v \end{bmatrix}$.

(c) Find u and v using the matrix algebra method. Perform computation by hand and show all steps.

(d) Find u and v using Cramer's rule.

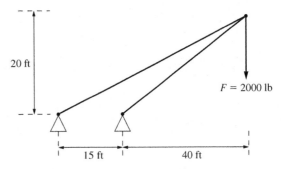

Figure P7.40   Finite element idealization of crane structure subject to loading.

CHAPTER 8 Derivatives in Engineering

8.1 INTRODUCTION

This chapter will discuss what a derivative is and why it is important in engineering. The concepts of maxima and minima along with the applications of derivatives to solve engineering problems in dynamics, electric circuits, and mechanics of materials are emphasized.

8.1.1 What Is a Derivative?

To explain what a derivative is, an engineering professor asks a student to drop a ball (shown in Fig. 8.1) from a height of $y = 1.0$ m to find the time when it impacts the ground. Using a high-resolution stopwatch, the student measures the time at impact as $t = 0.452$ s. The professor then poses the following questions:

(a) What is the average velocity of the ball?

(b) What is the speed of the ball at impact?

(c) How fast is the ball accelerating?

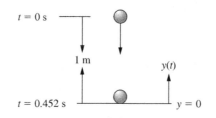

Figure 8.1   A ball dropped from a height of 1 meter.

Using the given information, the student provides the following answers:

(a) **Average Velocity, $\bar{v}$:** The average velocity is the total distance traveled per unit time, i.e.,

$$\bar{v} = \frac{\text{Total distance}}{\text{Total time}} = \frac{\Delta y}{\Delta t} = \frac{y_2 - y_1}{t_2 - t_1}$$

$$= -\frac{0 - 1.0}{0.452 - 0}$$

$$= -\frac{1.0}{0.452}$$

$$= -2.21 \text{ m/s}.$$

Note that the negative sign means the ball is moving in the negative y-direction.

(b) **Speed at Impact:** The student finds that there is not enough information to find the speed of ball when it impacts the ground. Using an ultrasonic motion detector in the laboratory, the student repeats the experiment and collects the data given in Table 8.1.

TABLE 8.1 Additional data collected from the dropped ball.

t, s	0	0.1	0.2	0.3	0.4	0.452
$y(t)$, m	1.0	0.951	0.804	0.559	0.215	0

The student then calculates the average velocity $\bar{v} = \Delta y / \Delta t$ in each interval. For example, in the interval $t = [0, 0.1]$, $\bar{v} = \dfrac{0.951 - 1.0}{0.1 - 0} = -0.490$ m/s. The average velocity in the remaining intervals is given in Table 8.2.

TABLE 8.2 Average velocity of the ball in different intervals.

Interval	[0, 0.1]	[0.1, 0.2]	[0.2, 0.3]	[0.3, 0.4]	[0.4, 0.452]
$\bar{v}$, m/s	−0.490	−1.47	−2.45	−3.44	−4.13

The student proposes an approximate answer of −4.13 m/s as the speed of impact with ground, but claims that he/she would need an infinite (∞) number of data points to get it exactly right, i.e.,

$$v(t = 0.452) = \lim_{t \to 0.452} \frac{y(0.452) - y(t)}{0.452 - t}.$$

The professor suggests that this looks like the **definition of a derivative**, i.e.,

$$v(t) = \lim_{\Delta t \to 0} \frac{y(t + \Delta t) - y(t)}{\Delta t} = \frac{dy}{dt}$$

where $\Delta t = 0.452 - t$.

The derivatives of some common functions in engineering are given below. Note that ω, a, n, c, c_1, and c_2 are constants and not functions of t.

TABLE 8.3 Some common derivatives used in engineering.

Function, $f(t)$	Derivative, $\dfrac{df(t)}{dt}$
$\sin(\omega t)$	$\omega\cos(\omega t)$
$\cos(\omega t)$	$-\omega\sin(\omega t)$
e^{at}	$a\,e^{at}$
t^n	$n\,t^{n-1}$
$c f(t)$	$c\dfrac{df(t)}{dt}$
c	0
$c_1 f_1(t) + c_2 f_2(t)$	$c_1\dfrac{df_1(t)}{dt} + c_2\dfrac{df_2(t)}{dt}$
$f(t)\cdot g(t)$	$f(t)\dfrac{dg(t)}{dt} + g(t)\dfrac{df(t)}{dt}$
$f(g(t))$	$\dfrac{df}{dg}\times\dfrac{dg(t)}{dt}$

The professor then suggests a quadratic curve fit of the measured data, which gives

$$y(t) = 1.0 - 4.905\,t^2.$$

The velocity at any time is thus calculated by taking the derivative as

$$v(t) = \frac{dy}{dt}$$

$$= \frac{d}{dt}\left(1.0 - 4.905\,t^2\right)$$

$$= \frac{d}{dt}(1.0) - 4.905\frac{d}{dt}\left(t^2\right)$$

$$= 0 - 4.905(2\,t)$$

$$= -9.81\,t \ \text{m/s}.$$

(c) The student is now asked to find the acceleration without taking any more data. The **acceleration is the rate of change of velocity**, i.e.,

$$a(t) = \lim_{\Delta t \to 0} \frac{\Delta v(t)}{\Delta t}$$

$$= \frac{dv(t)}{dt}$$

$$= \frac{d}{dt} \frac{dy(t)}{dt}$$

$$= \frac{d^2 y(t)}{dt^2}.$$

Thus, if $v(t) = -9.81\, t$, then

$$a(t) = \frac{d}{dt}(-9.81\, t)$$

$$= -9.81 \ \text{m/s}^2.$$

Hence, the **acceleration due to gravity is constant and is equal to -9.81 m/s^2**.

8.2   MAXIMA AND MINIMA

Suppose now that the ball is thrown upward with an initial velocity $v_o = 4.43$ m/s as shown in Fig. 8.2.

(a) How long does it take for the ball to reach its maximum height?
(b) What is the velocity at $y = y_{max}$?
(c) What is the **maximum** height y_{max} achieved by the ball?

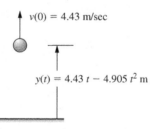

$v(0) = 4.43$ m/sec

$y(t) = 4.43\, t - 4.905\, t^2$ m

Figure 8.2   A ball thrown upward.

The professor suggests that the height of the ball is governed by the quadratic equation

$$y(t) = 4.43\, t - 4.905\, t^2 \ \ \text{m,} \tag{8.1}$$

and plotted as shown in Fig. 8.3.

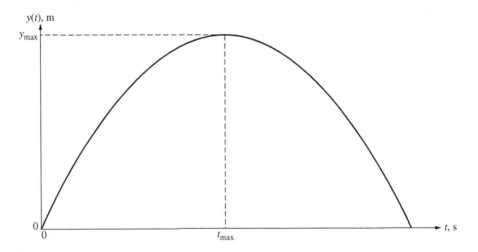

Figure 8.3 The height of the ball thrown upward.

Based on the definition of the derivative, the **velocity** $v(t)$ **at any time** t **is the slope of the line tangent to** $y(t)$ **at that instant**, as shown in Fig. 8.4. Therefore, at the time when $y = y_{max}$, the slope of the tangent line is zero (Fig. 8.5).

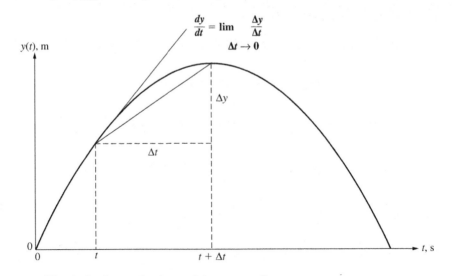

Figure 8.4 The derivative as the slope of the tangent line.

For the problem at hand, the velocity is given by

$$v(t) = \frac{dy(t)}{dt}$$

$$= \frac{d}{dt}\left(4.43\,t - 4.905\,t^2\right)$$

$$= 4.43 - 9.81\,t \text{ m/s}. \tag{8.2}$$

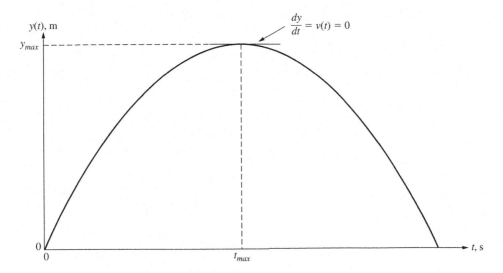

Figure 8.5   The slope of the tangent line at maximum height.

Given equation (8.2), the student answers the professor's questions as follows:

(a) **How long does it take to reach maximum height?** At the time of maximum height $t = t_{max}$, $v(t) = \dfrac{dy(t)}{dt} = 0$. Hence, setting $v(t)$ in equation (8.2) to zero gives

$$4.43 - 9.81 t_{max} = 0$$

$$t_{max} = \frac{4.43}{9.81}$$

or

$$t_{max} = 0.4515 \text{ s}.$$

Therefore, it takes 0.4515 s for the ball to reach the maximum height.

(b) **What is the velocity at $y = y_{max}$?**   Since the slope of the height at $t = t_{max}$ is zero, the velocity at $y = y_{max}$ is zero. The plot of the velocity, $v(t) = 4.43 - 9.81 t$ for times $t = 0$ to $t = 0.903$ s, is shown in Fig. 8.6. It can be seen that the velocity is maximum at $t = 0$ s (initial velocity = 4.43 m/s), reduces to 0 at $t = 0.4515$ s ($t = t_{max}$), and reaches a minimum value (-4.43 m/s) at $t = 0.903$ sec.

(c) **The maximum height:** The maximum height can now be obtained by substituting $t = t_{max} = 0.4515$ s in equation (8.1) for $y(t)$

$$y_{max} = y(t_{max})$$

$$= 4.43 \, (0.4515) - 4.905 \, (0.4515)^2$$

$$= 1.0 \text{ m}.$$

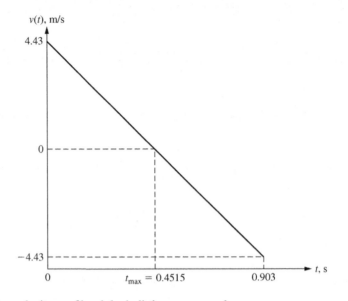

Figure 8.6   The velocity profile of the ball thrown upward.

It should be noted that the **derivative of a function is zero both at the points where the value of the function is maximum (maxima) and where the value of the function is minimum (minima).** So if the derivative is zero at both maxima and minima, how can one tell whether the value of the function found earlier is a maximum or a minimum? Consider the function shown in Fig. 8.7, which has local maximum and minimum values. As discussed earlier, the derivative of a function at a

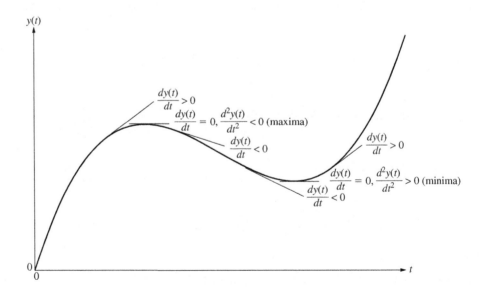

Figure 8.7   Plot of a function with local maximum and minimum values.

point is the slope of the tangent line at that point. At its maximum value, the derivative (slope) of the function shown in Fig. 8.7 changes from positive to negative. At its minimum value, the derivative (slope) of the function changes from negative to positive. In other words, the rate of change of the derivative (or the second derivative of the function) is negative at maxima and positive at minima. Therefore, to test for maxima and minima, the following rules apply:

At a Local Maximum:

$$\frac{dy(t)}{dt} = 0, \quad \frac{d^2y(t)}{dt^2} < 0$$

At a Local Minimum:

$$\frac{dy(t)}{dt} = 0, \quad \frac{d^2y(t)}{dt^2} > 0$$

To test whether the point where the slope (first derivative) of the trajectory of the ball thrown upward is zero is a maximum or a minimum, the student obtains the second derivative of the height as

$$\frac{d^2y(t)}{dt^2} = \frac{d}{dt}(4.43 - 9.81\,t) = -9.81 \ < \ 0.$$

Therefore, the point where the slope of the trajectory of the ball is zero is a maximum and thus the maximum height is 1.0 m.

In general, the procedure of finding the local maxima and minima of any function $f(t)$ is as follows:

(a) Find the derivative of the function with respect to t; in other words, find $f'(t) = \frac{df(t)}{dt}$.

(b) Find the solution of the equation $f'(t) = 0$; in other words, find the values of t where the function has a local maximum or a local minimum.

(c) To find which values of t gives the local maximum and which values of t gives the local minimum, determine the second derivative $\left(f''(t) = \frac{d^2f(t)}{dt^2}\right)$ of the function.

(d) Evaluate the second derivative at the values of t found in step (b). If the second derivative is negative $\left(\frac{d^2f(t)}{dt^2} < 0\right)$, the function has a local maximum for those values of t; however, if the second derivative is positive, the function has a local minimum for these values.

(e) Evaluate the function, $f(t)$, at the values of t found in step (b) to find the maximum and minimum values.

8.3   APPLICATIONS OF DERIVATIVES IN DYNAMICS

This section demonstrates the application of derivatives in determining the velocity and acceleration of an object if the position of the object is given. This section also demonstrates the application of derivatives in sketching plots of position, velocity, and acceleration.

8.3.1 Position, Velocity, and Acceleration

Suppose the position $x(t)$ of an object is defined by a linear function with parabolic blends, as shown in Fig. 8.8. This motion is similar to a vehicle starting from rest and accelerating with a maximum positive acceleration (parabolic position) to reach a constant speed, cruising at that constant speed (linear position), and then coming to stop with maximum braking (maximum negative acceleration, parabolic position).

As discussed previously, **velocity** $v(t)$ is the instantaneous rate of change of the position (i.e., the derivative of the position) and is given by

$$v(t) = lim_{\Delta t \to 0} \frac{\Delta x(t)}{\Delta t}$$

or

$$v(t) = \frac{dx(t)}{dt}.$$

Therefore, the velocity $v(t)$ is the slope of the position $x(t)$ as shown in Fig. 8.9. It can be seen from Fig. 8.9 that the object is starting from rest, moves at a linear velocity with positive slope until it reaches a constant velocity, cruises at that constant velocity, and comes to rest again after moving at a linear velocity with a negative slope.

Figure 8.8 Position of an object as a linear function with parabolic blends.

The **acceleration** $a(t)$ is the instantaneous rate of change of the velocity (i.e., the derivative of the velocity):

$$a(t) = \frac{dv(t)}{dt},$$

Figure 8.9 Velocity of the object moving as a linear function with parabolic blends.

or

$$a(t) = \frac{d^2 x(t)}{dt^2}.$$

Therefore, the acceleration $a(t)$ is the slope of the velocity $v(t)$, which is shown in Fig. 8.10. It can be seen from Fig. 8.10 that the object starts with maximum positive acceleration until it reaches a constant velocity, cruises with zero acceleration, and then comes to rest with maximum braking (constant negative acceleration).

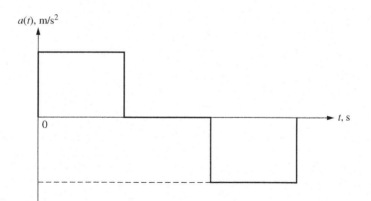

Figure 8.10 Acceleration of the object moving as a linear function with parabolic blends.

The following examples will provide some practice in taking basic derivatives using the formulas in Table 8.3.

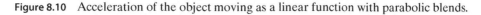

Example 8-1	The motion of the particle shown in Fig. 8.11 is defined by its position $x(t)$. Determine the position, velocity, and acceleration at $t = 0.5$ seconds if

(a) $x(t) = \sin(2\pi t)$ m
(b) $x(t) = 3t^3 - 4t^2 + 2t + 6$ m
(c) $x(t) = 20\cos(3\pi t) - 5t^2$ m

Figure 8.11   A particle moving in the horizontal direction.

Solution   (a) The velocity and acceleration of the particle can be obtained by finding the first and second derivatives of $x(t)$, respectively. Since

$$x(t) = \sin 2\pi t \text{ m},\tag{8.3}$$

the velocity is

$$v(t) = \frac{dx(t)}{dt}$$

$$= \frac{d}{dt}(\sin 2\pi t)$$

or

$$v(t) = 2\pi\cos 2\pi t \text{ m/s}.\tag{8.4}$$

The acceleration of the particle can now be found by differentiating the velocity as

$$a(t) = \frac{dv(t)}{dt}$$

$$= \frac{d}{dt}(2\pi\cos 2\pi t)$$

$$= (2\pi)\frac{d}{dt}(\cos 2\pi t)$$

$$= (2\pi)(-2\pi\sin 2\pi t)$$

or

$$a(t) = -4\pi^2 \sin 2\pi t \text{ m/s}^2.\tag{8.5}$$

The position, velocity, and acceleration of the particle at $t = 0.5$ s can now be calculated by substituting $t = 0.5$ in equations (8.3), (8.4), and (8.5) as

$$x(0.5) = \sin(2\pi(0.5)) = \sin\pi = 0 \text{ m}$$
$$v(0.5) = 2\pi\cos(2\pi(0.5)) = 2\pi\cos\pi = -2\pi \text{ m/s}$$
$$a(0.5) = -4\pi^2\sin(2\pi(0.5)) = -4\pi^2\sin\pi = 0 \text{ m/s}^2.$$

(b) The position of the particle is given by

$$x(t) = 3t^3 - 4t^2 + 2t + 6 \text{ m}.\tag{8.6}$$

The velocity of the particle can be calculated by differentiating equation (8.6) as

$$v(t) = \frac{dx(t)}{dt}$$

$$= \frac{d}{dt}(3t^3 - 4t^2 + 2t + 6)$$

$$= 3\frac{d}{dt}(t^3) - 4\frac{d}{dt}(t^2) + 2\frac{d}{dt}(t) + 6\frac{d}{dt}(1)$$

$$= 3(3t^2) - 4(2t) + 2(1) + 6(0)$$

or

$$v(t) = 9t^2 - 8t + 2 \text{ m/s}. \tag{8.7}$$

The acceleration of the particle can now be obtained by differentiating equation (8.7) as

$$a(t) = \frac{dv(t)}{dt}$$

$$= \frac{d}{dt}(9t^2 - 8t + 2)$$

$$= 9\frac{d}{dt}(t^2) - 8\frac{d}{dt}(t) + 2\frac{d}{dt}(1)$$

$$= 9(2t) - 8(1) + 2(0)$$

or

$$a(t) = 18t - 8 \text{ m/s}^2. \tag{8.8}$$

The position, velocity, and acceleration of the particle at $t = 0.5$ s can now be calculated by substituting $t = 0.5$ in equations (8.6), (8.7), and (8.8) as

$$x(0.5) = 3(0.5)^3 - 4(0.5)^2 + 2(0.5) + 6 = 6.375 \text{ m}$$
$$v(0.5) = 9(0.5)^2 - 8(0.5) + 2 = 0.25 \text{ m/s}$$
$$a(0.5) = 18(0.5) - 8 = 1.0 \text{ m/s}^2.$$

(c) The position of the particle is given by

$$x(t) = 20\cos(3\pi t) - 5t^2 \text{ m}. \tag{8.9}$$

The velocity of the particle can be calculated by differentiating equation (8.9) as

$$v(t) = \frac{dx(t)}{dt}$$

$$= \frac{d}{dt}(20\cos(3\pi t) - 5t^2)$$

$$= 20\frac{d}{dt}(\cos(3\pi t)) - 5\frac{d}{dt}(t^2)$$

$$= 20(-3\pi\sin(3\pi t)) - 5(2t)$$

or

$$v(t) = -60\,\pi\,\sin(3\,\pi t) - 10\,t \text{ m/s.} \tag{8.10}$$

The acceleration of the particle can now be obtained by differentiating equation (8.10) as

$$a(t) = \frac{dv(t)}{dt}$$

$$= \frac{d}{dt}\left(-60\,\pi\,\sin(3\,\pi t) - 10\,t\right)$$

$$= -60\,\pi\,\frac{d}{dt}\left(\sin(3\,\pi t)\right) - 10\,\frac{d}{dt}\,(t)$$

$$= -60\,\pi\,(3\,\pi\,\cos(3\,\pi t)) - 10\,(1).$$

or

$$a(t) = -180\,\pi^2\,\cos(3\,\pi t) - 10 \text{ m/s}^2 \tag{8.11}$$

The position, velocity, and acceleration of the particle at $t = 0.5$ s can now be calculated by substituting $t = 0.5$ in equations (8.9), (8.10), and (8.11) as

$$x(0.5) = 20\cos\left(3\,\pi(0.5)\right) - 5\,(0.5)^2$$

$$= 20\cos\left(\frac{3\,\pi}{2}\right) - 5\,(0.25)$$

$$= 0 - 1.25$$

$$= -1.25 \text{ m}$$

$$v(0.5) = -60\,\pi\,\sin\left(3\,\pi(0.5)\right) - 10\,(0.5)$$

$$= -60\,\pi\,\sin\left(\frac{3\,\pi}{2}\right) - 5$$

$$= 60\,\pi - 5$$

$$= 183.5 \text{ m/s}$$

$$a(0.5) = -180\,\pi^2\,\cos\left(3\,\pi(0.5)\right) - 10$$

$$= -180\,\pi^2\,\cos\left(\frac{3\,\pi}{2}\right) - 10$$

$$= -180\,\pi^2\,(0) - 10$$

$$= -10 \text{ m/s}^2.$$

The following example will illustrate how derivatives can be used to help sketch functions.

Example 8-2

The motion of a particle shown in Fig. 8.12 is defined by its position $y(t)$ as

$$y(t) = \frac{1}{3}t^3 - 5t^2 + 21t + 10 \text{ m.} \tag{8.12}$$

(a) Determine the value of the position and acceleration when the velocity is zero.
(b) Use the results of part (a) to sketch the graph of the position $y(t)$ for $0 \leq t \leq 9$ s.

Figure 8.12 The position of a particle in the vertical plane.

Solution

(a) The velocity of the particle can be calculated by differentiating equation (8.12) as

$$v(t) = \frac{dy(t)}{dt}$$

$$= \frac{d}{dt}\left(\frac{1}{3}t^3 - 5t^2 + 21t + 10\right)$$

$$= \frac{1}{3}\frac{d}{dt}(t^3) - 5\frac{d}{dt}(t^2) + 21\frac{d}{dt}(t) + 10\frac{d}{dt}(1)$$

$$= \frac{1}{3}(3t^2) - 5(2t) + 21(1) + 10(0)$$

or

$$v(t) = t^2 - 10t + 21 \text{ m/s.} \tag{8.13}$$

The time when the velocity is zero can be obtained by setting equation (8.13) equal to zero as

$$t^2 - 10t + 21 = 0. \tag{8.14}$$

The quadratic equation (8.14) can be solved using one of the methods discussed in Chapter 2. For example, factoring equation (8.14) gives

$$(t - 3)(t - 7) = 0. \tag{8.15}$$

The two solutions of equation (8.15) are given as:

$$t - 3 = 0 \quad \Rightarrow \quad t = 3 \text{ s}$$
$$t - 7 = 0 \quad \Rightarrow \quad t = 7 \text{ s}$$

Note that quadratic equation (8.14) can also be solved using the quadratic formula, which gives

$$t = \frac{10 \pm \sqrt{10^2 - 4(1)(21)}}{2(1)}$$

$$= \frac{10 \pm \sqrt{16}}{2}$$

$$= \frac{10 \pm 4}{2}$$

$$= \frac{10 - 4}{2}, \quad \frac{10 + 4}{2}$$

or

$$t = 3, \ 7 \text{ s}.$$

Therefore, the velocity is zero at both $t = 3$ s and $t = 7$ s. To evaluate the acceleration at these times, an expression for the acceleration is needed. The acceleration of the particle can be obtained by differentiating the velocity of the particle (equation (8.13)) as

$$a(t) = \frac{dv(t)}{dt}$$

$$= \frac{d}{dt}(t^2 - 10t + 21)$$

$$= \frac{d}{dt}(t^2) - 10\frac{d}{dt}(t) + 21\frac{d}{dt}(1)$$

or

$$a(t) = 2t - 10 \text{ m/s}^2. \tag{8.16}$$

The position and acceleration at time $t = 3$ s can be found by substituting $t = 3$ in equations (8.12) and (8.16) as

$$y(3) = \frac{1}{3}(3)^3 - 5(3)^2 + 21(3) + 10 = 37 \text{ m}$$

$$a(3) = 2(3) - 10 = -4 \text{ m/s}^2.$$

Similarly, the position and acceleration at $t = 7$ s can be found by substituting $t = 7$ in equations (8.12) and (8.16) as

$$y(7) = \frac{1}{3}(7)^3 - 5(7)^2 + 21(7) + 10 = 26.3 \text{ m}$$

$$a(7) = 2(7) - 10 = 4 \text{ m/s}^2.$$

(b) The results of part (a) can be used to sketch the graph of the position $y(t)$. It was shown in part (a) that the velocity of the particle is zero at $t = 3$ s and $t = 7$ s. Since the velocity is the derivative of the position, the derivative of the position at $t = 3$ s and $t = 7$ s is zero (i.e., the slope is zero). What this means is that the position $y(t)$ has a local minimum or maximum at $t = 3$ s and $t = 7$ s. To check whether $y(t)$ has a local minimum or maximum, the second derivative (acceleration) test is applied. Since the acceleration at $t = 3$ s is negative ($a(3) = -4$ m/s^2), the position $y(3) = 37$ m is a local maximum. Since the acceleration at $t = 7$ s is positive ($a(7) = 4$ m/s^2), the position $y(7) = 26.3$ m is a local minimum. This information, along with the positions of the particle at $t = 0$ ($y(0) = 10$ m) and $t = 9$ ($y(9) = 37$ m), can be used to sketch the position $y(t)$, as shown in Fig. 8.13.

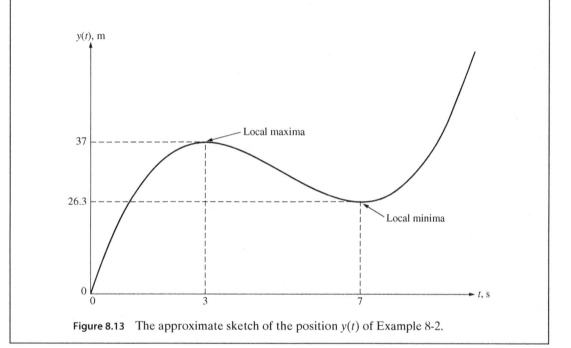

Figure 8.13 The approximate sketch of the position $y(t)$ of Example 8-2.

Derivatives are frequently used in engineering to help sketch functions for which no equation is given. Such is the case in the following example, which begins with a plot of the acceleration $a(t)$.

Example 8-3

The acceleration of a vehicle is measured as shown in Fig. 8.14. Knowing that the particle starts from rest at position $x = 0$ and travels a total of 16 m, sketch plots of the position $x(t)$ and velocity $v(t)$.

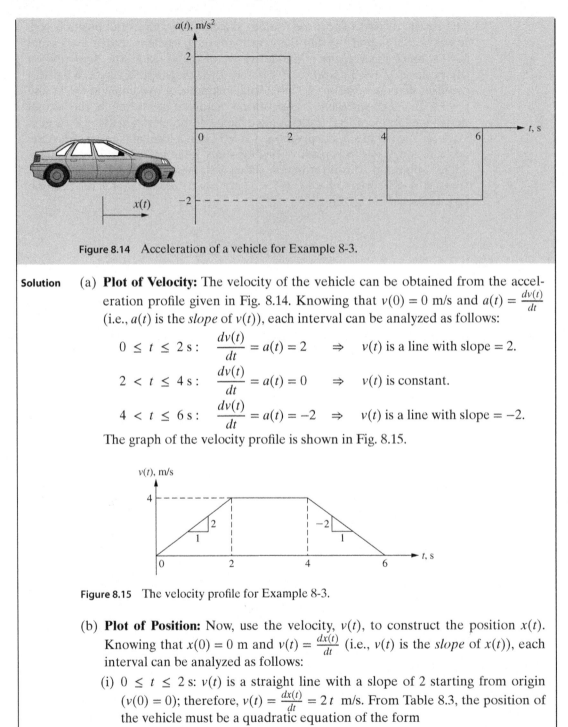

Figure 8.14   Acceleration of a vehicle for Example 8-3.

Solution   (a) **Plot of Velocity:** The velocity of the vehicle can be obtained from the acceleration profile given in Fig. 8.14. Knowing that $v(0) = 0$ m/s and $a(t) = \frac{dv(t)}{dt}$ (i.e., $a(t)$ is the *slope* of $v(t)$), each interval can be analyzed as follows:

$$0 \leq t \leq 2\,\text{s}: \quad \frac{dv(t)}{dt} = a(t) = 2 \quad \Rightarrow \quad v(t) \text{ is a line with slope} = 2.$$

$$2 < t \leq 4\,\text{s}: \quad \frac{dv(t)}{dt} = a(t) = 0 \quad \Rightarrow \quad v(t) \text{ is constant.}$$

$$4 < t \leq 6\,\text{s}: \quad \frac{dv(t)}{dt} = a(t) = -2 \quad \Rightarrow \quad v(t) \text{ is a line with slope} = -2.$$

The graph of the velocity profile is shown in Fig. 8.15.

Figure 8.15   The velocity profile for Example 8-3.

(b) **Plot of Position:** Now, use the velocity, $v(t)$, to construct the position $x(t)$. Knowing that $x(0) = 0$ m and $v(t) = \frac{dx(t)}{dt}$ (i.e., $v(t)$ is the *slope* of $x(t)$), each interval can be analyzed as follows:

(i) $0 \leq t \leq 2$ s: $v(t)$ is a straight line with a slope of 2 starting from origin ($v(0) = 0$); therefore, $v(t) = \frac{dx(t)}{dt} = 2\,t$ m/s. From Table 8.3, the position of the vehicle must be a quadratic equation of the form

$$x(t) = t^2 + C. \tag{8.17}$$

This can be checked by taking the derivative, for example, $v(t) = \frac{dx(t)}{dt} = \frac{d}{dt}(t^2 + C) = 2t$ m/s. Therefore, the equation of position given by (8.17) is correct. The value of C is obtained by evaluating equation (8.17) at $t = 0$ and substituting the value of $x(0) = 0$ as

$$x(0) = 0 + C$$
$$0 = 0 + C$$
$$C = 0.$$

Therefore, for $0 \leq t \leq 2$ s, $x(t) = t^2$ is a quadratic function with a positive slope (concave up) and $x(2) = 4$ m, as shown in Fig. 8.16.

(ii) $2 < t \leq 4$ s: $v(t)$ has a constant value of 4 m/s, for example, $v(t) = \frac{dx(t)}{dt} = 4$. Therefore, $x(t)$ is a straight line with a slope of 4 m/s starting with a value of 4 m at $t = 2$ s as shown in Fig. 8.16. Since the slope is 4 m/s, the position increases by 4 m every second. So during the two seconds between $t = 2$ and $t = 4$, its position increases by 8 m. And since the position at time $t = 2$ s was 4 m, its position at time $t = 4$ s will be 4 m + 8 m = 12 m. The equation of position for $2 < t \leq 4$ s can be written as

$$x(t) = 4 + 4(t - 2) = 4t - 4 \text{ m}.$$

(iii) $4 < t \leq 6$ s: $v(t)$ is a straight line with a slope of -2 m/s; therefore, $x(t)$ is a quadratic function with decreasing slope (concave down) starting at $x(4) = 12$ m and ending at $x(6) = 16$ m with zero slope. The resulting graph of the position is shown in Fig. 8.16.

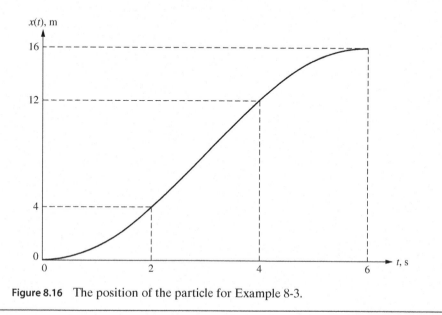

$x(t)$, m

Figure 8.16   The position of the particle for Example 8-3.

Example 8-4

The position of the cart moving on frictionless rollers shown in Fig. 8.17 is given by

$$x(t) = \cos(\omega t) \ m,$$

where $\omega = 2\pi$.

(a) Find the velocity of the cart.
(b) Show that the acceleration of the cart is given by $a(t) = -\omega^2 \cos(\omega t)$ m/s^2, where $\omega = 2\pi$.

Figure 8.17   A cart moving on frictionless rollers.

Solution

(a) The velocity $v(t)$ of the cart is obtained by differentiating the position $x(t)$ as

$$v(t) = \frac{dx(t)}{dt}$$

$$= \frac{d}{dt}\left(\cos(2\pi t)\right)$$

$$= -2\pi \sin(2\pi t) \ \text{m/s}.$$

(b) The acceleration $a(t)$ of the cart is obtained by differentiating the position $v(t)$ as

$$a(t) = \frac{dv(t)}{dt}$$

$$= \frac{d}{dt}\left(-2\pi \sin(2\pi t)\right)$$

$$= -2\pi \frac{d}{dt}\left(\sin(2\pi t)\right)$$

$$= -(2\pi)^2 \sin(2\pi t) \ \text{m/s}^2.$$

Note that the second derivative of $\sin(\omega t)$ or $\cos(\omega t)$ is the same function scaled by $-\omega^2$, for example

$$\frac{d^2}{dt^2}\sin(\omega t) = -\omega^2 \sin(\omega t)$$

$$\frac{d^2}{dt^2}\cos(\omega t) = -\omega^2 \cos(\omega t)$$

Example 8-5

An object of mass m moving at velocity v_0 impacts a cantilever beam (of length l and flexural rigidity EI) as shown in Fig. 8.18. The resulting displacement of the beam is given by

$$y(t) = \frac{v_0}{\omega} \sin \omega t \qquad (8.18)$$

where $\omega = \sqrt{\dfrac{3EI}{ml^3}}$ is the angular frequency of displacement. Find the following:

(a) The maximum displacement y_{max}.

(b) The values of the displacement and acceleration when the velocity is zero.

Figure 8.18   A mass impacting a cantilever beam.

Solution

(a) The maximum displacement can be found by first finding the time t_{max} when the displacement is maximum. This is done by equating the derivative of the displacement (or the velocity) to zero, (i.e., $v(t) = \dfrac{dy(t)}{dt} = 0$). Since v_0 and ω are constants, the derivative is given by

$$\begin{aligned} v(t) &= \frac{dy(t)}{dt} \\ &= \frac{v_0}{\omega} \frac{d}{dt} (\sin \omega t) \\ &= \frac{v_0}{\omega} (\omega \cos \omega t) \end{aligned}$$

or

$$v(t) = v_0 \cos \omega t \text{ m/s}. \qquad (8.19)$$

Equating equation (8.19) to zero gives

$$v_0 \cos \omega t_{max} = 0 \quad \Rightarrow \quad \cos \omega t_{max} = 0. \qquad (8.20)$$

The solutions of equation (8.20) are

$$\omega t_{max} = \frac{\pi}{2}, \frac{3\pi}{2}, \dots,$$

or

$$t_{max} = \frac{\pi}{2\omega}, \frac{3\pi}{2\omega}, \dots. \qquad (8.21)$$

Therefore, the displacement of the beam has local maxima or minima at the values of t given by equation (8.21). To find the time when the displacement is maximum, the second derivative rule is applied. The second derivative of the displacement is obtained by differentiating equation (8.19) as

$$\frac{d^2 y(t)}{dt^2} = v_o \frac{d}{dt}(\cos \omega t)$$

$$= v_0 (-\omega \sin \omega t)$$

or

$$\frac{d^2 y(t)}{dt^2} = -v_0 \omega \sin \omega t \ \text{m/s}^2. \qquad (8.22)$$

The value of the second derivative of the displacement at $\frac{\pi}{2\omega}$ is given by

$$\frac{d^2 y\left(\frac{\pi}{2\omega}\right)}{dt^2} = -v_0 \omega \sin\left(\frac{\pi}{2}\right) < 0.$$

Similarly, the value of the second derivative of the displacement at $\frac{3\pi}{2\omega}$ is given by

$$\frac{d^2 y\left(\frac{3\pi}{2\omega}\right)}{dt^2} = -v_0 \omega \sin\left(\frac{3\pi}{2}\right) > 0.$$

Therefore, the displacement is maximum at time

$$t_{max} = \frac{\pi}{2\omega}. \qquad (8.23)$$

The maximum displacement can be found by substituting $t_{max} = \frac{\pi}{2\omega}$ into equation (8.18) as

$$y_{max} = \frac{v_0}{\omega} \sin(\omega t_{max})$$

$$= \frac{v_0}{\omega} \sin \frac{\pi}{2}$$

$$= \frac{v_0}{\omega}$$

$$= \frac{v_0}{\sqrt{\dfrac{3EI}{ml^3}}}$$

or

$$y_{max} = v_0 \sqrt{\frac{ml^3}{3EI}}.$$

Note: The local maxima or minima of trigonometric functions can also be obtained without derivatives. The plot of the beam displacement given by equation (8.18) is shown in Fig. 8.19.

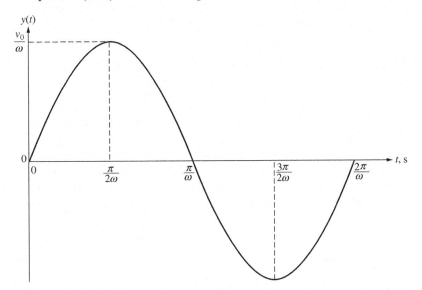

Figure 8.19   Displacement plot to find maximum value.

It can be seen from Fig. 8.19 that the maximum value of the beam displacement is simply the amplitude $y_{max} = \dfrac{v_0}{\omega} = \dfrac{v_0}{\sqrt{\dfrac{3\,E\,I}{m\,l^3}}}$ and the time where the displacement is maximum is given by

$$\omega t_{max} = \frac{\pi}{2}$$

or

$$t_{max} = \frac{\pi}{2\,\omega}.$$

(b) The position when the velocity is zero is simply the maximum value

$$y_{max} = v_0\,\sqrt{\frac{m\,l^3}{3\,E\,I}}. \qquad (8.24)$$

The acceleration found in part (a) is given by $a(t) = -v_0\,\omega\,\sin\omega t$. Therefore, the acceleration when the velocity is zero is given by

$$a(t_{max}) = -v_0\,\omega\,\sin\left(\omega\,t_{max}\right)$$
$$= -v_0\,\omega\,\sin\left(\frac{\pi}{2}\right)$$
$$= -v_0\,\omega$$
$$= -v_0\,\sqrt{\frac{3\,E\,I}{m\,l^3}}.$$

Note: $a(t) = -v_0\,\omega\sin\omega t = -\omega^2\left(\frac{v_0}{\omega}\sin\omega t\right) = -\omega^2\,y(t)$. Therefore, the accelera-tion is maximum when the displacement $y(t)$ is maximum. Since the second deriva-tive of a sinusoid is also a sinusoid of the same frequency (scaled by $-\omega^2$), this is a general result for harmonic motion of any system.

8.4 APPLICATIONS OF DERIVATIVES IN ELECTRIC CIRCUITS

Derivatives play a very important role in electric circuits. For example, the relation-ship between voltage and current for both the inductor and the capacitor is a deriva-tive relationship. The relationship between power and energy is also a derivative relationship. Before discussing the applications of derivatives in electric circuits, the relationship between different variables in circuit elements is discussed briefly here.

Consider a circuit element as shown in Fig. 8.20, where $v(t)$ is the voltage in volts (V) and $i(t)$ is the current in amperes (A). Note that the current always flows through the circuit element and the voltage is always across the element.

Figure 8.20 Voltage and current in a circuit element.

The voltage $v(t)$ is the rate of change of electric potential energy $w(t)$ (in joules (J)) per unit charge $q(t)$ (in coulomb (C)), that is, the voltage is the deriva-tive of the electric potential energy with respect to charge, written as

$$v(t) = \frac{dw}{dq} \ V.$$

The current $i(t)$ is the rate of change (i.e., derivative) of electric charge per unit time (t in s), written as

$$i(t) = \frac{dq(t)}{dt} \ A.$$

The power $p(t)$ (in watts (W)) is the rate of change (i.e., derivative) of electric energy per unit time, written as

$$p(t) = \frac{dw(t)}{dt} \ W.$$

Note that the power can be written as the product of voltage and current using the chain rule of derivatives:

$$p(t) = \frac{dw(t)}{dt} = \frac{dw}{dq} \times \frac{dq}{dt},$$

or

$$p(t) = v(t) \times i(t). \tag{8.25}$$

The chain rule of derivative is a rule for differentiating composition of functions; for example, if f is a function of g and g is a function of t, then the derivative of composite function $f(g(t))$ with respect to t can be written as

$$\frac{df}{dt} = \frac{df}{dg} \times \frac{dg}{dt}. \tag{8.26}$$

For example, the function $f(t) = \sin(2\pi t)$ can be written as $f(t) = \sin(g(t))$, where $g(t) = 2\pi t$. By the chain rule of equation (8.26),

$$\begin{aligned}
\frac{df}{dt} &= \frac{df}{dg} \times \frac{dg}{dt} \\
&= \frac{d}{dt}(\sin(g)) \times \frac{d}{dt}(2\pi t) \\
&= \cos(g) \times \frac{d}{dt}(2\pi t) \\
&= \cos(g) \times (2\pi) \\
&= 2\pi \cos(2\pi t).
\end{aligned}$$

The chain rule is also useful in differentiating the power of sinusoidal function such as $y_1(t) = \sin^2(2\pi t)$ or the power of the polynomial function such as $y_2(t) = (2t + 10)^2$. The derivatives of these functions are obtained as

$$\begin{aligned}
\frac{dy_1}{dt} &= \frac{d}{dt}\left((\sin(2\pi t))^2\right) \\
&= 2\,(\sin(2\pi t))^1 \times \frac{d}{dt}(\sin(2\pi t)) \\
&= 2\,\sin(2\pi t) \times (2\pi\,\cos(2\pi t)) \\
&= 4\pi\,\sin(2\pi t)\,\cos(2\pi t)
\end{aligned}$$

and

$$\begin{aligned}
\frac{dy_2}{dt} &= \frac{d}{dt}\left((2t + 10)^2\right) \\
&= 2\,(2t + 10)^1 \times \frac{d}{dt}(2t + 10) \\
&= 2\,(2t + 10) \times (2) \\
&= 4\,(2t + 10)
\end{aligned}$$

The following example will illustrate some of the derivative relationships discussed above.

Example
8-6

For a particular circuit element, the charge is

$$q(t) = \frac{1}{50} \sin 250\,\pi t \quad \text{C} \tag{8.27}$$

and the voltage supplied by the voltage source shown in Fig. 8.21 is

$$v(t) = 100 \sin 250\,\pi t \quad \text{V}. \tag{8.28}$$

Figure 8.21   Voltage applied to a particular circuit element.

Find the following quantities:

(a) The current, $i(t)$.

(b) The power, $p(t)$.

(c) The maximum power p_{max} delivered to the circuit element by the voltage source.

Solution   (a) **Current:** The current $i(t)$ can be determined by differentiating the charge $q(t)$ as

$$i(t) = \frac{dq(t)}{dt}$$

$$= \frac{d}{dt}\left(\frac{1}{50}\sin 250\,\pi t\right)$$

$$= \frac{1}{50}\,(250\,\pi\,\cos 250\,\pi t)$$

or

$$i(t) = 5\,\pi\,\cos 250\,\pi t \quad \text{A}. \tag{8.29}$$

(b) **Power:** The power $p(t)$ can be determined by multiplying the voltage given in equation (8.28) and the current calculated in equation (8.29) as

$$p(t) = v(t)\,i(t)$$

$$= (100\sin 250\,\pi t)\,(5\,\pi\cos 250\,\pi t)$$

$$= 500\,\pi(\sin 250\,\pi t)\,(\cos 250\,\pi t)$$

or

$$p(t) = 250\,\pi \sin 500\,\pi t \;\; \text{W}. \tag{8.30}$$

The power $p(t) = 250\,\pi \sin 500\,\pi t$ W in equation (8.30) is obtained by using the double-angle trigonometric identity $\sin(2\,\theta) = 2 \sin \theta \cos \theta$ or $(\sin 250\,\pi t)\,(\cos 250\,\pi t) = \dfrac{\sin(500\,\pi t)}{2}$, which gives $p(t) = 250\,\pi \sin 500\,\pi t$.

(c) **Maximum Power Delivered to the Circuit:** As discussed in Section 8.3, the maximum value of a trigonometric function such as $p(t) = 250\,\pi (\sin 500\,\pi t)$ can be found without differentiating the function and equating the result to zero. Since $-1\; \le \sin 500\,\pi t\; \le 1$, the power delivered to the circuit element is maximum when $\sin 500\,\pi t = 1$. Therefore,

$$p_{\max} = 250\,\pi\text{W},$$

which is simply the *amplitude* of the power.

8.4.1   Current and Voltage in an Inductor

The current-voltage relationship for an inductor element (Fig. 8.22) is given by

$$v(t) = L\,\frac{di(t)}{dt}, \tag{8.31}$$

where $v(t)$ is the voltage across the inductor in V, $i(t)$ is the current flowing through the inductor in A, and L is the inductance of the inductor in henry (H). Note that if the inductance is given in mH (1 mH (millihenry) $= 10^{-3}$ H), it must be converted to H before using it in equation (8.31).

Figure 8.22   Inductor as a circuit element.

Example 8-7

For the inductor shown in Fig. 8.22, if $L = 100$ mH and $i(t) = t\,e^{-3t}$ A.

(a) Find the voltage $v(t) = L\,\dfrac{di(t)}{dt}$.

(b) Find the value of the **current** when the voltage is zero.

(c) Use the results of (a) and (b) to sketch the current $i(t)$.

Solution   (a) The voltage $v(t)$ is determined as

$$v(t) = L \frac{di(t)}{dt}$$

$$= (100 \times 10^{-3}) \frac{di(t)}{dt}$$

or

$$v(t) = 0.1 \frac{di(t)}{dt},$$

where $\frac{di(t)}{dt} = \frac{d}{dt}(t\,e^{-3t})$. To differentiate the product of two functions (t and e^{-3t}), the product rule of differentiation (Table 8.3) is used:

$$\frac{d}{dt}(f(t)\,g(t)) = f(t)\frac{d}{dt}(g(t)) + g(t)\frac{d}{dt}(f(t)). \tag{8.32}$$

Substituting $f(t) = t$ and $g(t) = e^{-3t}$ in equation (8.32) gives

$$\frac{d}{dt}(t\,e^{-3t}) = (t)\left(\frac{d}{dt}(e^{-3t})\right) + \left(\frac{d}{dt}(t)\right)(e^{-3t})$$

$$= (t)(-3\,e^{-3t}) + (1)(e^{-3t})$$

$$= e^{-3t}(-3\,t + 1). \tag{8.33}$$

Therefore,

$$v(t) = 0.1\,e^{-3t}(-3\,t + 1)\ \text{V}. \tag{8.34}$$

(b) To find the current when the voltage is zero, first find the time t when the voltage is zero and then substitute this time in the expression for current. Setting equation (8.34) equal to zero gives

$$0.1\,e^{-3t}(-3\,t + 1) = 0.$$

Since e^{-3t} is never zero, it follows that $(-3\,t + 1) = 0$, which gives $t = 1/3$ sec. Therefore, the value of the current when the voltage is zero is determined by substituting $t = 1/3$ into the current $i(t) = t\,e^{-3t}$, which gives

$$i\left(\frac{1}{3}\right) = \left(\frac{1}{3}\right) e^{-3\left(\frac{1}{3}\right)}$$

$$= \frac{1}{3} e^{-1}$$

or

$$i = 0.123\ \text{A}.$$

(c) Since the voltage is proportional to the derivative of the current, the slope of the current is zero when the voltage is zero. Therefore, the current $i(t)$ is maximum ($i_{max} = 0.123$ A) at $t = 1/3$ s. Also, at $t = 0$, $i(0) = 0$ A. Using these values along with the values of the current at $t = 1$ s ($i(1\text{s}) = 0.0498$ A) and at $t = 2$ s ($i(2\text{s}) = 0.00496$ A), the approximate sketch of the current can be drawn as shown in Fig. 8.23. Note that $i = 0.123$ A **must** be a maximum (as opposed to a minimum) value, since it is the only location of zero slope and is greater than the values of $i(t)$ at either $t = 0$ or $t = 2$ s. Hence, no second derivative test is required!

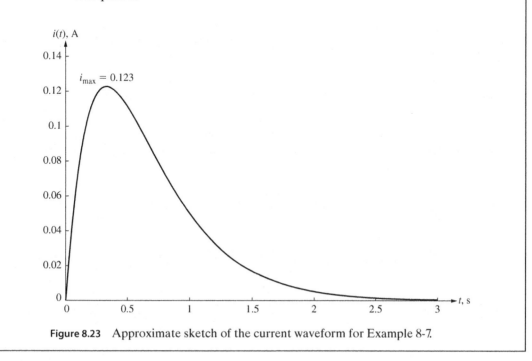

Figure 8.23   Approximate sketch of the current waveform for Example 8-7.

As seen in dynamics, derivatives are frequently used in circuits to sketch functions for which no equations are given, as illustrated in the following example:

Example 8-8

For the given input voltage (square wave) shown in Fig. 8.24, plot the current $i(t)$ and the power $p(t)$ if $L = 500$ mH. Assume $i(0) = 0$ A and $p(0) = 0$ W.

Figure 8.24   A square wave voltage applied to an inductor.

Solution   For an inductor, the current-voltage relationship is given by $v(t) = L \dfrac{di(t)}{dt}$. Since the voltage is known, the rate of change of the current is given by

$$(500 \times 10^{-3}) \frac{di(t)}{dt} = v(t)$$

$$\frac{di(t)}{dt} = \frac{1}{0.5} v(t) \qquad (8.35)$$

or

$$\frac{di(t)}{dt} = 2 v(t).$$

Therefore, the slope of the current is twice the applied voltage. Since $v(t) = \pm 9$ V (constant in each interval), the current waveform has a constant slope of ± 18 A/s, in other words, the current waveform is a straight line with a constant slope of ± 18 A/s. In the interval $0 \le t \le 2$, the current waveform is a straight line with a slope of 18 A/s that starts at 0 A ($i(0) = 0$ A). Therefore, the value of the current at $t = 2$ s is 36 A. In the interval, $2 < t \le 4$, the current waveform is a straight line starting at 36 A (at $t = 2$ s) with a slope of -18 A/s. Therefore, the value of the current at $t = 4$ s is 0 A. This completes one cycle of the current waveform. Since the value of the current at $t = 4$ s is 0 (the same as at $t = 0$ s) and the applied voltage between interval $4 < t \le 8$ is the same as the applied voltage between $0 \le t \le 4$, the waveform for the current from $4 \le t \le 8$ is the same as the waveform of the current from $0 \le t \le 4$. The resulting plot of $i(t)$ is shown in Fig. 8.25. Hence, when a square wave voltage is applied to an inductor, the resulting current is a

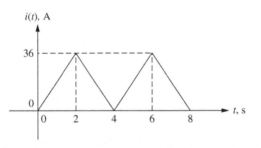

Figure 8.25   Sketch of the current waveform for Example 8-8.

triangular wave. Since $p(t) = v(t) i(t)$ and the voltage is ± 9 V, the power is given by $p(t) = (\pm 9) i(t)$ W. In the interval $0 \le t \le 2$, $v(t) = 9$ V, therefore, $p(t) = 9 i(t)$ W. The waveform of the power is a straight line starting at 0 W with a slope of $(9 \times 18) = 162$ W/s. Therefore, the power delivered to the inductor just before $t = 2$ sec is 324 W. Just after $t = 2$ sec, the voltage is -9 V and the current is 36 A. Thus, the power jumps down to $(-9)(36) = -324$ W. In the interval $2 < t \le 4$, $v(t) = -9$ V, and the current has a negative slope of -18 A/s. Therefore, the waveform for the power is a straight line starting at -324 W with a slope of 162 W/s. Thus, $p(4) = -324 + 2(162) = 0$ W. This completes one cycle of the power. Since the value of the power at $t = 4$ s is 0 (the same as at $t = 0$ s) and the applied voltage and current in the interval $4 \leq t \leq 8$ are the same as the voltage and current in the $0 \leq t \leq 4$, the waveform for the power from $4 \leq t \leq 8$ is the same as the waveform of the power from $0 \leq t \leq 4$. The resulting plot of $p(t)$ is shown in Fig. 8.26, and is typically referred to as a sawtooth curve.

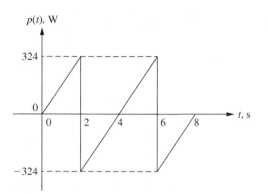

Figure 8.26 Sketch of the power for Example 8-8.

8.4.2 Current and Voltage in a Capacitor

The current-voltage relationship for a capacitive element (Fig. 8.27) is given by

$$i(t) = C \frac{dv(t)}{dt}, \tag{8.36}$$

where $v(t)$ is the voltage across the capacitor in V, $i(t)$ is the current flowing through the capacitor in A, and C is the capacitance of the capacitor in farad (F). Note that if the capacitance is given in μF (10^{-6} F), it must be converted to F before using it in equation (8.36).

Example 8-9

Consider the capacitive element shown in Fig. 8.27 with $C = 25 \ \mu F$ and $v(t) = 20 \, e^{-500t} \sin 5000 \, \pi t$ V. Find the current $i(t)$.

Figure 8.27 Capacitor as a circuit element.

Solution   The current $i(t)$ can be found by using equation (8.36) as

$$i(t) = C \, \frac{dv(t)}{dt}.$$

Substituting the value of C gives

$$i(t) = (25 \times 10^{-6}) \, \frac{dv(t)}{dt},$$

where $\dfrac{dv(t)}{dt} = \dfrac{d}{dt}(20 \, e^{-500t} \sin 5000 \, \pi t)$. To differentiate the product of the two functions e^{-500t} and $\sin 5000 \, \pi t$, the product rule of differentiation is required. Letting $f(t) = e^{-500t}$ and $g(t) = \sin 5000 \, \pi t$,

$$\frac{d}{dt}(20 \, e^{-500t} \sin 5000 \pi t) = 20 \times \left[e^{-500t} \frac{d}{dt}(\sin 5000 \, \pi t) \right.$$

$$\left. + (\sin 5000 \, \pi t) \frac{d}{dt}(e^{-500t}) \right]$$

$$= 20 \times [(e^{-500t})(5000\pi \cos 5000 \, \pi t)$$

$$+ (\sin 5000 \, \pi t)(-500 \, e^{-500t})]$$

$$= 10{,}000 \, e^{-500t} (10 \, \pi \cos 5000 \, \pi t - \sin 5000 \, \pi t).$$

Therefore,

$$i(t) = 25 \times 10^{-6} \left(10{,}000 \, e^{-500t} (10 \, \pi \cos 5000 \, \pi t - \sin 5000 \, \pi t) \right)$$

or

$$i(t) = 0.25 \, e^{-500t} (10 \, \pi \cos 5000 \, \pi t - \sin 5000 \, \pi t) \, \text{A}$$

Using the results of Chapter 6, this can also be written as $i(t) = 2.5 \, \pi e^{-500t} \sin(5000 \, \pi t + 92°) \, \text{A}$.

Example 8-10

The current shown in Fig. 8.28 is used to charge a capacitor with $C = 20 \, \mu\text{F}$. Knowing that $i(t) = \frac{dq(t)}{dt} = C \frac{dv(t)}{dt}$, plot the charge $q(t)$ stored in the capacitor and the corresponding voltage $v(t)$. Assume $q(0) = v(0) = 0$.

Figure 8.28   Charging of a capacitor.

Solution   (a) **Charge:** Since $i(t) = \dfrac{dq(t)}{dt}$, the slope of the charge $q(t)$ is given by each constant value of current in each interval, for example

$$0 \leq t \leq 1 \text{ ms}: \quad \frac{dq(t)}{dt} = \quad 4 \text{ C/s}$$

$$1 < t \leq 2 \text{ ms}: \quad \frac{dq(t)}{dt} = -2 \text{ C/s}$$

$$2 < t \leq 4 \text{ ms}: \quad \frac{dq(t)}{dt} = \quad 2 \text{ C/s}$$

$$4 < t \leq 5 \text{ ms}: \quad \frac{dq(t)}{dt} = -4 \text{ C/s}$$

$$5 < t \leq \infty \text{ ms}: \quad \frac{dq(t)}{dt} = \quad 0 \text{ C/s}.$$

Therefore, the plot of the charge $q(t)$ stored in the capacitor can be drawn as shown in Fig. 8.29. Note that since time t is in ms, the charge $q(t)$ is in mC.

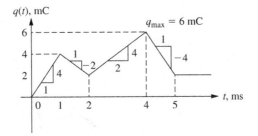

Figure 8.29   Charge on the capacitor in Example 8-10.

(b) **Voltage:** To find the voltage across the capacitor, the relationship between the charge and the voltage is first derived as:

$$i(t) = C\frac{dv(t)}{dt} = \frac{dq(t)}{dt} \quad \Rightarrow \quad \frac{dv(t)}{dt} = \frac{1}{C}\frac{dq(t)}{dt}.$$

Therefore, the derivative (slope) of the voltage $v(t)$ is equal to the derivative (slope) of the charge $q(t)$ multiplied by the reciprocal of the capacitance. Substituting the value of C gives

$$\frac{dv(t)}{dt} = \frac{1}{20 \times 10^{-6}}\frac{dq(t)}{dt},$$

or

$$\frac{dv(t)}{dt} = 50 \times 10^3 \frac{dq(t)}{dt}.$$

The plot of the voltage is thus the same as the plot of the charge with the ordinate scaled by 50×10^3, as shown in Fig. 8.30.

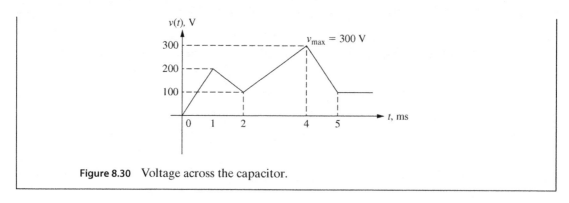

Figure 8.30 Voltage across the capacitor.

Note that while the time is still measured in ms, the voltage is measured in volts.

8.5 APPLICATIONS OF DERIVATIVES IN STRENGTH OF MATERIALS

In this section, the derivative relationship for beams under transverse loading conditions will be discussed. The locations and values of maximum deflections are obtained using derivatives, and results are used to sketch the deflection. This section also considers the application of derivatives to maximum stress under axial loading and torsion.

Consider a beam with elastic modulus E (lb/in.^{2} or N/m^2) and second moment of area I (in.^{4} or m^4) as shown in Fig. 8.31. The product EI is called the flexural rigidity and is a measure of how stiff the beam is. If the beam is loaded with a distributed transverse load $q(x)$ (lb/in. or N/m), the beam deflects in the y-direction with a deflection of $y(x)$ (in. or m) and a slope of $\theta(x) = \dfrac{dy(x)}{dx}$ in radians.

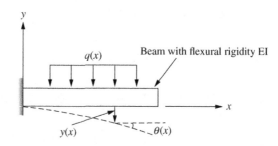

Figure 8.31 A beam loaded in the y-direction.

The internal moments and forces in the beam shown in Fig. 8.32 are given by the expressions

$$\textbf{Bending Moment: } M(x) = EI\,\frac{d\,\theta(x)}{dx} = EI\,\frac{d^2y(x)}{dx^2} \text{ lb-in. or N-m} \quad (8.37)$$

Shear Force: $$V(x) = \frac{d\,M(x)}{dx} = EI\,\frac{d^3 y(x)}{dx^3} \text{ lb or N} \qquad (8.38)$$

Distributed Load: $q(x) = -\dfrac{dV(x)}{dx} = -EI\,\dfrac{d^4 y(x)}{dx^4}$ lb/in. or N/m $\qquad (8.39)$

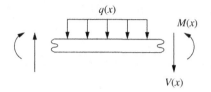

Figure 8.32 The internal forces in a beam loaded by a distributed load.

More detailed background on the above relations can be found in any book on strength of materials.

Example 8-11

Consider a cantilever beam of length l loaded by a force P at the free end, as shown in Fig. 8.33. If the deflection is given by

$$y(x) = \frac{P}{6\,EI}\left(x^3 - 3l\,x^2\right) \text{ m}, \qquad (8.40)$$

find the deflection and slope at the free end $x = l$.

Figure 8.33 Cantilever beam with an end load P.

Solution **Deflection:** The deflection of the beam at the free end can be determined by substituting $x = l$ in equation (8.40), which gives

$$y(l) = \frac{P}{6\,EI}\left[l^3 - 3l\,(l^2)\right]$$

$$= \frac{P}{6\,EI}\left(-2l^3\right)$$

or

$$y(l) = -\frac{Pl^3}{3\,EI} \text{ m}.$$

This classic result is used in a range of mechanical and civil engineering courses.

Slope: The slope of the deflection $\theta(x)$ can be found by differentiating the deflection $y(x)$ as

$$\theta(x) = \frac{dy(x)}{dx}$$

$$= \frac{d}{dx}\left[\frac{P}{6\,EI}\left(x^3 - 3\,l\,x^2\right)\right]$$

$$= \frac{P}{6\,EI}\left[\frac{d}{dx}(x^3) - 3\,l\,\frac{d}{dx}(x^2)\right]$$

$$= \frac{P}{6\,EI}\left[3\,(x^2) - 3\,l\,(2\,x)\right]$$

$$= \frac{P}{6\,EI}\left(3\,x^2 - 6\,l\,x\right)$$

or

$$\theta(x) = \frac{P}{2\,EI}\left(x^2 - 2\,l\,x\right) \text{ rad.} \tag{8.41}$$

Note that the parameters P, l, E, and I are all treated as constants.

The slope $\theta(x)$ at the free end can now be determined by substituting $x = l$ in equation (8.41) as

$$\theta(l) = \frac{P}{2\,EI}\left[l^2 - 2\,l\,(l)\right]$$

$$= \frac{P}{2\,EI}\left(-l^2\right)$$

or

$$\theta(l) = -\frac{P\,l^2}{2\,EI} \text{ rad.}$$

Note: It can be seen by inspection that both the deflection and the slope of deflection are maximum at the free end, for example:

$$y_{max} = -\frac{P\,l^3}{3\,EI}$$

$$\theta_{max} = -\frac{P\,l^2}{2\,EI}.$$

It can be seen that doubling the load P would increase the maximum deflection and slope by a factor of 2. However, doubling the length l would increase the maximum deflection by a factor of 8, and the maximum slope by a factor of 4!

Example 8-12

Consider a simply supported beam of length l subjected to a central load P, as shown in Fig. 8.34. For $0 \le x \le 1/2$, the deflection is given by

$$y(x) = \frac{P}{48\,EI}\left(4x^3 - 3\,l^2\,x\right) \text{ m.} \tag{8.42}$$

Determine the maximum defection y_{max}, as well as the slope at the end $x = 0$.

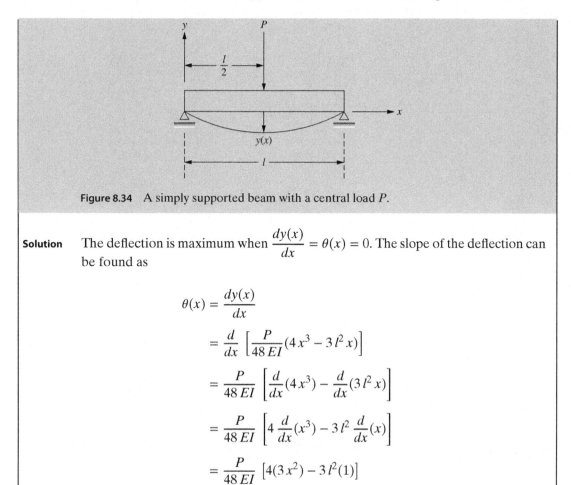

Figure 8.34 A simply supported beam with a central load P.

Solution The deflection is maximum when $\dfrac{dy(x)}{dx} = \theta(x) = 0$. The slope of the deflection can be found as

$$\theta(x) = \frac{dy(x)}{dx}$$

$$= \frac{d}{dx}\left[\frac{P}{48\,EI}(4x^3 - 3l^2 x)\right]$$

$$= \frac{P}{48\,EI}\left[\frac{d}{dx}(4x^3) - \frac{d}{dx}(3l^2 x)\right]$$

$$= \frac{P}{48\,EI}\left[4\frac{d}{dx}(x^3) - 3l^2\frac{d}{dx}(x)\right]$$

$$= \frac{P}{48\,EI}\left[4(3x^2) - 3l^2(1)\right]$$

$$= \frac{P}{48\,EI}\left(12x^2 - 3l^2\right)$$

or

$$\theta(x) = \frac{P}{16\,EI}\left(4x^2 - l^2\right). \tag{8.43}$$

To find the location of the maximum deflection, $\theta(x)$ is set to zero and the resulting equation is solved for the values of x as

$$\frac{P}{16\,EI}(4x^2 - l^2) = 0 \;\Rightarrow\; 4x^2 - l^2 = 0 \;\Rightarrow x = \pm\frac{l}{2}.$$

Since the deflection is given for $0 \le x \le l/2$, the deflection is maximum at $x = l/2$. The value of the maximum deflection can now be found by substituting $x = l/2$ in

equation (8.42) as

$$y_{max} = y\left(\frac{l}{2}\right)$$

$$= \frac{P}{48\,EI}\left[4\left(\frac{l}{2}\right)^3 - 3\,l^2\left(\frac{l}{2}\right)\right]$$

$$= \frac{P}{48\,EI}\left(\frac{l^3}{2} - \frac{3\,l^3}{2}\right)$$

or

$$y_{max} = -\frac{P\,l^3}{48\,EI}\ \text{m.}$$

Likewise, the slope $\theta(x)$ at $x = 0$ can be determined by substituting $x = 0$ in equation (8.43) as

$$\theta(0) = \frac{P}{16\,EI}\,(4 * 0 - l^2)$$

$$= \frac{P}{16\,EI}\,(-l^2)$$

or

$$\theta(0) = -\frac{P\,l^2}{16\,EI}\ \text{rad.}$$

Example 8-13

A simply supported beam of length l is subjected to a distributed load $q(x) = w_0 \sin\left(\frac{\pi x}{l}\right)$ as shown in Fig. 8.35. If the deflection is given by

$$y(x) = -\frac{w_0\,l^4}{\pi^4\,EI}\,\sin\left(\frac{\pi x}{l}\right)\ \text{m} \tag{8.44}$$

find the slope $\theta(x)$, the moment $M(x)$ and the shear force $V(x)$.

Figure 8.35   A simply supported beam with a sinusoidal load.

Solution   **Slope:** The slope of the deflection $\theta(x)$ is given by

$$\theta(x) = \frac{dy(x)}{dx}$$

$$= \frac{d}{dx}\left[-\frac{w_0\, l^4}{\pi^4\, EI}\sin\left(\frac{\pi x}{l}\right)\right]$$

$$= -\frac{w_0\, l^4}{\pi^4\, EI}\left[\frac{d}{dx}\left(\sin\left(\frac{\pi x}{l}\right)\right)\right]$$

$$= -\frac{w_0\, l^4}{\pi^4\, EI}\left[\frac{\pi}{l}\cos\left(\frac{\pi x}{l}\right)\right]$$

or

$$\theta(x) = -\frac{w_0\, l^3}{\pi^3\, EI}\cos\left(\frac{\pi x}{l}\right) \text{ rad.} \qquad (8.45)$$

Moment: By definition, the moment $M(x)$ is obtained by multiplying the derivative of the slope $\theta(x)$ by EI, or

$$M(x) = EI\frac{d\theta(x)}{dx} = EI\frac{d^2y(x)}{dx^2}.$$

Substituting equation (8.45) for $\theta(x)$ gives

$$M(x) = EI\frac{d}{dx}\left[-\frac{w_0\, l^3}{\pi^3\, EI}\cos\left(\frac{\pi x}{l}\right)\right]$$

$$= -\frac{w_0\, l^3}{\pi^3}\left[-\frac{\pi}{l}\sin\left(\frac{\pi x}{l}\right)\right]$$

or

$$M(x) = \frac{w_0\, l^2}{\pi^2}\sin\left(\frac{\pi x}{l}\right) \text{ N-m.} \qquad (8.46)$$

Shear Force: By definition, the shear force $V(x)$ is the derivative of the moment $M(x)$, or

$$V(x) = \frac{dM(x)}{dx} = EI\frac{d^3y(x)}{dx^3}.$$

Substituting equation (8.46) for $M(x)$ gives

$$V(x) = \frac{d}{dx}\left[\frac{w_0\, l^2}{\pi^2}\sin\left(\frac{\pi x}{l}\right)\right]$$

$$= \frac{w_0\, l^2}{\pi^2}\left[\frac{\pi}{l}\cos\left(\frac{\pi x}{l}\right)\right]$$

or

$$V(x) = \frac{w_0 l}{\pi} \cos\left(\frac{\pi x}{l}\right) \text{ N.} \tag{8.47}$$

The above answer can be checked by showing that $q(x) = -\dfrac{dV(x)}{dx}$ as

$$q(x) = -\frac{d}{dx}\left[\frac{w_0 l}{\pi} \cos\left(\frac{\pi x}{l}\right)\right]$$

$$= -\frac{w_0 l}{\pi}\left[-\frac{\pi}{l} \sin\left(\frac{\pi x}{l}\right)\right]$$

or

$$q(x) = w_0 \sin\left(\frac{\pi x}{l}\right) \text{ N/m,} \tag{8.48}$$

which matches the applied load in Fig. 8.35.

8.5.1 Maximum Stress under Axial Loading

In this section, the application of derivatives in finding maximum stress under axial loading is discussed. A normal stress σ results when a bar is subjected to an axial load P (through the centroid of the cross section), as shown in Fig. 8.36. The normal stress is given by

$$\sigma = \frac{P}{A}, \tag{8.49}$$

where A is the cross-sectional area of the section perpendicular to longitudinal axis of the bar. Therefore, the normal stress σ acts perpendicular to the cross section and has units of force per unit area (psi or N/m^2).

Figure 8.36 A rectangular bar under axial loading.

To find the stress on an oblique plane, consider an inclined section of the bar as shown in Fig. 8.37.

Figure 8.37 Inclined section of the rectangular bar.

The relationship between the cross-sectional area perpendicular to the longitudinal axis and the area of the inclined plane is given by

$$A_\theta \cos \theta = A$$

$$A_\theta = \frac{A}{\cos \theta}.$$

The force P can be resolved into components perpendicular to the inclined plane F and parallel to the inclined plane V. The free-body diagram of the forces acting on the oblique plane is shown in Fig. 8.38. Note that the resultant force in the axial direction must be equal to P to satisfy equilibrium.

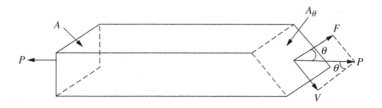

Figure 8.38   Free-body diagram.

The relationship among P, F, and V can be found by using the right triangle shown in Fig. 8.39 as

$$F = P \cos \theta$$
$$V = P \sin \theta$$

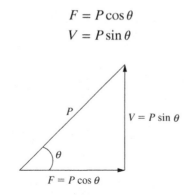

Figure 8.39   Triangle showing force P, F, and V.

The force perpendicular to the inclined cross section F produces a normal stress σ_θ (shown in Fig. 8.40) given by

$$\sigma_\theta = \frac{F}{A_\theta}$$

$$= \frac{P \cos \theta}{A/\cos \theta}$$

$$= \frac{P \cos^2 \theta}{A}. \qquad (8.50)$$

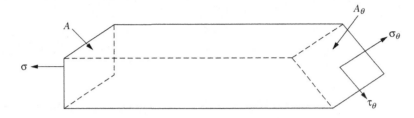

Figure 8.40 Normal and shear stresses acting on the inclined cross section.

The tangential force V produces a shear stress τ_θ given as

$$\tau_\theta = \frac{V}{A_\theta}$$

$$= \frac{P \sin \theta}{A/\cos \theta}$$

$$= \frac{P \sin \theta \cos \theta}{A}. \tag{8.51}$$

Substituting $\sigma = \dfrac{P}{A}$ from (8.49) into equations (8.50) and (8.51), the normal and shear stresses on the inclined cross section are given by

$$\sigma_\theta = \sigma \cos^2 \theta \tag{8.52}$$

$$\tau_\theta = \sigma \sin \theta \cos \theta. \tag{8.53}$$

In general, brittle materials like glass, concrete, and cast iron fail due to maximum values of σ_θ (normal stress). However, ductile materials like steel, aluminum, and brass fail due to maximum values of τ_θ (shear stress).

Example 8-14 Use derivatives to find the values of θ where σ_θ and τ_θ are maximum, and find their maximum values.

Solution (a) **First, find the derivative of σ_θ with respect to θ:** The derivative of σ_θ given by equation (8.52) is

$$\frac{d\sigma_\theta}{d\theta} = \sigma \frac{d}{d\theta}(\cos^2(\theta))$$

$$= \sigma \frac{d}{d\theta}(\cos \theta \cos \theta)$$

$$= \sigma (\cos \theta (-\sin \theta) + \cos \theta (-\sin \theta))$$

$$= -2 \sigma (\cos \theta \sin \theta). \tag{8.54}$$

(b) **Next, equate the derivative in equation (8.54) to zero** and solve the resulting equation for the value of θ between 0 and 90° where σ_θ is maximum. Therefore, $-2\cos\theta\sin\theta = 0$, which gives

$$\cos\theta = 0 \quad \Rightarrow \quad \theta = 90°$$

or

$$\sin\theta = 0 \quad \Rightarrow \quad \theta = 0°.$$

Therefore, $\theta = 0°$ and 90° are the critical points; in other words, at $\theta = 0°$ and 90°, σ_θ has a local maximum or minimum. To find the value of θ where σ_θ has a maximum, the second derivative test is performed. The second derivative of σ_θ is given by

$$\frac{d^2\sigma_\theta}{d\theta^2} = \frac{d}{d\theta}(-2\sigma\cos\theta\sin\theta)$$

$$= -2\sigma\frac{d}{d\theta}(\cos\theta\sin\theta)$$

$$= -2\sigma\left[\cos\theta\frac{d}{d\theta}(\sin\theta) + \frac{d}{d\theta}(\cos\theta)(\sin\theta)\right]$$

$$= -2\sigma\left[\cos\theta(\cos\theta) + (-\sin\theta)\sin\theta\right]$$

$$= -2\sigma(\cos^2\theta - \sin^2\theta)$$

or

$$\frac{d^2\sigma_\theta}{d\theta^2} = -2\sigma\cos 2\theta$$

where $\cos 2\theta = \cos^2\theta - \sin^2\theta$. For $\theta = 0°$, $\dfrac{d^2\sigma_\theta}{d\theta^2} = -2\sigma < 0$. So σ_θ has a **maximum value** at $\theta = 0°$.

For $\theta = 90°$, $\dfrac{d^2\sigma_\theta}{d\theta^2} = -2\sigma\cos(180°) = 2\sigma > 0$; therefore, σ_θ has a **minimum value** at $\theta = 90°$.

Maximum Value of σ_θ: Substituting $\theta = 0°$ in equation (8.52) gives

$$\sigma_{\max} = \sigma\cos^2(0°) = \sigma.$$

This means that the largest normal stress during axial loading is simply the applied stress σ!

Value of θ where τ_θ is maximum:

(a) **First, find the derivative of τ_θ with respect to θ:** The derivative of τ_θ given by equation (8.53) is given by

$$\frac{d\tau_\theta}{d\theta} = \frac{d}{d\theta}(\sigma \sin\theta \cos\theta)$$

$$= \sigma \frac{d}{d\theta}(\sin\theta \cos\theta)$$

$$= \sigma \frac{d}{d\theta}\left(\frac{\sin 2\theta}{2}\right)$$

$$= \frac{\sigma}{2}(2\cos 2\theta)$$

or

$$\frac{d\tau_\theta}{d\theta} = \sigma \cos 2\theta. \tag{8.55}$$

(b) **Next, equate the derivative in equation (8.55) to zero** and solve the resulting equation $\sigma \cos 2\theta = 0$ for the value of θ (between 0 and 90°) where τ_θ is maximum:

$$\cos 2\theta = 0 \quad \Rightarrow \quad 2\theta = 90° \quad \Rightarrow \quad \theta = 45°.$$

Therefore, τ_θ has a local maximum or minimum at $\theta = 45°$. To find whether τ_θ has a maximum or minimum at $\theta = 45°$, the second derivative test is performed. The second derivative of τ_θ is given by

$$\frac{d^2\tau_\theta}{d\theta^2} = \frac{d}{d\theta}(\sigma \cos 2\theta)$$

$$= \sigma(-\sin 2\theta)(2)$$

$$= -2\sigma \sin 2\theta$$

For $0 \leq \theta \leq 90°$ $\quad \Rightarrow \quad$ $0 \leq 2\theta \leq 180°$ $\quad \Rightarrow \quad$ $\sin 2\theta > 0$, therefore, $\dfrac{d^2\tau_\theta}{d\theta^2} <$ 0. Since the second derivative is negative, τ_θ has a maximum value at $\theta = 45°$.

Maximum Value of τ_θ: Substituting $\theta = 45°$ in equation (8.53) gives

$$\tau_{max} = \sigma \sin 45°, \cos 45°$$

$$= \sigma \left(\frac{\sqrt{2}}{2}\right)\left(\frac{\sqrt{2}}{2}\right)$$

or

$$\tau_{max} = \frac{\sigma}{2} \quad \text{at } \theta = 45°.$$

Thus, the maximum shear stress during axial loading is equal to half the applied normal stress, but at an angle of 45°. This is why a tensile test of a steel specimen results in failure at a 45° angle.

8.6   FURTHER EXAMPLES OF DERIVATIVES IN ENGINEERING

Example 8-15

The velocity of a skydiver jumping from a height of 12,042 ft is shown in Fig. 8.41.

(a) Find the equation of the velocity $v(t)$ for the five time intervals shown in Fig. 8.41.

(b) Knowing that $a(t) = \dfrac{dv}{dt}$, find the acceleration $a(t)$ of the skydiver for $0 \le t \le 301$ s.

(c) Use the results of part (b) to sketch the acceleration $a(t)$ for $0 \le t \le 301$ s.

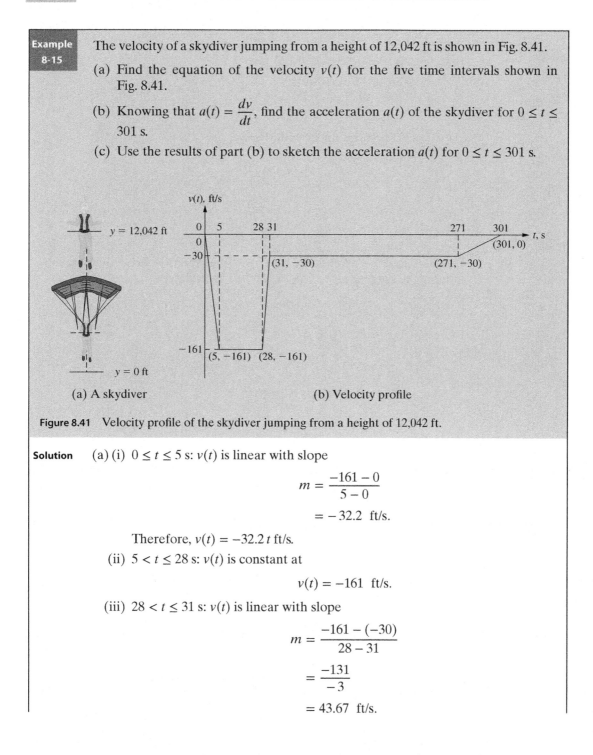

(a) A skydiver   (b) Velocity profile

Figure 8.41   Velocity profile of the skydiver jumping from a height of 12,042 ft.

Solution

(a) (i) $0 \le t \le 5$ s: $v(t)$ is linear with slope

$$m = \frac{-161 - 0}{5 - 0}$$

$$= -32.2 \ \text{ft/s}.$$

Therefore, $v(t) = -32.2\,t$ ft/s.

(ii) $5 < t \le 28$ s: $v(t)$ is constant at

$$v(t) = -161 \ \text{ft/s}.$$

(iii) $28 < t \le 31$ s: $v(t)$ is linear with slope

$$m = \frac{-161 - (-30)}{28 - 31}$$

$$= \frac{-131}{-3}$$

$$= 43.67 \ \text{ft/s}.$$

Therefore, $v(t) = 43.67\, t + b$ ft/s. The value of b (y-intercept) can be found by substituting the data point $(t,\ v(t)) = (31, -30)$, which gives

$$-30 = 43.67(31) + b \quad \Rightarrow \quad b = -1383.67$$

Therefore, $v(t) = 43.67\, t - 1383.67$ ft/s.

(iv) $31 < t \le 271$ s: $v(t)$ is constant at

$$v(t) = -30 \ \text{ft/s}.$$

(v) $271 < t \le 301$ s: $v(t)$ is linear with slope

$$
\begin{aligned}
m &= \frac{-30 - 0}{271 - 301} \\
&= \frac{-30}{-30} \\
&= 1 \ \text{ft/s}.
\end{aligned}
$$

Therefore, $v(t) = t + b$ ft/s. The value of b (y-intercept) can be found by substituting the data point $(t,\ v(t)) = (301, 0)$, which gives

$$0 = 1(301) + b \quad \Rightarrow \quad b = -301$$

Therefore, $v(t) = t - 301$ ft/s.

(b) (i) $0 \le t \le 5$ s:

$$
\begin{aligned}
a(t) &= \frac{dv(t)}{dt} \\
&= \frac{d}{dt}(-32.3\, t) \\
&= -32.2 \ \text{ft/s}^2.
\end{aligned}
$$

(ii) $5 < t \le 28$ s:

$$
\begin{aligned}
a(t) &= \frac{dv(t)}{dt} \\
&= \frac{d}{dt}(-161) \\
&= 0 \ \text{ft/s}^2.
\end{aligned}
$$

(iii) $28 < t \le 3$ s:

$$
\begin{aligned}
a(t) &= \frac{dv(t)}{dt} \\
&= \frac{d}{dt}(43.7\, t - 1383.67) \\
&= 43.7 \ \text{ft/s}^2.
\end{aligned}
$$

(iv) $31 < t \leq 271$ s:

$$a(t) = \frac{dv(t)}{dt}$$
$$= \frac{d}{dt}(-30)$$
$$= 0 \ \text{ft/s}^2.$$

(v) $271 < t \leq 301$ s:

$$a(t) = \frac{dv(t)}{dt}$$
$$= \frac{d}{dt}(t - 301)$$
$$= 1 \ \text{ft/s}^2.$$

(c) The acceleration of the skydiver found in part (b) can be drawn as shown in Fig. 8.42.

Figure 8.42   The acceleration of the skydiver.

A proposed highway traverses a hilltop bounded by uphill and downhill grades of 10% and −8%, respectively. These grades pass through benchmarks A and B located as shown in Fig. 8.43. With the origin of the coordinate axes (x, y) set at benchmark A, the engineer has defined the hilltop segment of the highway by a parabolic arc:

$$y(x) = a x^2 + b x \tag{8.56}$$

which is tangent to the uphill grade at the origin.

(a) Find the slope of the line for the uphill grade and the value of b for the parabolic arc.

(b) Find the equation of the line

$$\hat{y} = cx + d \qquad (8.57)$$

for the downhill grade.

(c) Given that at the downhill point of tangency ($\bar{x}$), both the elevation and the slope of the parabolic arc are equal to their respective values of the downhill line:

$$y(\bar{x}) = \hat{y}(\bar{x}) \qquad (8.58)$$

$$\frac{dy}{dx}\Big|_{x=\bar{x}} = \frac{d\hat{y}}{dx}\Big|_{x=\bar{x}} \qquad (8.59)$$

Determine the point of tangency ($\bar{x}, \bar{y}$) of the parabolic arc with the downhill grade. Also, compute its elevation.

(d) Find the equation of the parabolic arc.

Figure 8.43   Parabolic arc traversing highway hilltop.

Solution   (a) The initial slope of the parabolic arc is equal to the uphill grade, which is expressed as a decimal fraction of 0.1. Since the arc is tangent to the uphill grade at the origin, the initial slope is equal to the derivative of equation (8.56) evaluated at $x = 0$. The derivative of the parabolic arc is given by

$$\frac{dy}{dx} = \frac{d}{dx}(ax^2 + bx)$$
$$= 2ax + b. \qquad (8.60)$$

Therefore, the slope of the line for the uphill grade can be found by setting $x = 0$ in equation (8.60) as

$$\frac{dy}{dx}\Big|_{x=0} = b = 0.1. \qquad (8.61)$$

Hence, the initial slope of the arc is given by the coefficient b of the equation (8.56) and has the value $b = 0.1$.

(b) The slope of the line (equation (8.57)) for the downhill grade is given by $c = -0.08$. Therefore, the equation of the line can be written as

$$\hat{y} = -0.08\,x + d. \tag{8.62}$$

Since this line passes through benchmark B, the value of the y-intercept d can be found by substituting the data point $(200, -6)$ into equation (8.62) as

$$-6 = -0.08(200) + d$$
$$d = -6 + 16$$
$$d = 10.$$

Therefore, the equation of the line for the downhill slope is given by

$$\hat{y} = -0.08x + 10. \tag{8.63}$$

(c) Substituting $x = \bar{x}$ in equation (8.60) gives

$$\frac{dy}{dx}\Big|_{x=\bar{x}} = 2\,a\,\bar{x} + 0.1. \tag{8.64}$$

Evaluating the derivative of equation (8.63) at $x = \bar{x}$ gives

$$\frac{d\hat{y}}{dx}\Big|_{x=\bar{x}} = -0.08. \tag{8.65}$$

Substituting equations (8.64) and (8.65) into equation (8.59) gives

$$2\,a\,\bar{x} + 0.10 = -0.08$$
$$a\,\bar{x} = -0.09. \tag{8.66}$$

Evaluating equations (8.56) and (8.62) at $\bar{x}$ and substituting the results in equation (8.58) gives

$$a\,\bar{x}^2 + 0.1\,\bar{x} = -0.08\,\bar{x} + 10$$
$$\bar{x}\,(a\,\bar{x} + 0.18) = 10. \tag{8.67}$$

Now, substituting the value of $a\,\bar{x}$ from equation (8.66) into equation (8.67), the value of $\bar{x}$ can be found as

$$\bar{x}(-0.09 + 0.18) = 10$$
$$0.09\,\bar{x} = 10$$
$$\bar{x} = \frac{10}{0.09} \tag{8.68}$$
$$= 111.1 \text{ m.} \tag{8.69}$$

Thus, the point of tangency lies a horizontal distance of 111.1 m from benchmark A. Its elevation $\bar{y}$ is obtained by substituting this value of $\bar{x}$ into

equation (8.63) as

$$\bar{y} = \hat{y}(\bar{x}) = -0.08\,\bar{x} + 10$$

$$= -0.08(111.1) + 10$$

$$= 1.11 \text{ m}.$$

Therefore, the downhill point of tangency is given by $(111.1,\ 1.11)$ m.

(d) With the value of $\bar{x}$ known, the coefficient a for the parabolic arc can be obtained from equation (8.66) as

$$a\,\bar{x} = -0.09$$

$$a = \frac{-0.09}{111.1}$$

$$a = -0.00081 \text{ m}^{-1}. \tag{8.70}$$

The equation for the parabolic arc may now be written by substituting the values of a and b from equations (8.70) and (8.61) into equation (8.56) as

$$y = -0.00081\,x^2 + 0.1\,x.$$

PROBLEMS

8-1. A model rocket is fired from the roof of a 50 ft tall building as shown in Fig. P8.1. The height of the rocket is given by

$$y(t) = y(0) + v(0)\,t - \frac{1}{2}g\,t^2 \text{ ft}$$

Figure P8.1   A rocket fired from top of a building in problem P8-1.

where $y(t)$ is the height of the rocket at time t, $y(0) = H = 50$ ft is the initial height of the rocket, $v(0) = 150$ ft/s is the initial velocity of the rocket, and $g = 32.2$ ft/s^2 is the acceleration due to gravity. Find the following:

(a) Write the quadratic equation for the height $y(t)$ of the rocket.

(b) The velocity $v(t) = \frac{dy(t)}{dt}$.

(c) The acceleration $a(t) = \frac{dv(t)}{dt} = \frac{d^2 y(t)}{dt^2}$.

(d) The time required to reach the maximum height as well as the corresponding height y_{max}. Use your results to sketch $y(t)$.

8-2. Repeat problem P8-1 if $H = 15\,\text{m}$, $v(0) = 49$ m/s and $g = 9.8$ m/^{2}.

8-3. The height in the vertical plane of a ball thrown from the ground with an initial velocity of $v(0) = 250$ m/s satisfies the

relationship

$$y(t) = v(0)\,t - \frac{1}{2}g\,t^2 \text{ m.}$$

where $g = 9.8$ m/s^2 is the acceleration due to gravity.

Figure P8.3 A projectile in the vertical plane.

Find the following:

(a) Write the quadratic equation for the height $y(t)$ of the ball.

(b) The velocity $v(t) = \frac{dy(t)}{dt}$.

(c) The acceleration $a(t) = \frac{dv(t)}{dt} = \frac{d^2y(t)}{dt^2}$.

(d) The time required to reach the maximum height, as well as the corresponding height y_{max}. Use your results to sketch $y(t)$.

8-4. Repeat problem P8-3 if $v(0) = 161$ ft/s and $g = 32.2$ ft/s^2.

8-5. The motion of a particle moving in the horizontal direction as shown in Fig. P8.5 is described by its position $x(t)$. Determine the position, velocity, and acceleration at $t = 3.0$ s if

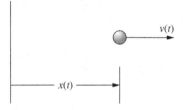

Figure P8.5 A particle moving in the horizontal direction.

(a) $x(t) = 4\cos\left(\frac{2}{3}\pi t\right) + 3\sin\left(\frac{3}{2}\pi t\right)$ m.

(b) $x(t) = 3\,t^5 - 5\,t^2 + \frac{7}{t} + 2\,\sqrt{t}$ m.

(c) $x(t) = 2\,e^{4t} + 3\,e^{-5t} + 2\,(e^t - 1)$ m.

8-6. The motion of a particle moving in the horizontal direction is described by its position $x(t)$. Determine the position, velocity, and acceleration at $t = 1.5$ s if

(a) $x(t) = 4\cos(5\,\pi t)$ m.

(b) $x(t) = 4\,t^3 - 6\,t^2 + 7\,t + 2$ m.

(c) $x(t) = 10\sin(10\,\pi t) + 5\,e^{3t}$ m.

8-7. The motion of a particle in the vertical plane is shown in Fig. P8.7. The height of the particle is given by

$$y(t) = t^3 - 12\,t^2 + 36\,t + 20 \text{ m.}$$

(a) Find the values of position and acceleration when the **velocity** is zero.

(b) Use your results in part (a) to sketch $y(t)$ for $0 \le t \le 9$ s.

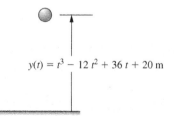

$$y(t) = t^3 - 12\,t^2 + 36\,t + 20 \text{ m}$$

Figure P8.7 Motion of a particle in the vertical plane.

8-8. The motion of a particle in the vertical plane is shown in Fig. P8.8. The height of the particle is given by

$$y(t) = 2\,t^3 - 15\,t^2 + 24\,t + 8 \text{ m}$$

(a) Find the values of the position and acceleration when the **velocity** is zero.

(b) Use your results in part (a) to sketch $y(t)$ for $0 \le t \le 4$ s.

$y(t) = 2\,t^3 - 15\,t^2 + 24\,t + 8$ m

Figure P8.8 Motion of a particle in the vertical plane.

8-9. The voltage across an inductor is given by $v(t) = L\frac{di(t)}{dt}$. If $i(t) = t^3 e^{-2t}$ A and $L = 0.125$ H,
  (a) Find the voltage, $v(t)$.
  (b) Find the value of the **current** when the voltage is zero.
  (c) Use the above information to sketch $i(t)$.

Figure P8.9 Voltage and current in an inductor.

8-10. Repeat problem P8-9 if $L = 0.25$ H and $i(t) = t^2 e^{-t}$ A.

8-11. The voltage across the inductor of Fig. P8.9 is given by $v(t) = L\frac{di(t)}{dt}$. Determine the voltage $v(t)$, the power $p(t) = v(t)\,i(t)$, and the maximum power transfered if the inductance is $L = 2$ mH and the current $i(t)$ is given by
  (a) $i(t) = 13\,e^{-200t}$ A.
  (b) $i(t) = 20\cos(2\,\pi 60\,t)$ A.

8-12. The current flowing through a capacitor shown in Fig. P8.12 is given by $i(t) = C\frac{dv(t)}{dt}$. If $C = 500\ \mu$F and $v(t) = 250\sin(200\,\pi t)$ V,

  (a) Find the current $i(t)$.
  (b) Find the power $p(t) = v(t)\,i(t)$ and its maximum value p_{max}.

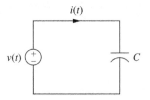

Figure P8.12 Current flowing through a capacitor.

8-13. Repeat problem P8-12 if $C = 40\ \mu$F and $v(t) = 500\cos(200\,\pi t)$ V.

8-14. The current flowing through the capacitor shown in Fig. P8.12 is given by $i(t) = C\frac{dv(t)}{dt}$. If $C = 2\ \mu$F and $v(t) = t^2 e^{-10t}$ V,
  (a) Find the current $i(t)$.
  (b) Find the value of the voltage when the **current** is zero.
  (c) Use the given information to sketch $v(t)$.

8-15. A vehicle starts from rest at position $x = 0$. The velocity of the vehicle for the next 8 seconds is shown in Fig. P8.15.
  (a) Knowing that $a(t) = \frac{dv}{dt}$, sketch the acceleration $a(t)$.
  (b) Knowing that $v(t) = \frac{dx}{dt}$, sketch the position $x(t)$ if the minimum position is $x = -16$ m and the final position is $x = 0$.

8-16. At time $t = 0$, a vehicle located at position $x = 0$ is moving at a velocity of 10 m/s. The velocity of the vehicle for the next 8 seconds is shown in Fig. P8.16.
  (a) Knowing that $a(t) = \frac{dv}{dt}$, sketch the acceleration $a(t)$.
  (b) Knowing that $v(t) = \frac{dx}{dt}$, sketch the position $x(t)$ if the maximum position is $x = 30$ m and the final position is $x = 10$ m.

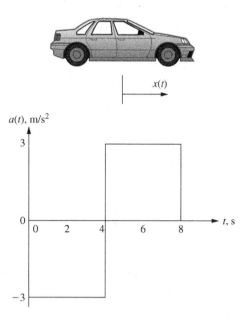

8-17. At time $t = 0$, a moving vehicle is located at position $x = 0$ and subjected to the acceleration $a(t)$ shown in Fig. P8.17.

Figure P8.15 Velocity of a vehicle for problem P8-15.

Figure P8.17 Acceleration of a vehicle for problem P8-17.

(a) Knowing that $a(t) = \frac{dv}{dt}$, sketch the velocity $v(t)$ if the initial velocity is 12 m/s.

(b) Knowing that $v(t) = \frac{dx}{dt}$, sketch the position $x(t)$ if the final position is $x = 48$ m.

8-18. A vehicle **starting from rest** at position $x = 0$ is subjected to the acceleration $a(t)$ shown in Fig. P8.18.

(a) Knowing that $a(t) = \frac{dv}{dt}$, sketch the velocity $v(t)$.

(b) Knowing that $v(t) = \frac{dx}{dt}$, sketch the position $x(t)$ if the final position is $x = 60$ m. Clearly indicate both its **maximum** and **final** values.

8-19. A vehicle **starting from rest** at position $x = 0$ is subjected to the following acceleration $a(t)$:

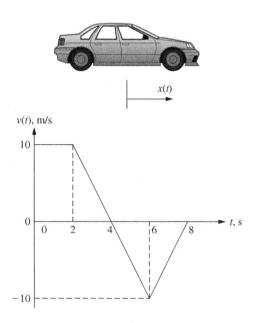

Figure P8.16 Velocity of a vehicle for problem P8-16.

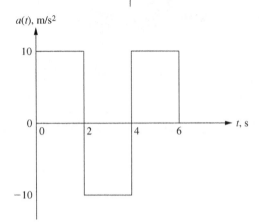

Figure P8.18   Acceleration of a vehicle for problem P8-18.

Figure P8.19   Acceleration of a vehicle for problem P8-19.

(a) Knowing that $a(t) = \frac{dv}{dt}$, sketch the velocity $v(t)$.

(b) Knowing that $v(t) = \frac{dx}{dt}$, sketch the position $x(t)$ if the final position is $x = 15$ m. Clearly indicate both its **maximum** and **final** values.

8-20. A vehicle **starting from rest** at position $x = 0$ is subjected to the following acceleration $a(t)$:

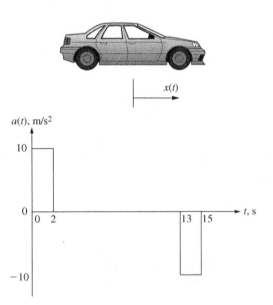

Figure P8.20   Acceleration of a vehicle for problem P8-20.

(a) Knowing that $a(t) = \frac{dv}{dt}$, sketch the velocity $v(t)$.

(b) Knowing that $v(t) = \frac{dx}{dt}$, sketch the position $x(t)$ if the final position is $x = 260$ m. Clearly indicate both its **maximum** and **final** values.

8-21. The voltage across an inductor is given in Fig. P8.21. Knowing that $v(t) = L \frac{di(t)}{dt}$ and $p(t) = v(t) i(t)$, sketch the graphs of $i(t)$ and $p(t)$. Assume $L = 2$ H, $i(0) = 0$ A, and $p(0) = 0$ W.

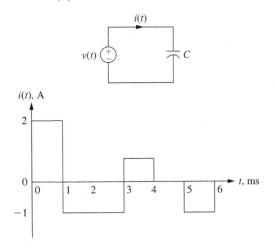

Figure P8.21 Voltage across an inductor for problem P8-21.

8-22. The voltage across an inductor is given in Fig. P8.22. Knowing that $v(t) = L\frac{di(t)}{dt}$ and $p(t) = v(t)\,i(t)$, sketch the graphs of $i(t)$ and $p(t)$. Assume $L = 0.25$ H, $i(0) = 0$ A, and $p(0) = 0$ W.

Figure P8.22 Voltage across an inductor for problem P8-22.

8-23. The voltage across an inductor is given in Fig. P8.23. Knowing that $v(t) = L\frac{di(t)}{dt}$ and $p(t) = v(t)\,i(t)$, sketch the graphs of $i(t)$ and $p(t)$. Assume $L = 2$ mH, $i(0) = 0$ A, and $p(0) = 0$ W.

Figure P8.23 Voltage across an inductor for problem P8-23.

8-24. The current applied to a capacitor is given in Fig. P8.24. Knowing that $i(t) = \frac{dq(t)}{dt} = C\frac{dv(t)}{dt}$, sketch the graphs of the stored charge $q(t)$ and the voltage $v(t)$. Assume $C = 250\ \mu$F, $q(0) = 0$ C, and $v(0) = 0$ V.

Figure P8.24 Current flowing through a capacitor for problem P8-24.

8-25. The current flowing through a capacitor is given in Fig. P8.25. Knowing that

$i(t) = \frac{dq(t)}{dt} = C \frac{dv(t)}{dt}$, sketch the graphs of the stored charge $q(t)$ and the voltage $v(t)$. Assume $C = 250 \, \mu F$, $q(0) = 0 \, C$, and $v(0) = 0 \, V$.

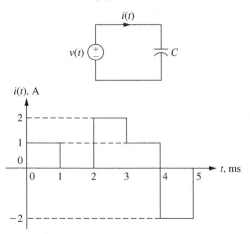

Figure P8.25 Current flowing through a capacitor for problem P8-25.

8-26. The current flowing through a 500 μF capacitor is given in Fig. P8.26. Knowing that $i(t) = C \frac{dv(t)}{dt}$, plot $v(t)$ for $0 \leq t \leq 4$ sec if $v(0) = -4 \, V$, $v(2) = 2 \, V$, $v(4) = 2 \, V$, and the maximum voltage is 4 V.

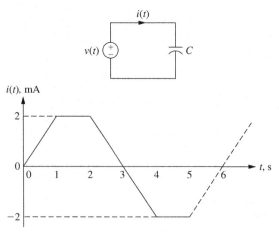

Figure P8.26 Current flowing through a capacitor for problem P8-26.

8-27. A simply supported beam is subjected to a load P as shown in Fig. P8.27. The deflection of the beam is given by

$$y(x) = \frac{P b x}{6 E I l}(x^2 + b^2 - l^2), \quad 0 \leq x \leq \frac{l}{2}$$

where EI is the flexural rigidity of the beam. Find the following:
   (a) The location and the value of maximum deflection y_{max}.
   (b) The value of the slope $\theta = \frac{dy(x)}{dx}$ at the end $x = 0$.

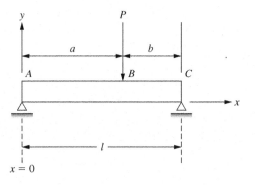

Figure P8.27 A simply supported beam for problem P8-27.

8-28. A simply supported beam is subjected to a load P at $x = \frac{L}{4}$ as shown in Fig. P8.28. The deflection of the beam is given by

$$y(x) = -\frac{P}{128 \, EI}(7 L^2 x - 16 x^3), \quad 0 \leq x \leq \frac{L}{4}$$

where EI is the flexural rigidity of the beam. Find the following:
   (a) The equation of the slope $\theta = \frac{dy(x)}{dx}$.
   (b) The location and magnitude of maximum deflection y_{max}.
   (c) Evaluate both the deflection and slope at $x = 0$ and $x = \frac{L}{4}$.
   (d) Use the results of parts (b) and (c) to sketch the deflection $y(x)$ for $0 \leq x \leq \frac{L}{4}$.

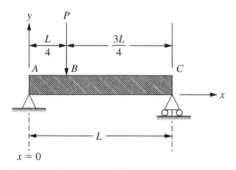

Figure P8.28 A simply supported beam for problem P8-28.

8-29. A simply supported beam is subjected to an applied moment M_o at its center $x = \frac{L}{2}$ as shown in Fig. P8.29. The deflection of the beam is given by

$$y(x) = \frac{M_o}{6\,EI\,L}\left(L^2 x - \frac{3\,L x^2}{4} - x^3\right),$$

$$0 \le x \le \frac{L}{2}$$

where EI is the flexural rigidity of the beam. Find the following:
(a) The equation for the slope $\theta = \frac{dy}{dx}$.
(b) The location and the value of maximum deflection y_{max}.
(c) Evaluate both the deflection and slope at the points A and B ($x = 0$ and $x = \frac{L}{2}$).
(d) Use the results of parts (a)–(c) to sketch the deflection $y(x)$ from $x = 0$ to $x = \frac{L}{2}$.

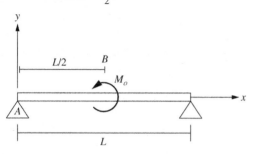

Figure P8.29 A simply supported beam for problem P8-29.

8-30. A simply supported beam is subjected to a sinusoidal distributed load, as shown in the Fig. P8.30. The deflection $y(x)$ of the beam is given by

$$y(x) = -\frac{w_o\,L^4}{16\,\pi^4\,EI}\sin\left(\frac{2\,\pi\,x}{L}\right)$$

where EI is the flexural rigidity of the beam. Find the following:

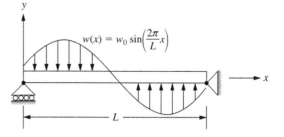

Figure P8.30 A simply supported beam for problem P8-30.

(a) The equation for the slope $\theta = \frac{dy}{dx}$.
(b) The value of the slope where the **deflection** is zero.
(c) The value of the deflection at the location where the **slope** is zero.
(d) Use the results of parts (b) and (c) to sketch the deflection $y(x)$.

8-31. Consider a beam under a linear distributed load and supported as shown in Fig. P8.31. The deflection of the beam is given by

$$y(x) = \frac{w_0}{120\,EI\,L}\left(-x^5 + 2\,L^2 x^3 - L^4 x\right)$$

where L is the length and EI is the flexural rigidity of the beam. Find the following:
(a) The location and value of the maximum deflection y_{max}.
(b) The value of the slope $\theta = \frac{dy(x)}{dx}$ at the ends $x = 0$ and $x = L$.
(c) Use your results in (a) and (b) to sketch the deflection $y(x)$.

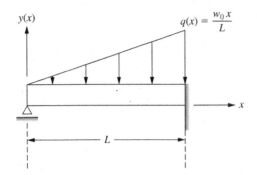

Figure P8.31   A beam under linear distributed load for problem P8-31.

8-32. A fixed-fixed beam is subjected to a sinusoidal distributed load, as shown in Fig. P8.32.

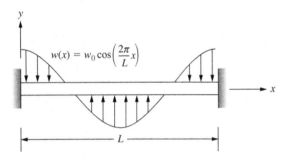

Figure P8.32   A fixed-fixed beam subjected to a sinusoidal distributed load.

The deflection $y(x)$ is given by

$$y(x) = -\frac{w_0 L^4}{16 \pi^4 EI} \left[1 - \cos\left(\frac{2\pi}{L} x \right) \right].$$

(a) Determine the equation for the slope $\theta = \frac{dy}{dx}$.

(b) Evaluate both the deflection and the slope at the points $x = 0$ and $x = L$.

(c) Determine both the location and value of the maximum deflection.

(d) Use your results of parts (b) and (c) to sketch the deflection $y(x)$, and clearly indicate both the location and value of the maximum deflection.

8-33. Consider the buckling of a pinned-pinned column under a compressive axial load P as shown in Fig. P8.33.

Figure P8.33   A pinned-pinned column under a compressive axial load.

The deflection $y(x)$ of the second buckling mode is given by

$$y(x) = -A \sin\left(\frac{2\pi x}{L} \right),$$

where A is an undetermined constant.

(a) Determine the equation for the slope $\theta = \frac{dy}{dx}$.

(b) Determine the value of the slope at the locations where the deflection is zero.

(c) Determine the value of the deflection at the locations where the slope is zero.

(d) Use your results of parts (b) and (c) to sketch the buckled deflection $y(x)$.

8-34. Consider the buckling of a pinned-fixed column under a compressive load P as shown in Fig. P8.34.

The deflection $y(x)$ of the buckled configuration is given by

$$y(x) = -A \left[\sin\left(\frac{4.4934}{L} x \right) + \frac{0.97616}{L} x \right]$$

where A is an undetermined constant.

(a) Determine the equation for the slope $\theta = \frac{dy}{dx}$.

Figure P8.34 A pinned-fixed column under a compressive load.

(b) Evaluate both the deflection and the slope at the points $x = 0$ and $x = L$.

(c) Determine the location and magnitude of the maximum deflection.

(d) Use your results of parts (b) and (c) to sketch the buckled deflection $y(x)$.

8-35. Consider a beam under a uniform distributed load and supported as shown in Fig. P8.35. The deflection of the beam is given by

$$y(x) = \frac{-w}{48 \, EI} (2x^4 - 5x^3 L + 3x^2 L^2)$$

where L is the length and EI is the flexural rigidity of the beam. Find the following:

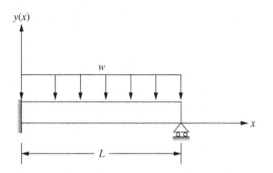

Figure P8.35 A beam under uniform distributed load for problem P8-35.

(a) The location and value of the maximum deflection y_{max}.

(b) The value of the slope $\theta = \dfrac{dy(x)}{dx}$ at the ends $x = 0$ and $x = L$.

(c) Use your results in (a) and (b) to sketch the deflection $y(x)$.

8-36. A cantilever beam is pinned at the end $x = L$, and subjected to an applied moment M_o as shown in Fig. P8.36. The deflection $y(x)$ of the beam is given by

$$y(x) = -\frac{M_o}{4 \, EI \, L} (x^3 - L x^2)$$

where L is the length and EI is the flexural rigidity of the beam. Find the following:

(a) The equation of the slope $\theta = \dfrac{dy(x)}{dx}$.

(b) The location and magnitude of the maximum deflection y_{max}.

(c) Evaluate both the deflection and slope at the points A and B ($x = 0$ and $x = L$).

(d) Use the results in parts (b) and (c) to sketch the deflection $y(x)$ and clearly indicate the location of the maximum deflection on the sketch.

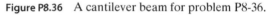

Figure P8.36 A cantilever beam for problem P8-36.

8-37. Consider a shaft subjected to an applied torque T, as shown in Fig. P8.37. The internal normal and shear stresses at the surface vary with the angle relative to the axis and are given by equations (8.71) and (8.72), respectively.

$$\sigma_\theta = \frac{32\,T}{\pi d^3}\sin\theta\cos\theta \qquad (8.71)$$

$$\tau_\theta = \frac{16\,T}{\pi d^3}(\cos^2\theta - \sin^2\theta) \qquad (8.72)$$

Find
(a) the angle θ where σ_θ is maximum
(b) the angle θ where τ_θ is maximum

Figure P8.37 Applied torque and internal forces in a shaft.

8-38. The velocity of a skydiver jumping from a height of 13,000 ft is shown in Fig. P8.38.
(a) Find the equation of the velocity $v(t)$ for the five time intervals shown in Fig. P8.38.
(b) Knowing that $a(t) = \dfrac{dv}{dt}$, sketch the acceleration $a(t)$ of the skydiver for $0 \le t \le 180$ s.
(c) Use the results of part (b) to sketch the height $y(t)$ for $0 \le t \le 180$ s.

8-39. A proposed highway traverses a hilltop bounded by uphill and downhill grades of 15% and −10%, respectively. These grades pass through benchmarks A and B located as shown in Fig. P8.39. With the origin of the coordinate axes (x, y) set at benchmark A, the engineer has defined the hilltop segment of the

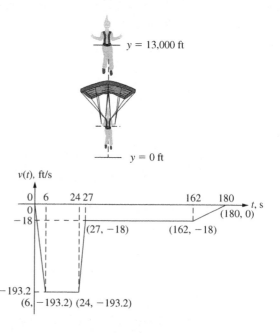

Figure P8.38 Velocity of a skydiver jumping from a height of 13,000 ft.

highway by a parabolic arc

$$y(x) = ax^2 + bx,$$

which is tangent to the uphill grade at the origin.
(a) Find the slope of the line for the uphill grade and the value of b for the parabolic arc.
(b) Find the equation of the line

$$\hat{y} = cx + d$$

for the downhill grade.
(c) Given that at the downhill point of tangency $(\bar{x})$, both the elevation and the slope of the parabolic arc are equal to their respective values of the downhill line, for example,

$$y(\bar{x}) = \hat{y}(\bar{x})$$

$$\frac{dy}{dx}\Big|_{x=\bar{x}} = \frac{d\hat{y}}{dx}_{x=\bar{x}}$$

Figure P8.39 Parabolic arc traversing highway hilltop.

determine the point of tangency $(\bar{x}, \bar{y})$ of the parabolic arc with the downhill grade. Also, compute its elevation.

(d) Find the equation of the parabolic arc.

Integrals in Engineering

This chapter will discuss what integration is and why engineers need to know it. It is important to point out that the objective of this chapter is not to teach techniques of integration, as discussed in a typical calculus course. Instead, the objective of this chapter is to expose students to the importance of integration in engineering and to illustrate its application to the problems covered in core engineering courses such as physics, statics, dynamics, and electric circuits.

9.1 INTRODUCTION: THE ASPHALT PROBLEM

An engineering co-op had to hire an asphalt contractor to widen the truck entrance to the corporate headquarters, as shown in Fig. 9.1. The asphalt extends 50 ft in the x- and y-directions and has a radius of 50 ft. Thus, the required asphalt is the area under the circular curve is given by

$$(x - 50)^2 + (y - 50)^2 = 2500. \tag{9.1}$$

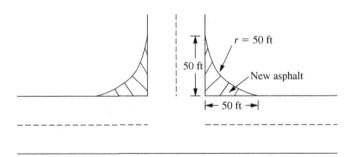

Figure 9.1 Driveway of corporate headquarters.

The asphalt company charges by the square foot and provides an estimate based on "eyeballing" the required area for new asphalt. The co-op asks a young engineer to estimate the area to make sure that the quote is fair. The young engineer proposes to estimate the area as a series of n inscribed rectangles as shown in Fig. 9.2. The area

A is given by

$$A \approx \sum_{i=1}^{n} f(x_i)\Delta x$$

where $\Delta x = \dfrac{50}{n}$ is the width of each rectangle and $f(x_i)$ is the height. The equation of the function $f(x)$ is obtained by solving equation (9.1) for y, which gives

$$y = f(x) = 50 - \sqrt{2500 - (x - 50)^2}. \qquad (9.2)$$

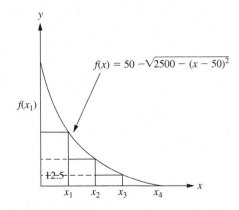

Figure 9.2 Division of asphalt area into n inscribed rectangles.

Suppose, for example, that $n = 4$, as shown in Fig. 9.3. Here $\Delta x = \dfrac{50}{4} = 12.5$ ft and the area can be estimated as

$$A \approx \sum_{i=1}^{4} f(x_i)\Delta x$$
$$= 12.5 * [f(x_1) + f(x_2) + f(x_3) + f(x_4)]. \qquad (9.3)$$

Figure 9.3 Calculation of area using four rectangles.

The values of $f(x_1) \dots f(x_4)$ are obtained by evaluating equation (9.2) at the corresponding values of x:

$$f(x_1) = f(12.5) = 16.93$$

$$f(x_2) = f(25.0) = 6.70$$

$$f(x_3) = f(37.5) = 1.59$$

$$f(x_4) = f(50.0) = 0.0.$$

Substituting these values in equation (9.3) gives

$$A \approx 12.5 \times (16.93 + 6.70 + 1.59 + 0)$$

$$= 315.4 \ \text{ft}^2.$$

This result clearly **underestimates** the actual value of the area. The young engineer claims he would need an ∞ number of rectangles to get it right, i.e.,

$$A = \lim_{n \to \infty} \sum_{i=1}^{n} f(x_i) \, \Delta x.$$

However, in comes the old engineer, who recognizes this as the definition of the *definite integral*, i.e.,

$$\lim_{n \to \infty} \sum_{i=1}^{n} f(x_i) \, \Delta x = \int_{a}^{b} f(x) \, dx. \tag{9.4}$$

In equation (9.4), $\lim_{n \to \infty} \sum_{i=1}^{n} f(x_i) \, \Delta x$ is the area under $f(x)$ between $x = a$ and $x = b$, while $\int_{a}^{b} f(x) \, dx$ is the definite integral of $f(x)$ between $x = a$ and $x = b$. In the case of the asphalt problem, $a = 0$ and $b = 50$ since these are the limits of the asphalt in the x-direction. Hence, the definite integral of a function over an interval is the area under the function over that same interval. The value of the integral is obtained from the fundamental theorem of calculus,

$$\int_{a}^{b} f(x) \, dx = [F(x)]_{a}^{b} = F(b) - F(a) \tag{9.5}$$

where $F(x)$ is an **antiderivative** of $f(x)$. If $F(x)$ is the antiderivative of $f(x)$, then $f(x)$ is the derivative of $F(x)$, i.e.,

$$f(x) = \frac{d}{dx} F(x). \tag{9.6}$$

Hence, evaluating the integral of a function amounts to finding its antiderivative (i.e., its derivative backward). Based on the knowledge of derivatives, the antiderivatives (integrals) of sin (x) and x^n can, for example, be written as

$$f(x) = \sin(x) \quad \Rightarrow \quad F(x) = -\cos(x) + C$$

$$f(x) = x^2 \quad \Rightarrow \quad F(x) = \frac{x^3}{3} + C$$

$$f(x) = x^n \quad \Rightarrow \quad F(x) = \frac{x^{n+1}}{n+1} + C,$$

$$\int \sin(x)\, dx = -\cos x + C$$

$$\int x^2\, dx = \frac{x^3}{3} + C$$

$$\int x^n\, dx = \frac{x^{n+1}}{n+1} + C.$$

The previous integrals are called *indefinite integrals* since there are no limits a and b. Since $F(x)$ is the antiderivative of $f(x)$,

$$F(x) = \int f(x)\, dx. \tag{9.7}$$

Equations (9.6) and (9.7) show that differentiation and integration are **inverse** operations; i.e.,

$$f(x) = \frac{d}{dx} F(x) = \frac{d}{dx} \int f(x)\, dx = f(x)$$

$$F(x) = \int f(x)\, dx = \int \frac{d}{dx} F(x)\, dx = F(x).$$

TABLE 9.1   The antiderivatives of some common functions in engineering.

Function, $f(x)$	Antiderivative, $F(x) = \int f(x)\, dx$
$\sin(\omega x)$	$-\frac{1}{\omega} \cos(\omega x) + C$
$\cos(\omega x)$	$\frac{1}{\omega} \sin(\omega x) + C$
e^{ax}	$\frac{1}{a} e^{ax} + C$
x^n	$\frac{1}{n+1} x^{n+1} + C$
$c f(x)$	$c \int f(x)\, dx$
$f_1(x) + f_2(x)$	$\int f_1(x)\, dx + \int f_2(x)\, dx$

Note that in Table 9.1, a, c, n, and ω are constants as they do not depend on x. With this background, the young engineer finds the total area as the sum of all elemental areas dA, as shown in Fig. 9.4.

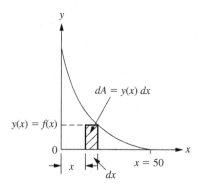

Figure 9.4   Asphalt area with elemental area dA.

The total area is thus

$$A = \int_0^{50} dA = \int_0^{50} y(x)\, dx.$$

Substituting the value of $y(x)$ from equation (9.2) gives

$$A = \int_0^{50} \left(50 - \sqrt{2500 - (x - 50)^2}\right) dx$$

$$= \int_0^{50} 50\, dx - \int_0^{50} \sqrt{2500 - (x - 50)^2}\, dx$$

$$= 50 \int_0^{50} dx - \int_0^{50} \sqrt{2500 - (x - 50)^2}\, dx$$

$$= 50\, [x]_0^{50} - \int_0^{50} \sqrt{2500 - (x - 50)^2}\, dx$$

$$= 50\,(50 - 0) - \int_0^{50} \sqrt{2500 - (x - 50)^2}\, dx$$

or

$$A = 2500 - I. \tag{9.8}$$

The integral $I = \int_0^{50} \sqrt{2500 - (x - 50)^2}\, dx$ is not easy to evaluate by hand. However, this integral can be evaluated using MATLAB (or other engineering programs), which gives $I = 625\,\pi$. Substituting I into equation (9.8) gives

$$A = 2500 - 625\,\pi$$
$$A = 536.5 \text{ ft}^2.$$

In comes the oldest engineer, who notes that the result can be calculated without calculus! The total area is simply the area of a square (of dimension 50×50) minus the area of a quarter-circle of radius $r = 50$ ft, as shown in Fig. 9.5. Therefore,

$$A = (50 \times 50) - \frac{1}{4} \left[\pi (50)^2 \right]$$

$$= 2500 - \frac{1}{4} (2500 \, \pi)$$

$$= 2500 - 625 \, \pi$$

$$A = 536.5 \text{ ft}^2.$$

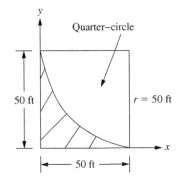

Figure 9.5 Calculation of area without calculus.

Indeed, one of the most important things about calculus in engineering is understanding when you really need to use it!

9.2 CONCEPT OF WORK

Work is done when a force is applied to an object to move it a certain distance. If the force F is constant, the work done is just the force times the distance, as shown in Fig. 9.6.

$$\boldsymbol{W = F \times d}$$

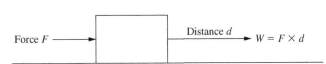

Figure 9.6 Force F moving an object a distance d.

If the object is moved by a constant force as shown in Fig. 9.7, the work done is the area under the force-displacement curve

$$W = F \times d$$
$$= F \times (x_2 - x_1)$$

where $d = x_2 - x_1$ is the distance moved.

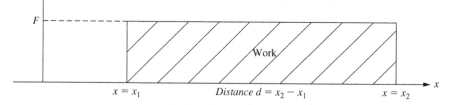

Figure 9.7 Work as area under a constant force curve.

If the force is not constant but is a function of x, as shown in Fig. 9.8, the area under the curve (i.e., the work) must be determined by integration:

$$W = \int_{x_1}^{x_2} F(x)\,dx$$

$$= \int_0^d F(x)\,dx. \tag{9.9}$$

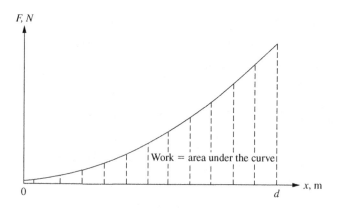

Figure 9.8 Work as area under a variable force curve.

Calculations used to find the work done by a variable force (equation (9.9)) are demonstrated in the following examples.

Example 9-1

The work done on the block shown in Fig. 9.7 is defined by equation (9.9). If $d = 1.0$ m, find the work done for the following forces:

(a) $f(x) = 2x^2 + 3x + 4$ N

(b) $f(x) = 2\sin\left(\dfrac{\pi}{2}x\right) + 3\cos\left(\dfrac{\pi}{2}x\right)$ N

(c) $f(x) = 4\,e^{\pi x}$ N

Solution   (a)

$$W = \int_0^d f(x)\,dx$$

$$= \int_0^1 (2x^2 + 3x + 4)\,dx$$

$$= 2\int_0^1 x^2\,dx + 3\int_0^1 x\,dx + 4\int_0^1 1\,dx$$

$$= 2\left[\frac{x^3}{3}\right]_0^1 + 3\left[\frac{x^2}{2}\right]_0^1 + 4\,(x)_0^1$$

$$= \frac{2}{3}(1-0) + \frac{3}{2}(1-0) + 4\,(1-0)$$

$$= \frac{2}{3} + \frac{3}{2} + 4$$

$$= \frac{37}{6}$$

or

$$W = 6.17 \text{ N-m.}$$

(b)

$$W = \int_0^d f(x)\,dx$$

$$= \int_0^1 \left[2\sin\left(\frac{\pi}{2}x\right) + 3\cos\left(\frac{\pi}{2}x\right)\right]\,dx$$

$$= 2\int_0^1 \sin\left(\frac{\pi}{2}x\right)\,dx + 3\int_0^1 \cos\left(\frac{\pi}{2}x\right)\,dx$$

$$= 2\left[-\frac{\cos\left(\frac{\pi}{2}x\right)}{\frac{\pi}{2}}\right]_0^1 + 3\left[\frac{\sin\left(\frac{\pi}{2}x\right)}{\frac{\pi}{2}}\right]_0^1$$

$$= 2\left[-\frac{2}{\pi}\cos\left(\frac{\pi}{2}x\right)\right]_0^1 + 3\left[\frac{2}{\pi}\sin\left(\frac{\pi}{2}x\right)\right]_0^1$$

$$= -\frac{4}{\pi}\left[\cos\left(\frac{\pi}{2}x\right)\right]_0^1 + \frac{6}{\pi}\left[\sin\left(\frac{\pi}{2}x\right)\right]_0^1$$

$$= -\frac{4}{\pi}\left[\cos\left(\frac{\pi}{2}\right) - \cos(0)\right] + \frac{6}{\pi}\left[\sin\left(\frac{\pi}{2}\right) - \sin(0)\right]$$

$$= -\frac{4}{\pi}(0 - 1) + \frac{6}{\pi}(1 - 0)$$

$$= \frac{4}{\pi} + \frac{6}{\pi} = \frac{10}{\pi}$$

or

$$W = 3.18 \text{ N-m.}$$

(c)

$$W = \int_0^d f(x)\, dx$$

$$= \int_0^1 (4\, e^{\pi x})\, dx$$

$$= 4 \int_0^1 e^{\pi x}\, dx$$

$$= 4 \left[\frac{1}{\pi}\, e^{\pi x} \right]_0^1$$

$$= \frac{4}{\pi}\, \left[e^{\pi} - e^{0} \right]$$

$$= \frac{4}{\pi}\, \left[e^{\pi} - 1 \right]$$

or

$$W = 28.2 \text{ N-m.}$$

Note: In all three cases, the distance moved by the object is 1.0 m, but the work (energy) expended by the force is completely different!

9.3 APPLICATION OF INTEGRALS IN STATICS

9.3.1 Center of Gravity (Centroid)

The centroid, or center of gravity, of an object is a point within the object that represents the average location of its mass. For example, the centroid of a two-dimensional object bounded by a function $y = f(x)$ is given by a point $G = (\bar{x}, \bar{y})$ as shown in Fig. 9.9. The average x-location $\bar{x}$ of the material is given by

$$\bar{x} = \frac{\sum \bar{x}_i A_i}{\sum A_i} \tag{9.10}$$

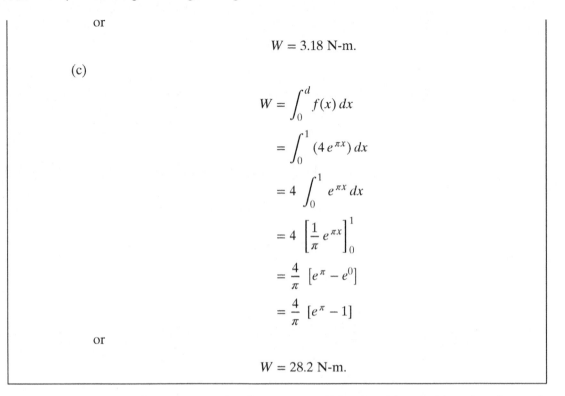

Figure 9.9 Centroid of a two-dimensional object.

while the average y-location $\bar{y}$ of the area is given by

$$\bar{y} = \frac{\sum \bar{y}_i A_i}{\sum A_i}. \tag{9.11}$$

To evaluate the summation in equations (9.10) and (9.11), consider a rectangular element of the area of width dx and centroid $(\bar{x}_i, \bar{y}_i) = \left(x, \frac{y}{2}\right)$, as shown in Fig. 9.10. Now, if $\bar{x}_i = x$, $\bar{y}_i = \frac{y(x)}{2}$, and $A_i = dA = y(x)\,dx$ is the elemental area, equations (9.10) and (9.11) can be written as

$$\bar{x} = \frac{\sum \bar{x}_i A_i}{\sum A_i} = \frac{\int x\,dA}{\int dA} = \frac{\int x\,y(x)\,dx}{\int y(x)\,dx} \tag{9.12}$$

and

$$\bar{y} = \frac{\sum \bar{y}_i A_i}{\sum A_i} = \frac{\int \dfrac{y(x)}{2}\,dA}{\int dA} = \frac{\dfrac{1}{2}\int (y(x))^2\,dx}{\int y(x)\,dx}. \tag{9.13}$$

Figure 9.10 Two-dimensional object with elemental area dA.

Example 9-2

Centroid of Triangular Section: Consider a triangular section of width b and height h, as shown in Fig. 9.11. Find the location of the centroid.

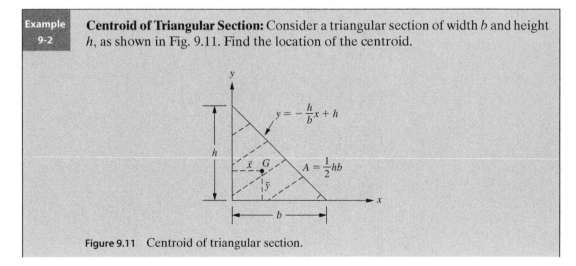

Figure 9.11 Centroid of triangular section.

Solution   The two-dimensional triangular section shown in Fig. 9.11 is the area bounded by the line

$$y(x) = -\frac{h}{b}x + h. \tag{9.14}$$

The area of the section is given by

$$A = \int_0^b y(x)\,dx$$

or

$$A = \frac{1}{2}bh, \tag{9.15}$$

which is simply the area of the triangle. The above result can also be obtained by integrating $y(x) = -\frac{h}{b}x + h$ with respect to x from 0 to b. Using the information in equations (9.14) and (9.15), the x-location of the centroid is calculated from equation (9.12) as

$$\bar{x} = \frac{\int x\,y(x)\,dx}{\int y(x)\,dx}$$

$$= \frac{\int_0^b x\left(-\frac{h}{b}x + h\right)dx}{\frac{1}{2}bh}$$

$$= 2\frac{\int_0^b \left(-\frac{h}{b}x^2 + hx\right)dx}{bh}$$

$$= \left(\frac{2}{bh}\right)\left(-\frac{h}{b}\left[\frac{x^3}{3}\right]_0^b + h\left[\frac{x^2}{2}\right]_0^b\right)$$

$$= \left(\frac{2}{bh}\right)\left(-\frac{h}{3b}(b^3 - 0) + \frac{h}{2}(b^2 - 0)\right)$$

$$= \left(\frac{2}{bh}\right)\left(-\frac{hb^2}{3} + \frac{hb^2}{2}\right)$$

$$= \left(\frac{2}{bh}\right)\left(\frac{hb^2}{6}\right)$$

or

$$\bar{x} = \frac{b}{3}.$$

Similarly, the y-location of the centroid is calculated from equation (9.13) as

$$\bar{y} = \frac{\frac{1}{2} \int y^2(x)\, dx}{\int y(x)\, dx}$$

$$= \left(\frac{1}{2}\right)\left(\frac{2}{bh}\right) \int_0^b \left(-\frac{h}{b}x + h\right)^2 dx$$

$$= \left(\frac{1}{hb}\right) \int_0^b \left[\left(\frac{h^2}{b^2}\right)x^2 - \left(\frac{2h}{b}\right)xh + h^2\right] dx$$

$$= \left(\frac{1}{hb}\right)\left(\frac{h^2}{b^2}\left[\frac{x^3}{3}\right]_0^b - \left(\frac{2h^2}{b}\right)\left[\frac{x^2}{2}\right]_0^b + h^2\,(x)_0^b\right)$$

$$= \left(\frac{1}{hb}\right)\left(\frac{h^2}{3b^2}(b^3 - 0) - \frac{h^2}{b}(b^2 - 0) + h^2\,(b - 0)\right)$$

$$= \left(\frac{1}{hb}\right)\left(\frac{h^2\,b}{3} - h^2\,b + h^2\,b\right)$$

$$= \left(\frac{1}{bh}\right)\left(\frac{h^2\,b}{3}\right)$$

or

$$\bar{y} = \frac{h}{3}.$$

Thus, the centroid of a triangular section of width b and height h is given by $(\bar{x},\, \bar{y}) = \left(\frac{b}{3}, \frac{h}{3}\right)$. In the above example, the coordinates of the centroid were found by using vertical rectangles. These coordinates can also be calculated using the horizontal rectangles, as shown in Fig. 9.12.

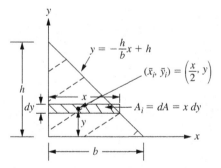

Figure 9.12 Evaluation of centroid using horizontal elemental areas.

The area of the horizontal element is given by

$$A_i = dA = x\,dy = g(y)dy$$

where $x = g(y) = -\dfrac{b}{h} y + b$ is obtained by solving the equation of the line for x. Therefore, the elemental area of the horizontal rectangle is given by

$$A_i = dA = \left(-\frac{b}{h} y + b\right) dy.$$

The y-coordinate of the triangular section can be calculated as

$$\bar{y} = \frac{\sum \bar{y}_i A_i}{\sum A_i} = \frac{\int y \, dA}{A} \qquad (9.16)$$

or

$$\bar{y} = \frac{\int_0^h y \left(-\dfrac{b}{h} y + b\right) dy}{\dfrac{1}{2} b h}$$

$$= \frac{2}{bh} \int_0^h \left(-\frac{b}{h} y^2 + b y\right) dy$$

$$= \frac{2}{bh} \left[\left(-\frac{b}{h}\right) \int_0^h y^2 \, dy + [b] \int_0^h y \, dy\right]$$

$$= \frac{2}{bh} \left[\left(-\frac{b}{3h}\right) [y^3]_0^h + \left(\frac{b}{2}\right) [y^2]_0^h\right]$$

$$= \frac{2}{bh} \left[\left(-\frac{b}{3h}\right) (h^3 - 0) + \left(\frac{b}{2}\right) (h^2 - 0)\right]$$

$$= \frac{2}{bh} \left[-\frac{b h^2}{3} + \frac{b h^2}{2}\right]$$

which gives

$$\bar{y} = \frac{h}{3}.$$

This is the same result previously found using the vertical rectangles!

Example 9-3

The geometry of a cooling fin is described by the shaded area bounded by the parabola shown in Fig. 9.13.

(a) If the equation of the parabola is $y(x) = -x^2 + 4$, determine the height h and the width b of the fin.

(b) Determine the area of the cooling fin by integration with respect to x.

(c) Determine the x-coordinate of the centroid by integration with respect to x.

(d) Determine the y-coordinate of the centroid by integration with respect to x.

Figure 9.13 Geometry of a cooling fin.

Solution   (a) The equation of the parabola describing the cooling fin is given by

$$y(x) = -x^2 + 4. \tag{9.17}$$

The height h of the cooling fin can be found by substituting $x = 0$ in equation (9.17) as

$$h = y(0) = -0^2 + 4 = 4''.$$

The width b of the fin can be obtained by setting $y(x) = 0$, which gives

$$y(x) = -x^2 + 4 = 0 \;\Rightarrow\; x^2 = 4 \;\Rightarrow\; x = \pm 2.$$

Since the width of the fin must be positive, it follows that $b = 2''$.

(b) The area A of the fin is calculated by integrating equation (9.17) from 0 to b as

$$A = \int_0^b y(x)\, dx$$

$$= \int_0^2 (-x^2 + 4)\, dx$$

$$= \left[-\frac{x^3}{3} + 4x \right]_0^2$$

$$= \left[\left(-\frac{2^3}{3} + 4\,(2) \right) - (0 + 0) \right]$$

or

$$A = \frac{16}{3} \text{ in.}^2.$$

(c) The x-coordinate of the centroid can be found using the vertical rectangles as illustrated in Fig. 9.14. By definition,

$$\bar{x} = \frac{\sum \bar{x}_i A_i}{\sum A_i}$$

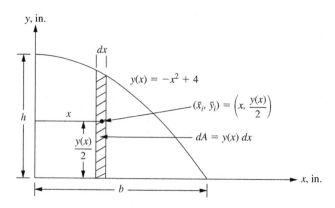

Figure 9.14 Determination of centroid using vertical rectangles.

where $\bar{x}_i = x$ and $A_i = dA = y(x)\, dx$. Thus,

$$\bar{x} = \frac{\int xy(x)\, dx}{A}$$

$$= \frac{\int_0^2 x\,(-x^2 + 4)\, dx}{\dfrac{16}{3}}$$

$$= \frac{3}{16} \int_0^2 (-x^3 + 4\,x)\, dx$$

$$= \frac{3}{16} \left[-\frac{x^4}{4} + 4\,\frac{x^2}{2} \right]_0^2$$

$$= \frac{3}{16} \left[\left(-\frac{2^4}{4} + 2\,(2^2) \right) - (0 + 0) \right]$$

or

$$\bar{x} = \frac{12}{16} \text{ in.}$$

Therefore, $\bar{x} = \dfrac{3}{4}$ in.

(d) Similarly, the y-coordinate of the centroid can be determined by integration with respect to x as

$$\bar{y} = \frac{\sum \bar{y}_i A_i}{\sum A_i}$$

where $\bar{y}_i = \dfrac{y(x)}{2}$ and $A_i = y(x)\,dx$. Thus,

$$\bar{y} = \frac{\dfrac{1}{2}\displaystyle\int y^2(x)\,dx}{A}$$

$$= \frac{\dfrac{1}{2}\displaystyle\int_0^2 (-x^2+4)^2\,dx}{\dfrac{16}{3}}$$

$$= \frac{3}{32}\int_0^2 (x^4 - 8x^2 + 16)\,dx$$

$$= \frac{3}{32}\left[\frac{x^5}{5} - 8\frac{x^3}{3} + 16x\right]_0^2$$

$$= \frac{3}{32}\left[\left(\frac{2^5}{5} - 8\frac{2^3}{3} + 16\,(2)\right) - (0+0+0)\right]$$

$$= \frac{3}{32}\left[\frac{32}{5} - \frac{64}{3} + 32\right]$$

$$= 3\left[\frac{1}{5} - \frac{2}{3} + 1\right]$$

$$= 3\left[\frac{3}{15} - \frac{10}{15} + \frac{15}{15}\right]$$

$$= \frac{24}{15}$$

or

$$\bar{y} = \frac{8}{5}.$$

Therefore, $\bar{y} = \dfrac{8}{5}$ in.

9.3.2 Alternate Definition of the Centroid

If the origin of the x-y coordinate system is located at the centroid as shown in Fig. 9.15, then the x- and y-coordinates of the centroid are given by

$$\bar{x} = \frac{\int x\,dA}{A} = 0 \tag{9.18}$$

$$\bar{y} = \frac{\int y\,dA}{A} = 0. \tag{9.19}$$

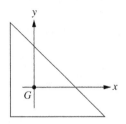

Figure 9.15 Triangular section with origin at centroid.

Hence, an alternative definition of the centroid is the location of the origin such that $\int x\, dA = \int y\, dA = 0$ (i.e., there is no first moment about the origin). As shown in Fig. 9.16, this means that the first moment of the area about both the x- and y-axes is zero:

$$\text{First moment of area about } x\text{-axis} = M_x = \int y\, dA = 0$$

$$\text{First moment of area about } y\text{-axis} = M_y = \int x\, dA = 0.$$

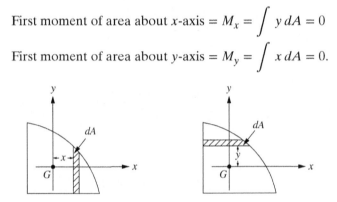

Figure 9.16 First moment of area.

The above definition of the centroid is illustrated for a rectangular section in the following example.

Example 9-4
Show that the coordinates of the centroid of the rectangle in Fig. 9.17 are $\bar{x} = \bar{y} = 0$.

Figure 9.17 Rectangular section.

Solution The first moment of area about the y-axis can be calculated using vertical rectangles, as shown in Fig. 9.18, which gives

$$\int x\, dA = \int_{-\frac{b}{2}}^{\frac{b}{2}} x\, h\, dx$$

$$= h \left[\frac{x^2}{2} \right]_{-\frac{b}{2}}^{\frac{b}{2}}$$

$$= \frac{h}{2} \left(\frac{b^2}{4} - \frac{b^2}{4} \right)$$

$$= 0.$$

Hence, $\bar{x} = \dfrac{\int x\, dA}{A} = 0.$

Figure 9.18 x-coordinate of the centroid using vertical rectangles.

Similarly, the first moment of area about the x-axis can be calculated using horizontal rectangles, as shown in Fig. 9.19, which gives

$$\int y\, dA = \int_{-\frac{h}{2}}^{\frac{h}{2}} y\, b\, dy$$

$$= b \left[\frac{y^2}{2} \right]_{-\frac{h}{2}}^{\frac{h}{2}}$$

$$= \frac{b}{2} \left(\frac{h^2}{4} - \frac{h^2}{4} \right)$$

$$= 0.$$

Hence, $\bar{y} = \dfrac{\int y\, dA}{A} = 0.$

Figure 9.19 y-coordinate of the centroid using horizontal rectangles.

9.4 DISTRIBUTED LOADS

In this section, integrals are used to find the resultant force due to a distributed load, as well as the location of that force required for statically equivalent loading. These are among the primary applications of integrals in statics.

9.4.1 Hydrostatic Pressure on a Retaining Wall

Consider a retaining wall of height h and width b that is subjected to a hydrostatic pressure from fluid of density ρ (Fig. 9.20). The pressure acting on the wall satisfies the linear equation

$$p(y) = \rho g y$$

where g is the acceleration due to gravity.

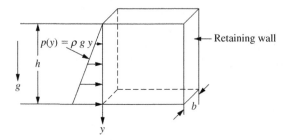

Figure 9.20 Hydrostatic force acting on the rectangular retaining wall.

The resultant force acting on the wall is calculated by adding up (i.e., integrating) all the differential forces dF shown in Fig. 9.22. Since pressure is force/unit area, the differential force is found by multiplying the value of the pressure at any depth y by an elemental area of the wall as

$$dF = p(y)\,dA$$

where $dA = b\,dy$.

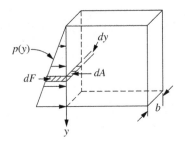

Figure 9.21 Forces acting on the retaining wall.

The resultant force acting on the wall is obtained by integration as

$$F = \int dF$$

$$= \int_0^h p(y)\, b\, dy$$

$$= \int_0^h \rho g y b\, dy$$

$$= \rho g b \int_0^h y\, dy$$

$$= \rho g b \left[\frac{y^2}{2} \right]_0^h$$

or

$$F = \frac{\rho g b h^2}{2}.$$

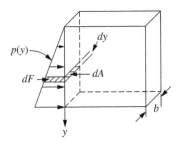

Figure 9.22 Forces acting on the retaining wall.

Note that $\int_0^h p(y)\, b\, dy = b \int_0^h p(y)\, dy$ is simply the width b times the area of the triangle shown in Fig. 9.23. Therefore, the resultant force can be obtained from the area under the distributed load. Since this is a triangular load, the area can be calculated without using integration, and is given by

$$A = \frac{1}{2}(\rho g h)(h) = \frac{\rho g h^2}{2}.$$

$p(y)$

h

$\rho g h$

y

Figure 9.23 Area under hydrostatic pressure.

The resultant force is obtained by multiplying the area with the width b, which gives

$$F = b \frac{\rho g h^2}{2} = \frac{\rho g b h^2}{2}.$$

9.4.2 Distributed Loads on Beams: Statically Equivalent Loading

Fig. 9.24 shows a simply supported beam with a distributed load applied over the entire length L. The distributed load $w(x)$ varies in intensity with position x and has units of force per unit length (lb/ft or N/m). The goal is to replace the distributed load with a statically equivalent point load. As concluded in the previous section, the equivalent load R is the area under the distributed load, and is given by

$$R = \int_0^L w(x)\, dx. \tag{9.20}$$

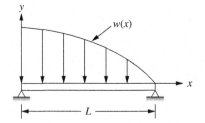

y

$w(x)$

x

L

Figure 9.24 Distributed load on a simply supported beam.

The equivalent load R and location l are shown in Fig. 9.25.

To find the location of the statically equivalent force, the resultant load shown in Fig. 9.25 must have the same **moment** about every point as the distributed load

Figure 9.25 Beam with an equivalent point load.

shown in Fig. 9.24. For example, the moment (force times distance) about the point $x = 0$ must be the same for both the distributed and equivalent loads. The moment M_0 for the distributed load can be calculated by summing moments due to elemental loads dw, as shown in Fig. 9.26. Hence,

$$M_0 = \int xdw$$

$$= \int_0^L xw(x)\, dx. \tag{9.21}$$

Figure 9.26 Beam with a small elemental load.

The moment M_0 about the $x = 0$ point of the equivalent load R is given by

$$M_0 = Rl,$$

or

$$M_0 = l \int_0^L w(x)\, dx. \tag{9.22}$$

Equating the two moments in equations (9.21) and (9.22) gives

$$l \int_0^L w(x)\, dx = \int_0^L xw(x)\, dx$$

or

$$l = \frac{\int_0^L xw(x)\, dx}{\int_0^L w(x)\, dx}. \tag{9.23}$$

Equation (9.23) is identical to equation (9.12) but with $w(x)$ instead of $y(x)$. Hence, it can be concluded that $l = \bar{x}$, which is the x-coordinate of the centroid of the area under the load! Thus, for the purpose of statics, a distributed load can always be replaced by its resultant force acting at its centroid.

Example 9-5   Find the magnitude and location of the statically equivalent point load for the beam of Fig. 9.27. Use your results to find the reactions at the support (Fig. 9.28).

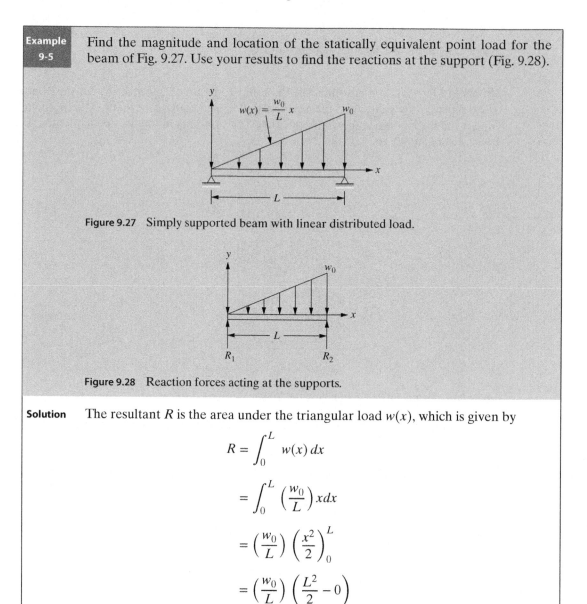

Figure 9.27   Simply supported beam with linear distributed load.

Figure 9.28   Reaction forces acting at the supports.

Solution   The resultant R is the area under the triangular load $w(x)$, which is given by

$$R = \int_0^L w(x)\, dx$$

$$= \int_0^L \left(\frac{w_0}{L}\right) x\, dx$$

$$= \left(\frac{w_0}{L}\right) \left(\frac{x^2}{2}\right)_0^L$$

$$= \left(\frac{w_0}{L}\right) \left(\frac{L^2}{2} - 0\right)$$

or

$$R = \frac{1}{2} w_0 L.$$

Note that the above result is just the area under the triangle defined by $w(x)$. The location l of the statically equivalent load is the x-coordinate of the centroid of the area under $w(x)$ and can be calculated using equation (9.23) as

$$l = \frac{\int_0^L x\,w(x)\,dx}{\int_0^L w(x)\,dx}$$

$$= \frac{\int_0^L x\left(\frac{w_0}{L}x\right)dx}{\frac{1}{2}w_0 L}$$

$$= \frac{2}{L^2}\int_0^L x^2\,dx$$

$$= \frac{2}{L^2}\left(\frac{x^3}{3}\right)_0^L$$

$$= \frac{2}{3L^2}(L^3 - 0)$$

or

$$l = \frac{2L}{3}.$$

Note that this is two-thirds of the way from the base of the triangular load (i.e., the centroid of the triangle). Hence, the location l could have been determined without any further calculus.

The statically equivalent loading is shown in Fig. 9.29.

Figure 9.29 Statically equivalent loading of the beam.

For equilibrium, the sum of the forces in the y-direction must be zero, i.e.,

$$R_1 + R_2 = R$$

or

$$R_1 + R_2 = \frac{1}{2}w_0 L. \tag{9.24}$$

Also, the sum of moments about any point on the beam must be zero. Taking the moments about $x = 0$ gives

$$R_2 \, L - \left(\frac{1}{2} w_0 \, L\right) \frac{2L}{3} = 0$$

which gives

$$R_2 \, L = \frac{w_0 \, L^2}{3}$$

or

$$R_2 = \frac{w_0 \, L}{3}. \tag{9.25}$$

Substituting equation (9.25) in equation (9.24) gives

$$R_1 + \frac{w_0 \, L}{3} = \frac{w_0 \, L}{2}.$$

Solving for R_1 gives

$$R_1 = \frac{w_0 \, L}{2} - \frac{w_0 \, L}{3}$$

or

$$R_1 = \frac{w_0 \, L}{6}. \tag{9.26}$$

9.5 APPLICATIONS OF INTEGRALS IN DYNAMICS

It was discussed in Chapter 8 that if the position of a particle moving in the x-direction as shown in Fig. 9.30 is given by $x(t)$, the velocity $v(t)$ is the first derivative of the position, and the acceleration is the first derivative of the velocity (second derivative of the position):

$$v(t) = \frac{dx(t)}{dt}$$

$$a(t) = \frac{dv(t)}{dt} = \frac{d^2x(t)}{dt^2}.$$

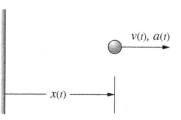

Figure 9.30 A particle moving in the horizontal direction.

Now, if the acceleration $a(t)$ of the particle is given, both the velocity $v(t)$ and the position $x(t)$ can be determined by integrating with respect to t. Beginning with

$$\frac{dv(t)}{dt} = a(t) \tag{9.27}$$

integrating both sides between $t = t_0$ and any time t gives

$$\int_{t_0}^{t} \frac{dv(t)}{dt} \, dt = \int_{t_0}^{t} a(t) \, dt. \tag{9.28}$$

By definition, $\int_{t_0}^{t} \frac{dv(t)}{dt} \, dt = [v(t)]_{t_0}^{t}$, so that equation (9.28) can be written as

$$[v(t)]_{t_0}^{t} = \int_{t_0}^{t} a(t) \, dt$$

which gives

$$v(t) - v(t_0) = \int_{t_0}^{t} a(t) \, dt$$

or

$$\boldsymbol{v(t) = v(t_0) + \int_{0}^{t} a(t) \, dt.} \tag{9.29}$$

Thus, the velocity of the particle at any time t is equal to the velocity at $t = t_0$ (initial velocity) plus the integral of the acceleration from $t = t_0$ to the time t. Now, given the velocity $v(t)$, the position $x(t)$ can be determined by integrating with respect to t. Beginning with

$$\frac{dx(t)}{dt} = v(t) \tag{9.30}$$

integrating both sides between $t = t_0$ and any time t gives

$$\int_{t_0}^{t} \frac{dx(t)}{dt} \, dt = \int_{t_0}^{t} v(t) \, dt. \tag{9.31}$$

By definition, $\int_{t_0}^{t} \frac{dx(t)}{dt} \, dt = [x(t)]_{t_0}^{t}$, so that equation (9.31) can be written as

$$[x(t)]_{t_0}^{t} = \int_{t_0}^{t} v(t) \, dt$$

which gives

$$x(t) - x(t_0) = \int_{t_0}^{t} v(t) \, dt$$

or

$$\boldsymbol{x(t) = x(t_0) + \int_{t_0}^{t} v(t) \, dt.} \tag{9.32}$$

Thus, the position of the particle at any time t is equal to the position at $t = t_0$ (initial position) plus the integral of the velocity from $t = t_0$ to the time t.

Example 9-6

A ball is dropped from a height of 1.0 m at $t = t_0 = 0$, as shown in Fig. 9.31. Find $v(t)$, $y(t)$, and the time it takes for the ball to hit the ground.

Figure 9.31   A ball dropped from a height of 1 m.

Solution

Since the ball is dropped from rest at time $t = 0$ sec, $v(0) = 0$ m/sec. Substituting $t_0 = 0$, $a(t) = -9.81$ m/s^2, and $v(0) = 0$ in equation (9.29), the velocity at any time t can be obtained as

$$v(t) = 0 + \int_0^t -9.81 \, dt$$

$$= -9.81 \, [t]_0^t$$

$$= -9.81 \, (t - 0)$$

or

$$v(t) = -9.81 \, t \text{ m/s.}$$

Now, substituting $v(t)$ into equation (9.32), the position $y(t)$ of the ball at any time t can be obtained as

$$y(t) = y(0) + \int_0^t -9.81 \, t \, dt$$

$$= y(0) - 9.81 \left[\frac{t^2}{2} \right]_0^t$$

$$= y(0) - \frac{9.81}{2} \, (t^2 - 0)$$

$$y(t) = y(0) - 4.905 \, t^2.$$

Since the initial height is $y(0) = 1$ m, the position of the ball at any time t is given by

$$y(t) = 1.0 - 4.905 \, t^2 \text{ m.}$$

The time to impact is obtained by setting $y(t) = 0$ as

$$1.0 - 4.905\, t_{\text{impact}}^2 = 0,$$

which gives

$$4.905\, t_{\text{impact}}^2 = 1.$$

Solving for t_{impact} gives

$$t_{\text{impact}} = \sqrt{\frac{1.0}{4.905}}$$

or

$$t_{\text{impact}} = 0.452 \text{ s}.$$

Example 9-7

Suppose that a ball is thrown upward from ground level with an initial velocity $v(0) = v_0 = 4.43$ m/s, as shown in Fig. 9.32. Find $v(t)$ and $y(t)$.

$v_0 = 4.43$ m/s $\qquad$ $y(t)$

$a = -9.81 \text{ m/s}^2$

Figure 9.32 A ball thrown upward with an initial velocity.

Solution

Substituting $t_0 = 0$, $v(0) = 4.43$ m/s, and $a(t) = -9.81$ m/s^2 into equation (9.29), the velocity of the ball at any time t is given by

$$v(t) = 4.43 + \int_0^t -9.81\, dt$$

$$= 4.43 - 9.81\, [t]_0^t$$

$$= 4.43 - 9.81\, (t - 0)$$

or

$$v(t) = 4.43 - 9.81\, t \text{ m/s}. \qquad (9.33)$$

Now substituting the velocity $v(t)$ into equation (9.32), the position of the ball is given as

$$y(t) = y(0) + \int_0^t (4.43 - 9.81\, t)\, dt$$

$$= y(0) + 4.43\, [t]_0^t - 9.81 \left[\frac{t^2}{2}\right]_0^t$$

$$= y(0) + 4.43\, (t - 0) - \frac{9.81}{2}\, (t^2 - 0)$$

or

$$y(t) = y(0) + 4.43\,t - 4.905\,t^2.$$

Since the initial position is $y(0) = 0$ m, the position of the ball at any time t is given by

$$y(t) = 4.43\,t - 4.905\,t^2 \text{ m}.$$

Example 9-8

A stone is thrown from the top of a 50 m high building with an initial velocity of 10 m/s, as shown in Fig. 9.33.

Figure 9.33   A stone thrown from the top of a building.

Knowing that the velocity is

$$v(t) = v(0) + \int_0^t a(t)\,dt \qquad (9.34)$$

and the position is

$$y(t) = y(0) + \int_0^t v(t)\,dt \qquad (9.35)$$

(a) Find and plot the velocity $v(t)$.
(b) Find and plot the position $y(t)$.
(c) Determine both the time and the velocity when the stone hits the ground.

Solution

(a) The velocity of the stone can be calculated by substituting $v(0) = 10$ m/s and $a(t) = -9.81$ m/s^2 into equation (9.34) as

$$v(t) = v(0) + \int_0^t a(t)\,dt$$

$$= 10 + \int_0^t -9.81\,dt$$

$$= 10 - 9.81\,[t]_0^t$$

$$= 10 - 9.81\,(t - 0)$$

or

$$v(t) = 10 - 9.81 \, t \text{ m/s}. \tag{9.36}$$

The plot of the velocity is a straight line with y-intercept $v_o = 10$ m/s and slope -9.81 m/s^2 as shown in Fig. 9.34.

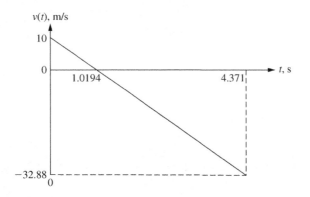

Figure 9.34   Velocity of the stone.

(b) The position of the stone can be calculated by substituting $y(0) = 50$ m and $v(t)$ from equation (9.36) into equation (9.35) as

$$y(t) = y(0) + \int_0^t v(t) \, dt$$

$$= 50 + \int_0^t (10 - 9.81 \, t) \, dt$$

$$= 50 + 10 \, [t]_0^t - 9.81 \left[\frac{t^2}{2} \right]_0^t$$

$$= 50 + 10 \, (t - 0) - \frac{9.81}{2} \, (t^2 - 0)$$

or

$$y(t) = 50 + 10 \, t - 4.905 \, t^2 \text{ m}. \tag{9.37}$$

The plot of the position is as shown in Fig. 9.35. The maximum height can be determined by setting $\dfrac{dy}{dt} = v(t) = 0$, which gives $v(t) = 10 - 9.81 \, t = 0$. Solving for t gives $t_{max} = 1.0194$ sec. The maximum height is thus

$$y_{max} = 50 + 10 \, (1.0194) - 4.905 \, (1.0194)^2$$

or

$$y_{max} = 55.097 \text{ m}.$$

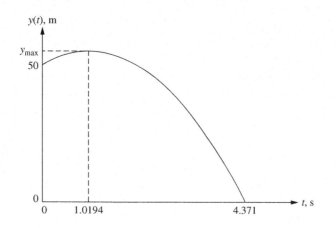

Figure 9.35   Position of the stone.

(c) The time it takes for the stone to hit the ground can be calculated by setting the position $y(t)$ equal to zero as

$$y(t) = 50 + 10t - 4.905\, t^2 = 0$$

or

$$t^2 - 2.039\, t - 10.194 = 0. \tag{9.38}$$

The quadratic equation (9.38) can be solved by using one of the methods described in Chapter 2. For example, we can complete the square as

$$t^2 - 2.039\, t = 10.194$$

$$t^2 - 2.039\, t + \left(\frac{2.039}{2}\right)^2 = 10.194 + \left(\frac{2.039}{2}\right)^2$$

$$\left(t - \frac{2.039}{2}\right)^2 = (\pm\sqrt{11.233})^2$$

$$t - 1.0194 = \pm\, 3.3516$$

$$t = 1.0194 \pm 3.3516$$

or

$$t = 4.371, -2.332 \text{ s}. \tag{9.39}$$

Since the negative time ($t = -2.332$ s) in equation (9.39) is not a possible solution, it takes $t = 4.371$ s for the stone to hit the ground. The velocity when the stone hits the ground is obtained by evaluating $v(t)$ at $t = 4.371$, which gives

$$v(4.371) = 10 - 9.81\,(4.371) = -32.88 \text{ m/s}.$$

9.5.1 Graphical Interpretation

The velocity $v(t)$ can be determined by integrating the acceleration. It follows that the change in velocity can be determined from the **area** under the graph of $a(t)$, as shown in Fig. 9.36.

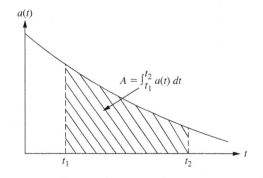

Figure 9.36 Velocity as an area under the acceleration graph.

This can be shown by considering the definition of acceleration, $a(t) = \frac{dv(t)}{dt}$. Integrating both sides from time t_1 to time t_2, we get

$$\int_{t_1}^{t_2} a(t)\, dt = \int_{t_1}^{t_2} \frac{dv(t)}{dt}\, dt$$

$$= [v(t)]_{t_1}^{t_2}$$

or

$$\int_{t_1}^{t_2} a(t)\, dt = v_2 - v_1.$$

In other words, the area under $a(t)$ between t_1 and t_2 equals the change in $v(t)$ between t_1 and t_2. The change in velocity $v_2 - v_1$ can be added to the initial velocity v_1 at time t_1 to obtain the velocity at time t_2, as shown in Fig. 9.37.

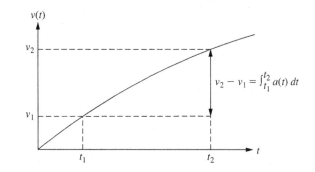

Figure 9.37 Change in velocity from time t_1 to t_2.

Similarly, the position $x(t)$ can be determined by integrating velocity. It follows that the change in position can be determined from the **area** under the graph of $v(t)$, as shown in Fig. 9.38.

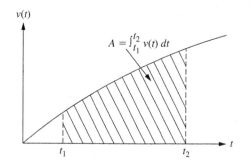

Figure 9.38 Position as an area under the velocity graph.

This can be shown by considering the definition of velocity, $v(t) = \frac{dx(t)}{dt}$. Integrating both sides from time t_1 to time t_2 gives

$$\int_{t_1}^{t_2} v(t)\, dt = \int_{t_1}^{t_2} \frac{dx(t)}{dt}\, dt$$

$$= [x(t)]_{t_1}^{t_2}$$

or

$$\int_{t_1}^{t_2} v(t)\, dt = x_2 - x_1.$$

In other words, the area under $v(t)$ between t_1 and t_2 equals the change in $x(t)$ between t_1 and t_2. The change in position $x_2 - x_1$ can be added to the initial position x_1 at time t_1 to obtain the position at time t_2, as shown in Fig. 9.39.

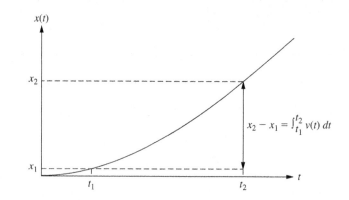

Figure 9.39 Change in position from time t_1 to t_2.

Example
9-9

The acceleration of a vehicle is measured as shown in Fig. 9.40. Knowing that the vehicle starts from rest at position $x = 0$, sketch the velocity $v(t)$ and position $x(t)$ using integrals.

Figure 9.40 Acceleration of a vehicle for Example 9-9.

Solution (a) **Velocity:** Knowing $v(0) = 0$ and $v(t) - v(t_0) = \int_{t_0}^{t} a(t)\, dt, 0 \leq t \leq 2$:
$a(t) = 2$ m/s^2 = constant. Therefore, $v(t)$ is a straight line with a slope of 2 m/s^2. Also, the change in velocity is

$$v_2 - v_0 = \int_0^2 a(t)\, dt$$

or

$$v_2 - v_0 = \text{area under } a(t) \text{ between 0 and 2 s.}$$

Thus,

$$v_2 - v_0 = (2)\,(2) = 4 \quad \text{(area of a rectangle)},$$

which gives

$$v_2 = v_0 + 4.$$

Since $v_0 = 0$,

$$v_2 = 0 + 4 = 4 \text{ m/s}.$$

$2 < t \leq 4$: Since $a(t) = 0$ m/s^2, $v(t)$ is constant. Also,

$$v_4 - v_2 = \int_2^4 a(t)\, dt$$

$$= \text{area under } a(t) \text{ between 2 and 4 s}$$

$$= 0.$$

Thus,

$$v_4 = v_2 + 0$$

$$= 4 + 0$$

or

$$v_4 = 4 \text{ m/s}.$$

$4 < t \leq 6$: $a(t) = -2 \text{ m/s}^2 = $ constant. Therefore, $v(t)$ is a straight line with a slope of -2 m/s^2. Also, the change in $v(t)$ is

$$v_6 - v_4 = \int_4^6 a(t)\, dt$$

$$= \text{area under } a(t) \text{ between 4 and 6}$$

$$= (-2)(2) \quad \text{(area of a rectangle)}$$

or

$$v_6 - v_4 = -4.$$

Thus,
$$v_6 = v_4 - 4$$
$$= 4 - 4$$

or

$$v_6 = 0 \text{ m/s}.$$

The graph of the velocity obtained above is as shown in Fig. 9.41.

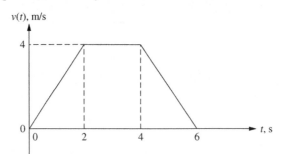

Figure 9.41 Velocity of the vehicle for Example 9-9.

(b) **Position:** Now use $v(t)$ to sketch $x(t)$ knowing that $x(0) = 0$ and $x(t) - x(t_0) = \int_{t_0}^t v(t)\, dt$.

$0 \leq t \leq 2$: $v(t)$ is a linear function (straight line) with a slope of 2 m/s. Therefore, $x(t)$ is a quadratic function with increasing slope (concave up). Also, the change in $x(t)$ is

$$x_2 - x_0 = \int_0^2 v(t)\, dt$$

$$= \text{area under } v(t) \text{ between 0 and 2}$$

$$= \frac{1}{2}(2)(4) \quad \text{(area of a triangle)}$$

or

$$x_2 - x_0 = 4.$$

Thus,

$$x_2 = x_0 + 4.$$

Since $x_0 = 0$,

$$x_2 = 0 + 4$$

or

$$x_2 = 4 \text{ m}.$$

$2 < t \leq 4:$ $v(t)$ has a constant value of 4 m/s. Therefore, $x(t)$ is a straight line with a slope of 4 m/s. Also, the change in $x(t)$ is

$$x_4 - x_2 = \int_2^4 v(t)\, dt$$

$$= \text{area under } v(t) \text{ between 2 and 4}$$

$$= (2)(4) \quad \text{(area of a rectangle)}$$

or

$$x_4 - x_2 = 8.$$

Thus,

$$x_4 = x_2 + 8$$

$$= 4 + 8$$

or

$$x_4 = 12 \text{ m}.$$

$4 < t \leq 6:$ $v(t)$ is a linear function (straight line) with a slope of -2 m/s. Therefore, $x(t)$ is a quadratic function with a decreasing slope (concave down). Also, the change in position is

$$x_6 - x_4 = \int_4^6 v(t)\, dt$$

$$= \text{area under } v(t) \text{ between 4 and 6}$$

$$= \frac{1}{2}(2)(4) \quad \text{(area of a triangle)}$$

or

$$x_6 - x_4 = 4.$$

Thus,

$$x_6 = x_4 + 4$$

$$= 12 + 4$$

or

$$x_6 = 16 \text{ m}.$$

The graph of the position obtained above is shown in Fig. 9.42.

Figure 9.42 Position of the particle for Example 9-7.

APPLICATIONS OF INTEGRALS IN ELECTRIC CIRCUITS

9.6.1 Current, Voltage, and Energy Stored in a Capacitor

In this section, integrals are used to obtain the voltage across a capacitor when a current is passed through it (charging and discharging of a capacitor), as well as the total stored energy.

Example 9-10

For $t \geq 0$, a current $i(t) = 24\,e^{-40t}$ mA is applied to a $3\,\mu$ F capacitor, as shown in Fig. 9.43.

(a) Given that $i(t) = C\,\dfrac{dv(t)}{dt}$, find the voltage $v(t)$ across the capacitor.

(b) Given that $p(t) = v(t)\,i(t) = \dfrac{dw(t)}{dt}$, find the stored energy $w(t)$ and show that $w(t) = \dfrac{1}{2}\,C\,v^2(t)$.

Assume that the capacitor is initially completely discharged; i.e., the initial voltage across the capacitor and initial energy stored are zero.

Figure 9.43 Current applied to a capacitor.

Solution   (a) Given $i(t) = C\dfrac{dv(t)}{dt}$, it follows that

$$\frac{dv(t)}{dt} = \frac{1}{C}\,i(t). \qquad (9.40)$$

Integrating both sides gives

$$\int_0^t \frac{dv(t)}{dt}\,dt = \int_0^t \frac{1}{C}\,i(t)\,dt$$

$$[v(t)]_0^t = \frac{1}{C}\int_0^t i(t)\,dt$$

$$v(t) - v(0) = \frac{1}{C}\int_0^t i(t)\,dt$$

or

$$v(t) = v(0) + \frac{1}{C}\int_0^t i(t)\,dt. \qquad (9.41)$$

Substituting $v(0) = 0$ V, $C = 3 \times 10^{-6}$ F, and $i(t) = 0.024\,e^{-40t}$ A into equation (9.41), the voltage across the capacitor at any time t is given by

$$v(t) = \frac{1}{3.0 \times 10^{-6}}\int_0^t 0.024\,e^{-40t}\,dt$$

$$= \frac{0.024}{3.0 \times 10^{-6}}\int_0^t e^{-40t}\,dt$$

$$= 8000\left[-\frac{1}{40}\,e^{-40t}\right]_0^t$$

$$= -\frac{8000}{40}\left(e^{-40t} - e^0\right)$$

$$= -200\left(e^{-40t} - 1\right)$$

or

$$v(t) = 200\left(1 - e^{-40t}\right) \text{ V}.$$

(b) By definition, the power supplied to a capacitor is given by

$$p(t) = \frac{dw(t)}{dt} \qquad (9.42)$$

Integrating both sides of equation (9.42) gives

$$\int_0^t \frac{dw(t)}{dt}\,dt = \int_0^t p(t)\,dt$$

$$[w(t)]_0^t = \int_0^t p(t)\,dt$$

$$w(t) - w(0) = \int_0^t p(t)\,dt$$

or

$$w(t) = w(0) + \int_0^t p(t)\, dt. \tag{9.43}$$

Since the energy stored in the capacitor at time $t = 0$ is zero, $w(0) = 0$. The power $p(t)$ is given by

$$p(t) = v(t)\, i(t)$$
$$= 200\,(1 - e^{-40t})\,(0.024\, e^{-40t})$$
$$= (200)(0.024)\, e^{-40t} - (200\, e^{-40t})(0.024\, e^{-40t})$$

or

$$p(t) = 4.8\, e^{-40t} - 4.8\, e^{-80t}\ \text{W}. \tag{9.44}$$

Substituting $w(0) = 0$ and $p(t)$ from equation (9.44) into equation (9.43) gives

$$w(t) = 0 + \int_0^t (4.8\, e^{-40t} - 4.8\, e^{-80t})\, dt$$

$$= 4.8 \left[-\frac{1}{40} e^{-40t} \right]_0^t - 4.8 \left[-\frac{1}{80} e^{-80t} \right]_0^t$$

$$= -\frac{4.8}{40}(e^{-40t} - 1) + \frac{4.8}{80}(e^{-80t} - 1)$$

$$= -0.12\, e^{-40t} + 0.12 + 0.06\, e^{-80t} - 0.06$$

or

$$w(t) = 0.06\, e^{-80t} - 0.12\, e^{-40t} + 0.06\ \text{J}. \tag{9.45}$$

To show that $w(t) = \frac{1}{2} C v^2(t)$, the quantity $\frac{1}{2} C v^2(t)$ can be calculated as

$$\frac{1}{2} C v^2(t) = \frac{1}{2}(3 \times 10^{-6})(200 - 200\, e^{-40t})^2$$

$$= (1.5 \times 10^{-6})\left[200^2 - 2(200)(200)e^{-40t} + (200e^{-40t})^2\right]$$

$$= (1.5 \times 10^{-6})(4 \times 10^4 - 8 \times 10^4\, e^{-40t} + 4 \times 10^4\, e^{-80t})$$

$$= 0.06\, e^{-80t} - 0.12\, e^{-40t} + 0.06\ \text{J}. \tag{9.46}$$

Comparison of equations (9.45) and (9.46) reveals that $w(t) = \frac{1}{2} C v^2(t)$.

Example 9-11

The current $i(t)$ shown in Fig. 9.44 is applied to a $20\,\mu\text{F}$ capacitor. Sketch the voltage $v(t)$ across the capacitor knowing that $i(t) = C\dfrac{dv(t)}{dt}$, or $v(t) = v(t_0) + \dfrac{1}{C}\int_{t_0}^t i(t)\, dt$. Assume that the initial voltage across the capacitor is zero (i.e., $v(0) = 0$ V).

Figure 9.44 Current applied to a capacitor.

Solution   Since $C = 20\,\mu\text{F}$,

$$\frac{1}{C} = \frac{1}{20 \times 10^{-6}}$$

$$= \frac{10^6}{20}$$

$$= \frac{10^2 \times 10^4}{20}$$

or

$$\frac{1}{C} = 50{,}000\ \text{F}^{-1}.$$

The voltage during each time interval can now be calculated as follows:

(a) $0 \le t \le 1$ ms: $t_0 = 0$, $v(t_0) = v_0 = 0$ V, and $i(t) = 4$ A $=$ constant. Therefore, $v(t) = \dfrac{1}{C} \displaystyle\int i(t)\,dt$ is a straight line with positive slope. The voltage at time $t = 1$ ms can be calculated as

$$v_1 = v_0 + 50{,}000 \int_0^{0.001} 4\,dt$$

$$= 0 + 200{,}000\ [\,t\,]_0^{0.001}$$

$$= 200{,}000\ (0.001 - 0)$$

or

$$v_1 = 200\ \text{V}.$$

The change in voltage across the capacitor between 0 and 1 ms can also be calculated without evaluating the integral (i.e., from geometry) as

$$v_1 - v_0 = \frac{1}{C} \times \text{area under the current between 0 and 1 ms}$$

$$= 50{,}000\,[(4)\,(0.001)] \quad \text{(area of a rectangle)}$$

or

$$v_1 - v_0 = 200\,\text{V}.$$

Therefore, $v_1 = v_0 + 200 = 200\,\text{V}.$

(b) $1\,\text{ms} < t \le 2\,\text{ms}$: $t_0 = 1\,\text{ms}$, $v(t_0) = v_1 = 200\,\text{V}$, and $i(t) = -2\,\text{A} = \text{constant}.$

Therefore, $v(t) = \dfrac{1}{C} \displaystyle\int i(t)\,dt$ is a straight line of negative slope. The voltage at $t = 2\,\text{ms}$ can be calculated as

$$
\begin{aligned}
v_2 &= v_1 + 50{,}000 \int_{0.001}^{0.002} (-2)\,dt \\
&= 200 - 100{,}000\,[\,t\,]_{0.001}^{0.002} \\
&= 200 - 100{,}000\,(0.002 - 0.001) \\
&= 200 - 100
\end{aligned}
$$

or

$$v_2 = 100\,\text{V}.$$

The change in voltage across the capacitor between 1 ms and 2 ms can also be calculated from geometry as

$$
\begin{aligned}
v_2 - v_1 &= \frac{1}{C} \times \text{area under the current waveform between 1 and 2 ms} \\
&= 50{,}000\,[\,(-2)\,(0.001)\,] \\
v_2 - v_1 &= -100\,\text{V}
\end{aligned}
$$

Therefore, $v_2 = v_1 - 100 = 200 - 100 = 100\,\text{V}.$

(c) $2\,\text{ms} < t \le 4\,\text{ms}$: $t_0 = 2\,\text{ms}$, $v(t_0) = v_2 = 100\,\text{V}$, and $i(t) = 2\,\text{A} = \text{constant}.$
Therefore, $v(t) = \frac{1}{C} \int i(t)\,dt$ is a straight line with positive slope. The voltage at $t = 4\,\text{ms}$ can be calculated as

$$
\begin{aligned}
v_4 &= v_2 + 50{,}000 \int_{0.002}^{0.004} 2\,dt \\
&= 100 + 100{,}000\,[t]_{0.002}^{0.004} \\
&= 100 + 100{,}000\,(0.004 - 0.002) \\
&= 100 + 100{,}000\,(0.002) \\
&= 100 + 200
\end{aligned}
$$

or

$$v_4 = 300\,\text{V}.$$

The change in voltage across the capacitor between 2 ms and 4 ms can also be calculated from geometry as

$$
\begin{aligned}
v_4 - v_2 &= \frac{1}{C} \times \text{area under the current waveform between 2 and 4 ms} \\
&= 50{,}000\,[\,(2)\,(.002)\,]
\end{aligned}
$$

or

$$v_4 - v_2 = 200 \text{ V}.$$

Therefore, $v_4 = v_2 + 200 = 100 + 200 = 300 \text{ V}.$

(d) 4 ms $< t \le$ 5 ms: $t_0 = 4$ ms, $v(t_0) = v_4 = 300$ V, and $i(t) = -4$ A = constant.
Therefore, $v(t) = \frac{1}{C} \int i(t) \, dt$ is a straight line with negative slope. The voltage at $t = 5$ ms can be calculated as

$$v_5 = v_4 + 50,000 \int_{0.004}^{0.005} (-4) \, dt$$

$$= 300 - 200,000 \, [t]_{0.004}^{0.005}$$

$$= 300 - 200,000 \, (0.005 - 0.004)$$

$$= 300 - 200,000 \, (0.001)$$

$$= 300 - 200$$

or

$$v_5 = 100 \text{ V}.$$

The change in voltage across the capacitor between 4 ms and 5 ms can also be calculated from geometry as

$$v_5 - v_4 = \frac{1}{C} \times \text{area under the current waveform between 4 and 5 ms}$$

$$= 50,000 \, [\, (-4) \, (.001) \,]$$

or

$$v_5 - v_4 = -200 \text{ V}$$

Therefore, $v_5 = v_4 - 200 = 300 - 200 = 100 \text{ V}.$

The sketch of the voltage across the capacitor is shown in Fig. 9.45.

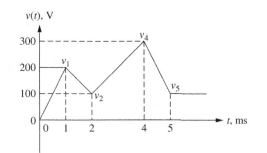

Figure 9.45 Voltage across the capacitor in Example 9-11.

**Example
9-12** The sawtooth current $i(t)$ shown in Fig. 9.46 is applied to a 0.5 F capacitor. Sketch the voltage $v(t)$ across the capacitor knowing that $i(t) = C \dfrac{dv(t)}{dt}$ or $v(t) = \dfrac{1}{C} \displaystyle\int i(t)\,dt$. Assume that the capacitor is completely discharged at $t = 0$ (i.e., $v(0) = 0$ V).

Figure 9.46 Sawtooth current applied to a capacitor.

Solution The voltage across the capacitor during different time intervals can be calculated as follows:

(a) $0 \le t \le 1$ s: $i(t)$ is a straight line; therefore, $v(t) = \frac{1}{C} \int i(t)\,dt$ is a quadratic function. The change in voltage across the capacitor between the time interval 0 to 1 s can be calculated as

$$v_1 - v_0 = \frac{1}{C} \int_0^t i(t)\,dt$$

$$= \frac{1}{0.5} \text{ (area under the current waveform between 0 and 1 s)}$$

$$= 2 \left[\frac{1}{2}(1)(-2) \right] \quad \text{(area of a triangle)}$$

or

$$v_1 - v_0 = -2.$$

Solving for v_1 gives,

$$v_1 = v_0 + (-2)$$
$$= 0 - 2$$

or

$$v_1 = -2 \text{ V.}$$

Also, since $\frac{dv(t)}{dt} = \frac{1}{C} i(t)$ is the slope of $v(t)$ at time t, the slope of the voltage at $t = 0$ is $\frac{dv}{dt} = \frac{1}{0.5}(-2) = -4$ V/s.

At $t = 1$ s, the slope of the voltage is $\frac{dv}{dt} = \frac{1}{0.5}(0) = 0$ V/s.

Therefore, $v(t)$ is a decreasing quadratic function starting at $v(0) = 0$ V and ending at -2 V, with a zero slope at $t = 1$ s.

(b) $1 < t \leq 2$ s: $i(t)$ is a straight line; therefore, $v(t) = \frac{1}{C} \int i(t)\, dt$ is a quadratic
function. The change in voltage across the capacitor during the time interval
1 to 2 s can be calculated as

$$v_2 - v_1 = \frac{1}{0.5} \text{ (area under the current waveform between 1 and 2 s)}$$

$$= 2 \left[\frac{1}{2}(1)(2) \right] \quad \text{(area of a triangle)}$$

or

$$v_2 - v_1 = 2.$$

Solving for v_2 gives

$$v_2 = v_1 + 2$$
$$= -2 + 2$$

or

$$v_2 = 0 \text{ V}.$$

Also, since $\frac{dv(t)}{dt} = \frac{1}{C} i(t)$ is the slope of $v(t)$,

$$\frac{dv(t)}{dt} = 0 \text{ V/s at } t = 1 \text{ s},$$

$$\frac{dv(t)}{dt} = \frac{1}{0.5}(2) = 4 \text{ V/s at } t = 2 \text{ s}.$$

Therefore, $v(t)$ is an increasing quadratic function starting with a zero slope at
$v_1 = -2$ V and ending at $v_2 = 0$ V with a slope of 4 V/s.

Since the voltage at $t = 2$ s is zero and the current supplied to the capacitor between
2 and 4 s is the same as the current applied to the capacitor from 0 to 2 s, the voltage
across the capacitor between 2 and 4 s is identical to the voltage between 0 and
2 s. The same is true for remaining intervals. A sketch of the voltage across the
capacitor is shown in Fig. 9.47.

Figure 9.47   Voltage across a capacitor subjected to a sawtooth current.

9.7 CURRENT AND VOLTAGE IN AN INDUCTOR

In this section, integrals are used to find the current flowing through an inductor when it is connected across a voltage source.

Example 9-13

A voltage $v(t) = 10\cos(10t)$ V is applied across a 100 mH inductor, as shown in Fig. 9.48.

$i(t)$

$v(t)$ $L = 100$ mH

Figure 9.48 Voltage applied to an inductor.

(a) Suppose the initial current flowing through the inductor is $i(0) = 10$ A. Knowing that $v(t) = L\frac{di(t)}{dt}$, integrate both sides of the equation to determine the current $i(t)$. Also, plot the current for $0 \le t \le \pi/5$ s.

(b) Given your results in part (a), find the power $p(t) = v(t)i(t)$ supplied to the inductor. If the initial energy stored in the inductor is $w(0) = 5$ J, find the stored energy

$$w(t) = w(0) + \int_0^t p(t)\,dt, \tag{9.47}$$

and show that $w(t) = \frac{1}{2}L i^2(t)$.

Solution

(a) The voltage/current relationship for an inductor is given by

$$L\frac{di(t)}{dt} = v(t)$$

or

$$\frac{di(t)}{dt} = \frac{1}{L}v(t). \tag{9.48}$$

Integrating both sides of equation (9.48) from an initial time t_0 to time t gives

$$\int_{t_0}^t \frac{di(t)}{dt} = \frac{1}{L}\int_{t_0}^t v(t)\,dt$$

$$[i(t)]_{t_0}^t = \frac{1}{L}\int_{t_0}^t v(t)\,dt$$

$$i(t) - i(t_0) = \frac{1}{L}\int_{t_0}^t v(t)\,dt$$

or

$$i(t) = i(t_0) + \frac{1}{L} \int_{t_0}^{t} v(t)\, dt. \tag{9.49}$$

Substituting $t_0 = 0$, $L = 0.1$ H, $i(0) = 10$ A, and $v(t) = 10\cos(10\,t)$ V in equation (9.49) gives

$$i(t) = 10 + \frac{1}{0.1} \int_{0}^{t} 10\cos(10\,t)\, dt$$

$$= 10 + 100 \left[\frac{1}{10} \sin(10\,t) \right]_{0}^{t}$$

$$= 10 + 10\, (\, \sin 10\,t - 0\,)$$

or

$$i(t) = 10 + 10\sin 10\,t \text{ A}. \tag{9.50}$$

The current $i(t)$ obtained in equation (9.50) is a periodic function with frequency $\omega = 10$ rad/s. The period is

$$T = \frac{2\,\pi}{\omega}$$

$$= \frac{2\,\pi}{10}$$

or

$$T = \frac{\pi}{5} \text{ s}.$$

Thus, the plot of $i(t)$ is simply the sinusoid $10\sin(10\,t)$ shifted upward by 10 amps as shown in Fig. 9.49.

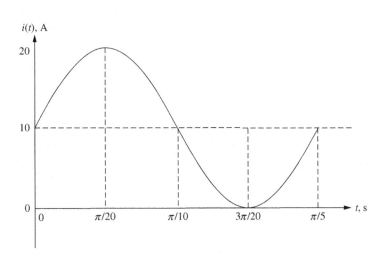

Figure 9.49   Current flowing through the inductor in Example 9-13.

(b) The power $p(t)$ supplied to the inductor is given by

$$p(t) = v(t)\, i(t)$$
$$= (10\cos(10t))\,(10 + 10\sin(10t))$$
$$= 100\cos(10t) + 100\sin(10t)\cos(10t)$$
$$= 100\cos(10t) + 50\,(2\sin(10t)\cos(10t))$$
$$= 100\cos(10t) + 50\sin(20t)$$

or

$$p(t) = 100\,(\cos(10t) + 0.5\sin(20t))\ \text{W}. \tag{9.51}$$

The energy stored in the inductor is given by

$$w(t) = w(0) + \int_0^t p(t)\, dt. \tag{9.52}$$

Substituting $w(0) = 5$ J and $p(t)$ calculated in equation (9.51) gives

$$w(t) = 5 + \int_0^t 100\,(\cos(10t) + 0.5\sin(20t))\, dt$$

$$= 5 + 100\left[\frac{\sin(10t)}{10}\right]_0^t + 50\left[-\frac{\cos 20t}{20}\right]_0^t$$

$$= 5 + 10\,(\sin(10t) - 0) - 2.5\,(\cos(20t) - 1)$$

or

$$w(t) = 7.5 + 10\sin(10t) - 2.5\cos(20t)\ \text{J}. \tag{9.53}$$

To show that $w(t) = \frac{1}{2}L\,i^2(t)$, the quantity $\frac{1}{2}L\,i^2(t)$ can be calculated as

$$\frac{1}{2}L\,i^2(t) = \frac{1}{2}\,(0.1)\,(10 + 10\sin(10t))^2$$

$$= 0.05\left[10^2 + 2(10)(10)\sin(10t) + (10\sin(10t))^2\right]$$

$$= 0.05\,(100 + 200\sin(10t) + 100\sin^2(10t)).$$

Noting that $\sin^2(10t) = \left(\dfrac{1 - \cos(20t)}{2}\right)$, we get

$$\frac{1}{2}L\,i^2(t) = 5 + 10\sin(10t) + 5\left(\frac{1 - \cos(20t)}{2}\right)$$

$$= 5 + 10\sin(10t) + 2.5 - 2.5\cos(20t)$$

$$= 7.5 + 10\sin(10t) - 2.5\cos(20t)\ \text{J}, \tag{9.54}$$

which is the same as equation (9.53).

Example 9-14   A voltage $v(t)$ is applied to a 500 mH inductor as shown in Fig. 9.50. Knowing that $v(t) = L\frac{di(t)}{dt}$ (or $i(t) = \frac{1}{L}\int v(t)\,dt$), plot the current $i(t)$ using integrals. Assume the initial current flowing through the inductor is zero (i.e., $i(0) = 0$ A).

Figure 9.50   Voltage applied to an inductor.

Solution   Using equation (9.49), the current $i(t)$ flowing through the inductor during each time interval can be determined as follows:

(a) $0 \leq t \leq 2$ s: $v(t) = 9$ V = constant. Therefore, $i(t) = \frac{1}{L}\int v(t)\,dt$ is a straight line with positive slope. Also, the change in current is

$$i_2 - i_0 = \frac{1}{L}\int_0^2 v(t)\,dt$$

$$= \frac{1}{0.5}(\text{area under the voltage waveform between 0 and 2 s})$$

$$= \frac{1}{0.5}\,[\,(2)\,(9)\,]$$

or

$$i_2 - i_0 = 36 \text{ A}.$$

Solving for i_2 is

$$i_2 = i_0 + 36$$
$$= 0 + 36$$

or

$$i_2 = 36 \text{ A}.$$

Note that the equation for the current flowing through the inductor at any time t between 0 and 2 s can also be calculated as

$$i(t) = i(0) + \frac{1}{L}\int_0^t v(t)\,dt$$

$$= 0 + \frac{1}{0.5}\int_0^t 9\,dt$$

$$= (2)\,(9)[t]_0^t$$

or

$$i(t) = 18\,t.$$

(b) $2 < t \le 4\,\text{s}$: $v(t) = -9\,\text{V} = \text{constant}$. Therefore, $i(t) = \frac{1}{L} \int v(t)\,dt$ is a straight line with a negative slope. Also, the change in current is

$$i_4 - i_2 = \frac{1}{0.5}(\text{area under the voltage waveform between 2 and 4 s})$$

$$= \frac{1}{0.5}\,[\,(2)\,(-9)\,]$$

or

$$i_4 - i_2 = -36\ \text{A}.$$

Solving for i_4 gives

$$i_4 = i_2 - 36$$
$$= 36 - 36$$

or

$$i_4 = 0\ \text{A}.$$

Note that the equation for the current flowing through the inductor at any time t between 2 and 4 s can also be calculated as

$$i(t) = i(2) + \frac{1}{L}\int_2^t v(t)\,dt$$

$$= 36 + \frac{1}{0.5}\int_2^t -9\,dt$$

$$= 36 - 18\,[t]_2^t$$

$$= 36 - 18\,(t - 2)$$

or

$$i(t) = -18\,t + 72.$$

Since the current at $t = 4$ s is zero (the same as at $t = 0$) and the voltage applied to the inductor between 4 and 8 s is the same as the voltage from 0 to 4 s, the current flowing through the inductor between 4 and 8 s is identical to the current between 0 and 4 s. The resulting current waveform (triangle wave) is shown in Fig. 9.51.

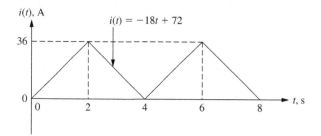

Figure 9.51 Current flowing through the inductor in Example 9-14.

9.8 | **FURTHER EXAMPLES OF INTEGRALS IN ENGINEERING**

Example 9-15

A biomedical engineer measures the velocity profiles of a belted and unbelted occupant during a 35 mph ($\approx$ 16 m/s) frontal collison, as shown in Fig. 9.52.

(a) Knowing that $x(t) = x(0) + \int_0^t v(t)dt$, find and plot the displacement $x(t)$ of the belted occupant for time 0 to 50 ms. Assume that the initial displacement at $t = 0$ is 0 m (i.e., $x(0) = 0$ m).

(b) Find and plot the displacement $x(t)$ of the unbelted occupant for time 0 to 50 ms. Assume $x(0) = 0$ m.

(c) Based on the results of parts (a) and (b), how much farther did the unbelted occupant travel compared to the belted occupant?

Figure 9.52   Velocities of the belted and unbelted occupants during frontal collison.

Solution

(a) The displacement of the belted occupant can be calculated using the expression

$$x(t) = x(0) + \int_0^t v(t)dt$$

(i) for $0 \le t \le 0.01$ s

$$x(t) = 0 + \int_0^t 16\cos(50\,\pi t)dt = 16\left[\frac{\sin(50\,\pi t)}{50\,\pi}\right]_0^t = \frac{16}{50\,\pi}\left[\sin(50\,\pi t) - 0\right]$$

$$= \frac{8}{25\,\pi}\sin(50\,\pi t)\ \text{m}. \tag{9.55}$$

Therefore, $x(0.01) = \frac{8}{25\,\pi}\sin\left(\frac{\pi}{2}\right) = 0.102$ m.

(ii) $0.01 \leq t \leq 0.05$ s

$$x(t) = 0.102 + \int_0^t 0 \, dt$$

$$= 0.102 \text{ m.} \tag{9.56}$$

The displacement of the belted occupant during a frontal collison is shown in Fig. 9.53.

Figure 9.53 Displacement of the belted occupant during a frontal collison.

(b) The displacement of the unbelted occupant for $0 \leq t \leq 0.05$ s can be calculated as

$$x(t) = 0 + \int_0^t 16 \cos(10 \pi t) dt$$

$$= 16 \left[\frac{\sin(10 \pi t)}{10 \pi} \right]_0^t$$

$$= \frac{16}{10 \pi} [\sin(10 \pi t) - 0]$$

$$= \frac{8}{5 \pi} \sin(10 \pi t) \text{ m.} \tag{9.57}$$

Therefore, $x(0.05) = \frac{8}{5 \pi} \sin\left(\frac{\pi}{2}\right) = 0.509$ m. The displacement of the unbelted occupant is shown in Fig. 9.54.

(c) To find how much farther the unbelted occupant traveled during collison as compared to the belted occupant, the total distance traveled by the belted occupant in 50 ms is subtracted from the total distance traveled by the unbelted occupant as

$$\Delta x = x_{\text{unbelted}}(0.05) - x_{\text{belted}}(0.05)$$

$$= 0.509 - 0.102$$

$$= 0.407 \text{ m.}$$

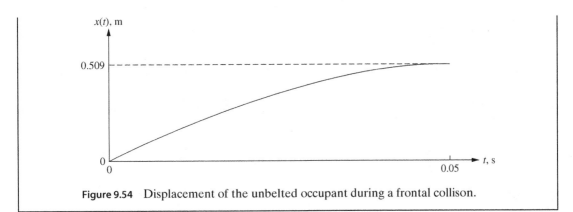

Figure 9.54 Displacement of the unbelted occupant during a frontal collison.

Example 9-16

A biomedical engineer is evaluating an energy-absorbing aviation seat on a vertical deceleration tower, as shown in Fig. 9.55. The acceleration profile of the drop cage is described by

$$a(t) = 500 \sin(40 \pi t) \text{ m/s}^2.$$

(a) Knowing that $v(t) = v(0) + \int_0^t a(t)dt$, find and plot the velocity $v(t)$ of the drop cage. Assume the drop cage starts from rest at $t = 0$ s.

(b) What is the impact velocity v_{impact} of the drop cage if it takes 25 ms to hit the ground?

(c) The total impulse I is equal to the change in momentum. For example $I = \Delta p = p_f - p_i = mv_{impact} - mv_0$, where m is the mass of the system, v_{impact} is the final velocity, v_0 is the initial velocity, p is the momentum, and I is the impulse. Find the total impulse after 25 ms. Assume that the total mass of the drop cage, seat, and crash test dummy is 1,000 kg.

Figure 9.55 Energy-absorbing aviation seat.

Solution   (a) The velocity of the cage can be calculated as

$$v(t) = v(o) + \int_0^t a(t)dt$$

$$= 0 + \int_0^t 500 \sin(40\,\pi t)dt$$

$$= 500 \left[-\frac{\cos(40\,\pi t)}{40\,\pi} \right]_0^t$$

$$= -\frac{500}{40\,\pi} [\cos(40\,\pi t) - 1]$$

$$= \frac{25}{2\,\pi} [1 - \cos(40\,\pi t)] \text{ m/s.} \tag{9.58}$$

(b) The velocity of the cage when it impacts the ground can be found by substituting $t = 25$ ms $= 0.025$ s in equation (9.58) as

$$v_{\text{impact}} = \frac{25}{2\,\pi} [1 - \cos(40\,\pi(0.025))]$$

$$= \frac{25}{2\,\pi} [1 - \cos(\pi)]$$

$$= \frac{25}{2\,\pi} [1 - (-1)]$$

$$= \frac{50}{2\,\pi}$$

$$= 7.96 \text{ m/s.}$$

(c) The total impulse I can be found as

$$I = m\,v_{\text{impact}} - m\,v_0$$
$$= (1000)(7.96) - (1000)(0)$$
$$= 7960 \frac{\text{kg m}}{\text{s}}.$$

Example 9-17

A civil engineer designs a building overhang to withstand a parabolic snow loading per unit length $p(x) = \hat{p}\left(1 - \frac{x}{L}\right)^2$, as shown in Fig. 9.56.

(a) Compute the resulting force $V = \int_0^L p(x)dx$.

(b) Compute the corresponding moment $M = \int_0^L xp(x)dx$.

(c) Locate the centroid of the loading $\bar{x} = \dfrac{\int_o^L xp(x)\,dx}{\int_0^L p(x)\,dx} = \dfrac{M}{V}$.

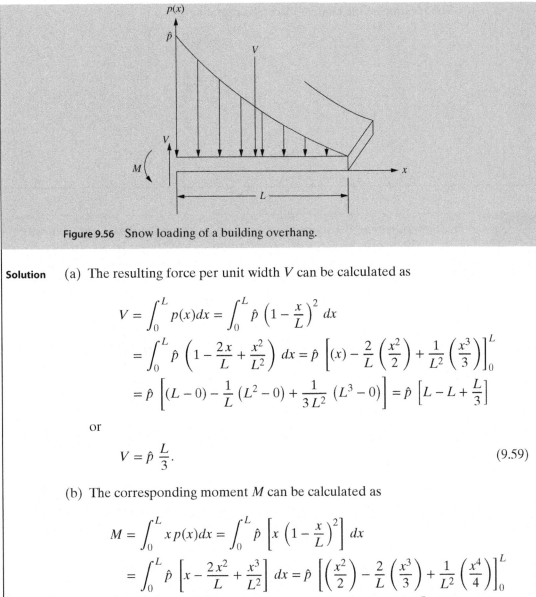

Figure 9.56 Snow loading of a building overhang.

Solution   (a) The resulting force per unit width V can be calculated as

$$V = \int_0^L p(x)dx = \int_0^L \hat{p}\left(1 - \frac{x}{L}\right)^2 dx$$

$$= \int_0^L \hat{p}\left(1 - \frac{2x}{L} + \frac{x^2}{L^2}\right)dx = \hat{p}\left[(x) - \frac{2}{L}\left(\frac{x^2}{2}\right) + \frac{1}{L^2}\left(\frac{x^3}{3}\right)\right]_0^L$$

$$= \hat{p}\left[(L - 0) - \frac{1}{L}\left(L^2 - 0\right) + \frac{1}{3L^2}\left(L^3 - 0\right)\right] = \hat{p}\left[L - L + \frac{L}{3}\right]$$

or

$$V = \hat{p}\,\frac{L}{3}. \tag{9.59}$$

(b) The corresponding moment M can be calculated as

$$M = \int_0^L x\,p(x)dx = \int_0^L \hat{p}\left[x\left(1 - \frac{x}{L}\right)^2\right]dx$$

$$= \int_0^L \hat{p}\left[x - \frac{2x^2}{L} + \frac{x^3}{L^2}\right]dx = \hat{p}\left[\left(\frac{x^2}{2}\right) - \frac{2}{L}\left(\frac{x^3}{3}\right) + \frac{1}{L^2}\left(\frac{x^4}{4}\right)\right]_0^L$$

$$= \hat{p}\left[\left(\frac{1}{2}\right)\left(L^2 - 0\right) - \frac{2}{3L}\left(L^3 - 0\right) + \frac{1}{4L^2}\left(L^4 - 0\right)\right]$$

$$= \hat{p}\left[\frac{L^2}{2} - \frac{2L^2}{3} + \frac{L^2}{4}\right]$$

or

$$M = \hat{p}\,\frac{L^2}{12}.$$

(c) The location of the centroid $\bar{x}$ can be found as

$$\bar{x} = \frac{M}{V}$$

$$= \frac{\dfrac{\hat{p}\,L^2}{12}}{\dfrac{\hat{p}\,L}{3}}$$

or $\qquad \bar{x} = \dfrac{L}{4}.$

Example 9-18

A building overhang is subjected to a triangular snow loading $p(x)$, as shown in Fig. 9.57. The overhang is constructed from two metal face plates separated by a nonmetallic core of thickness h. This sandwich beam construction deforms primarily due to shear deformation with deflection $y(x)$ satisfying

$$y(x) = \frac{1}{h\,G} \int_0^x \left[\int_0^x p(x)\,dx - V \right] dx, \qquad (9.60)$$

where $V = p_0 \dfrac{L}{2}$, $p(x) = p_0 \left(1 - \dfrac{x}{L}\right)$, and G is the shear modulus.

(a) Evaluate (9.60) for the deflection $y(x)$.

(b) Find the location and value of maximum deflection.

(a) Cantilever overhang.   (b) Sandwich construction.

Figure 9.57 Triangular snow loading of sandwich-constructed building overhang.

Solution

(a) Substituting $V = p_0 \dfrac{L}{2}$ and $p(x) = p_0 \left(1 - \dfrac{x}{L}\right)$ into equation (9.60) gives

$$y(x) = \frac{1}{h\,G} \int_0^x \left[\int_0^x p_0 \left(1 - \frac{x}{L}\right) dx - p_0 \frac{L}{2} \right] dx$$

$$= \frac{1}{h\,G} \int_0^x \left[p_0 \left[x - \frac{x^2}{2\,L} \right]_0^x - p_0 \frac{L}{2} \right] dx$$

$$= \frac{1}{hG} \int_0^x \left[p_0 \left\{ \left(x - \frac{x^2}{2L} \right) - (0 - 0) \right\} - p_0 \frac{L}{2} \right] dx$$

$$= \frac{p_0}{hG} \int_0^x \left[x - \frac{x^2}{2L} - \frac{L}{2} \right] dx$$

$$= \frac{p_0}{hG} \left[\frac{x^2}{2} - \frac{x^3}{6L} - \frac{L}{2} x \right]_0^x$$

$$= \frac{p_0}{hG} \left[\left(\frac{x^2}{2} - \frac{x^3}{6L} - \frac{L}{2} \right) - (0 - 0 - 0) \right]$$

$$= \frac{p_0 x}{2hG} \left(x - \frac{x^2}{3L} - L \right)$$

or

$$y(x) = -\frac{p_0 x}{2hG} \left[L - x \left(1 - \frac{x}{3L} \right) \right]. \tag{9.61}$$

(b) The location of the maximum value of the deflection can be found by equating the derivative of $y(x)$ to zero as

$$\frac{dy(x)}{dx} = 0$$

$$\frac{d}{dx} \left(-\frac{p_0 x}{2hG} \left[L - x \left(1 - \frac{x}{3L} \right) \right] \right) = 0$$

$$-\frac{p_0}{2hG} \left(L - 2x + \frac{3x^2}{3L} \right) = 0$$

$$x^2 - 2xL + L^2 = 0$$

$$(x - L)^2 = 0.$$

Therefore, the deflection is maximum at $x = L$. The value of the maximum deflection can now be obtained by substituting $x = L$ in equation (9.61) as

$$y(L) = -\frac{p_0 L}{2hG} \left[L - L \left(1 - \frac{L}{3L} \right) \right]$$

$$= -\frac{p_0 L}{2hG} \left(\frac{L}{3} \right)$$

$$= -\frac{p_0 L^2}{6hG}.$$

Therefore, the maximum value of deflection is $\dfrac{p_0 L^2}{6hG}$.

PROBLEMS

9-1. The profile of a gear tooth shown in Fig. P9.1(a) is approximated by the quadratic equation $y(x) = -\frac{4k}{l} x(x - l)$.

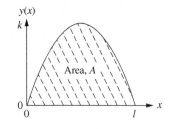

$y(x)$
k

Area, A

0
0 l x

(a) The profile of a gear tooth.

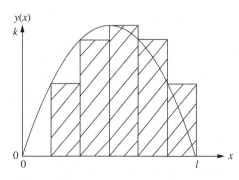

$y(x)$
k

0
0 l x

(b) The gear tooth area inscribed by six rectangles.

Figure P9.1   Area of a gear tooth for problem P9-1.

(a) Estimate the area A using six rectangles of equal width ($\Delta x = l/6$), as shown in Fig. 9.1(b).

(b) Calculate the exact area by evaluating the definite integral, $A = \int_0^l y(x)\, dx$.

9-2. The profile of a gear tooth shown in Fig. P9.2 is approximated by the trigonometric equation $y(x) = \frac{k}{2}\left(1 - \cos\left(\frac{2\pi x}{l}\right)\right)$.

(a) Estimate the area A using eight rectangles of equal width

$\Delta x = l/8$, i.e.,

$$A = \sum_{i=1}^{8} y(x_i)\,\Delta x.$$

(b) Calculate the exact area by integration, i.e.,

$$A = \int_0^l y(x)\, dx.$$

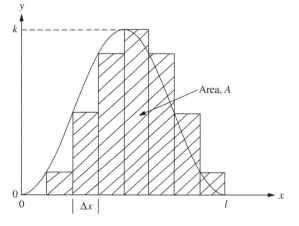

y
k

Area, A

0
0 $|\ \Delta x\ |$ l x

Figure P9.2   Profile of a gear tooth for problem P9-2.

9-3. The velocity of an object as a function of time is shown in Fig. P9.3. The acceleration is constant during the first 4 seconds of motion, so the velocity is a linear function of time with $v(t) = 0$ at $t = 0$ and $v(t) = 100$ ft/s at $t = 4$ s. The velocity is constant during the last 6 s.

(a) Estimate the total distance covered (the area, A) under the velocity curve using five rectangles of equal width ($\Delta t = 10/5 = 2$ s).

(b) Now estimate the total distance covered using 10 rectangles of equal width.

(c) Calculate the exact area under the velocity curve; i.e., find the total distance traveled by evaluating the definite integral $\Delta x = \int_0^{10} v(t)\, dt$.

(d) Calculate the exact area by adding the area of the triangle and the area of the rectangle bounded by the velocity curve.

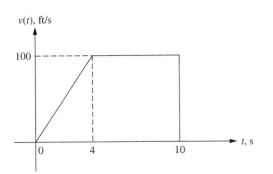

$v(t)$, ft/s

100

0   4   10 t, s

Figure P9.3   The velocity of an object.

9-4. A particle is accelerated along a curved path of length l under the action of an applied force $f(x)$, as shown in Fig. P9.4. The total work done on the particle is

$$W = \int_0^l f(x)\, dx \quad \text{N-m}.$$

If $l = 4.0$ m, determine the work done for
(a) $f(x) = 8x^3 + 6x^2 + 4x + 2$ N.
(b) $f(x) = 4e^{-2x}$ N.
(c) $f(x) = 1 - 2\sin^2\left(\frac{\pi x}{2}\right)$ N.
  Hint: Use a trigonometric identity.

$f(x)$ x

Figure P9.4   A particle moving on a curved path.

9-5. A particle is accelerated along a curved path of length $l = 3.0$ m under the action of an applied force $f(x)$, as shown in Fig. P9.4. The total work done on the particle is

$$W = \int_0^l f(x)\, dx.$$

Determine the work done for
(a) $f(x) = 2x^4 + 3x^3 + 4x - 1$ N.
(b) $f(x) = e^{-2x}(1 + e^{4x})$ N.
(c) $f(x) = 2\sin\left(\frac{\pi x}{l}\right) + 3\cos\left(\frac{\pi x}{l}\right)$ N.

9-6. When a variable force is applied to an object, it travels a distance of 5 m. The total work done on the object is given by

$$W = \int_0^5 f(x)\, dx \quad \text{N-m}.$$

Determine the work done if the force is given by
(a) $f(x) = (x + 1)^3$ N.
(b) $f(x) = 10\sin\left(\frac{\pi}{10}x\right)\cos\left(\frac{\pi}{10}x\right)$ N.
  Hint: Use the double angle formula.

9-7. A triangular area is bounded by a straight line in the x-y plane, as shown in Fig. P9.7(a).
(a) Find the equation of the line $y(x)$.
(b) Find the area A by integration,
$$A = \int_0^b y(x)\, dx.$$
(c) Find the centroid G by integration with vertical rectangles, as shown in Fig. P9.7(b); in other words, find
$$\bar{x} = \frac{\displaystyle\int x\, dA}{A} = \frac{\displaystyle\int x\, y(x)\, dx}{A}$$
and
$$\bar{y} = \frac{\displaystyle\int \frac{y}{2}\, dA}{A} = \frac{\frac{1}{2}\displaystyle\int (y(x))^2\, dx}{A}.$$
(d) Now solve for x as a function of y, and recalculate the y-coordinate of the centroid G by integration with

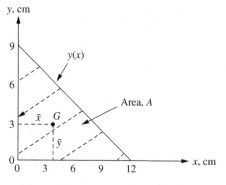

(a) Area bounded by a straight line.

(b) Vertical rectangles.

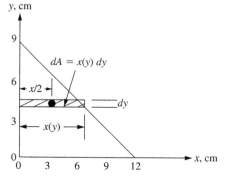

(c) Horizontal rectangles.

Figure P9.7 Centroid of a triangular cross section.

horizontal rectangles, as shown in Fig. P9.7(c), i.e.,

$$\bar{y} = \frac{\int y \, dA}{A} = \frac{\int y \, x(y) \, dy}{A}.$$

9-8. An area in the x-y plane is bounded by the curve $y = k x^n$ and the line $x = b$, as shown in Fig. P9.8.
(a) Determine the area A by integration with respect to x.
(b) Determine the coordinates of the centroid G by integration with respect to x.
(c) Evaluate your answer to part (b) for the case $n = 1$.

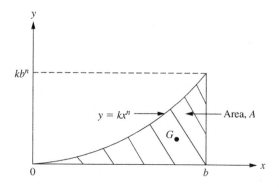

Figure P9.8 Area bounded by a curved surface.

9-9. Consider the shaded area under the line $y(x)$, as illustrated in Fig. P9.9.
(a) Find the equation of the line $y(x)$.

(b) Determine the area under the line by integration with respect to x.

(c) Determine the x-coordinate of the centroid by integration with respect to x.

(d) Determine the y-coordinate of the centroid by integration with respect to x.

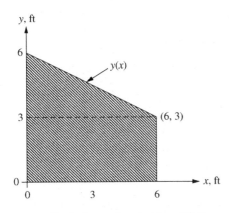

Figure P9.9 Shaded area for problem P9-9.

9-10. The geometry of a cooling fin is defined by the shaded area that is bounded by the parabola $y(x) = -x^2 + 16$, as illustrated in Fig. P9.10.

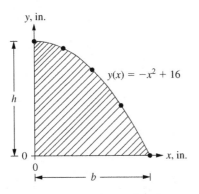

Figure P9.10 Geometry of a cooling fin.

(a) Given the above equation for $y(x)$, determine the height h and width b of the fin.

(b) Determine the area of the cooling fin by integration with respect to x.

(c) Determine the x-coordinate of the centroid by integration with respect to x.

(d) Determine the y-coordinate of the centroid by integration with respect to x.

9-11. Repeat problem P9-10 if the shaded area of the cooling fin is $y(x) = -x^2 + 4$.

9-12. Repeat problem P9-10 if the shaded area of the cooling fin is $y(x) = 9 - x^2$.

9-13. Repeat problem P9-10 if the geometry of a cooling fin is described by the shaded area shown in Fig. P9.13.

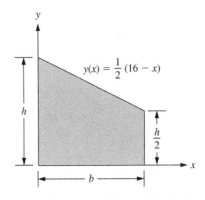

Figure P9.13 Geometry of the cooling fin for problem P9-13.

9-14. The cross section of an airfoil is described by the shaded area that is bounded by a cubic equation $y(x) = -x^3 + 9x$, as shown in Fig. P9.14.

(a) Given the above equation for $y(x)$, determine the height h and width b of the fin.

(b) Determine the area of the cooling fin by integration with respect to x.

(c) Determine the x-coordinate of the centroid by integration with respect to x.

(d) Determine the y-coordinate of the centroid by integration with respect to x.

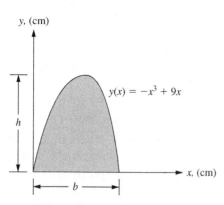

y, (cm)

$y(x) = -x^3 + 9x$

h

b

x, (cm)

Figure P9.14 Geometry of an airfoil.

9-15. The geometry of a decorative heat-sink fin of height 2 cm and length b (in cm) is approximated, as shown in Fig. P9.15.

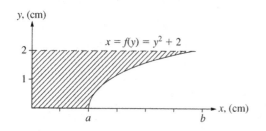

y, (cm)

$x = f(y) = y^2 + 2$

2

1

a

b

x, (cm)

Figure P9.15 Geometry of an airfoil.

(a) Determine the values of a and b for the heat-sink fin.

(b) Determine the area of the cooling fin by integration with respect to y.

(c) Determine the x-coordinate of the centroid by integration with respect to y.

(d) Determine the y-coordinate of the centroid by integration with respect to y.

9-16. The geometry of a gear tooth is approximated by the following quadratic equation, as shown in Fig. P9.16.

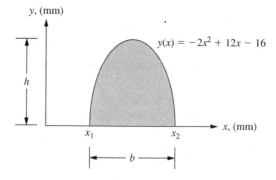

y, (mm)

$y(x) = -2x^2 + 12x - 16$

h

x_1

x_2

x, (mm)

b

Figure P9.16 Geometry of an airfoil.

(a) Determine the height h of the tooth (i.e., the maximum value of $y(x)$).

(b) Determine the coordinates x_1 and x_2 where $y(x) = 0$, and calculate the width b.

(c) Determine the area of the gear tooth by integration with respect to x.

(d) Determine the x-coordinate of the centroid by integration with respect to x.

(e) Determine the y-coordinate of the centroid by integration with respect to x.

9-17. A cantilever beam is subjected to a quadratic distributed load, as shown in Fig. P9.17.

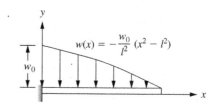

y

$w(x) = -\dfrac{w_0}{l^2}(x^2 - l^2)$

w_0

x

Figure P9.17 Cantilever beam subjected to a quadratic distributed load.

(a) Determine the total resultant force,

$$R = \int_0^l w(x)\,dx.$$

(b) Determine the x-location of the resultant R; i.e., determine the centroid of the area under the distributed load,

$$\bar{x} = \frac{\displaystyle\int_0^l x\,w(x)\,dx}{R}.$$

9-18. A simply supported beam is subjected to a quadratic distributed load, as shown in Fig. P9.18.

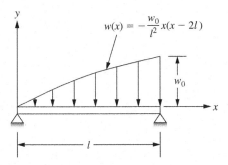

$$w(x) = -\frac{w_0}{l^2}x(x - 2l)$$

Figure P9.18 Simply supported beam subjected to a quadratic distributed load.

(a) Determine the total resultant force,
$$R = \int_0^l w(x)\,dx.$$

(b) Determine the x-location of the resultant R; i.e., determine the centroid of the area under the distributed load,

$$\bar{x} = \frac{\displaystyle\int_0^l x\,w(x)\,dx}{R}.$$

9-19. Determine the velocity $v(t)$ and the position $y(t)$ of a vehicle that starts from rest at position $y(0) = 0$ and is subjected to the following accelerations:
(a) $a(t) = 40\,t^3 + 30\,t^2 + 20\,t + 10$ m/s^2.
(b) $a(t) = 4\sin 2t\cos 2t$ m/s^2.
    Hint: Use a trigonometric identity.

9-20. A particle starts from rest at a position $x(0) = 0$. Find the velocity $v(t)$ and position $x(t)$ if the particle is subjected to the following accelerations:
(a) $a(t) = 6\,t^3 - 4\,t^2 + 7\,t - 8$ m/s^2.
(b) $a(t) = 5\,e^{-5t} + \frac{\pi}{4}\cos\left(\frac{\pi}{4}\,t\right)$ m/s^2.

9-21. The acceleration of an automobile is measured as shown in Fig. P9.21. If the automobile starts from rest at position $x(0) = 0$, sketch the velocity $v(t)$ and position $x(t)$ of the automobile.

Figure P9.21 The acceleration of an automobile.

9-22. A vehicle starting from rest at position $x(0) = 0$ is subjected to the acceleration shown in Fig. P9.22.
(a) Plot the velocity $v(t)$ of the vehicle, and clearly indicate both its maximum and final values.
(b) Given your results of part (a), plot the position $x(t)$ of the vehicle, and clearly indicate both its maximum and final values.

9-23. A vehicle starting from rest at a position $x(0) = 0$ is subjected to the acceleration shown in Fig. P9.23.
(a) Plot the velocity $v(t)$ of the vehicle, and clearly indicate both its maximum and final values.

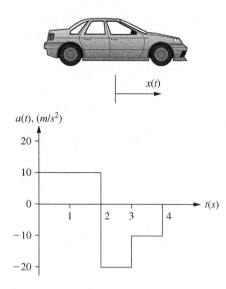

Figure P9.22 A vehicle subjected to a given acceleration.

Figure P9.23 A vehicle subjected to a given acceleration.

(b) Given your results of part (a), plot the position $x(t)$ of the vehicle, and clearly indicate both its maximum and final values.

9-24. A vehicle starting from rest at a position $x(0) = 0$ is subjected to the acceleration shown in Fig. P9.24.

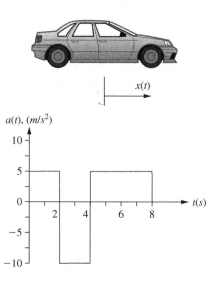

Figure P9.24 A vehicle subjected to a given acceleration.

(a) Plot the velocity $v(t)$ of the vehicle, and clearly indicate both its maximum and final values.

(b) Given your results of part (a), plot the position $x(t)$ of the vehicle, and clearly indicate both its maximum and final values.

9-25. The current flowing in a resistor is given by

$$i(t) = (t-1)^2 \, \text{A}.$$

Knowing that $i(t) = \frac{dq(t)}{dt}$, find the equation of the charge of (t) if $q(0) = 0$.

9-26. The RLC circuit shown in Fig. P9.26 has $R = 10 \, \Omega$, $L = 2 \, \text{H}$, and $C = 0.5 \, \text{F}$. If the current $i(t)$ flowing through the circuit is $i(t) = 10 \sin(240 \pi t)$ A, find the voltage $v(t)$ supplied by the voltage source, which is given by:

$$v(t) = iR + L\frac{di(t)}{dt} + \frac{1}{C}\int_0^t i(t)\,dt$$

Figure P9.26   A series RLC circuit.

9-27. An input voltage $v_{in} = 10\sin(10t)$ V is applied to an Op–Amp circuit as shown in Fig. P9.27.

Figure P9.27   An Op–Amp circuit for problem P9-27.

(a) The relationship between the input and output of the Op–Amp is given by

$$v_{in} = -0.1\frac{dv_o(t)}{dt}. \qquad (9.62)$$

Suppose that the initial output voltage of the Op–Amp is zero. Integrate both sides of equation (9.62) to determine the voltage $v_o(t)$. Also, sketch the output voltage $v_o(t)$ for one cycle.

(b) Suppose that the instantaneous power absorbed by the capacitor is $p(t) = 100(1 - \cos(10t))\sin(10t)$ mW, and that the initial stored energy is $w(0) = 0$. Knowing that $p(t) = \frac{dw(t)}{dt}$, integrate both sides of the equation to determine the total stored energy $w(t)$.

9-28. Repeat problem P9-27 if $v_{in} = 10\,e^{-10t}$ V.

9-29. Repeat problem P9-27 if $v_{in} = 10(1 - \cos(100t))$ V.

9-30. An input voltage $v_{in} = 5\cos(20t)$ V is applied to an Op–Amp circuit as shown in Fig. P9.30. The relationship between the input and output is given by

$$v_O = -\left(2v_{in} + 5\int_0^t v_{in}(t)\right)dt. \qquad (9.63)$$

Determine the voltage $v_o(t)$. Also, sketch the output voltage $v_o(t)$ for one cycle.

Figure P9.30   An Op–Amp circuit for problem P9-30.

9-31. A current $i(t) = 10\,e^{-10t}$ mA is applied to a $100\,\mu F$ capacitor, as shown in Fig. P9.31.
(a) Suppose the initial voltage across the capacitor is $v_o = 10$ V. Knowing that $i(t) = C\frac{dv_o(t)}{dt}$, integrate both sides of the equation to determine

the voltage $v_o(t)$. Evaluate the voltage at times $t = 0$, 0.1, 0.2, and 0.5 s, and use your results to sketch $v(t)$.

(b) Suppose the stored power is $p(t) = 0.2\,e^{-20t}$ W, and that the initial stored energy is $w(0) = 0.005$ J. Integrate both sides of the equation $p(t) = \frac{dw}{dt}$ to determine the total stored energy $w(t)$.

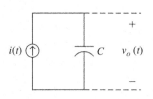

Figure P9.31   A current applied to a capacitor.

9-32. For the circuit shown in Fig. P9.32, the voltage is $v(t) = 5\cos(5t)$ V, the current is $i(t) = 10\sin(5t)$ A, and the total power is $p(t) = 25\sin(10t)$ W. If the initial stored energy is $w(0) = 0$ J, determine the total stored energy, $w(t) = w(0) + \int_0^t p(t)\,dt$, and plot one cycle of $w(t)$.

Figure P9.32   A voltage applied to an inductor.

9-33. The sawtooth current $i(t)$ shown in Fig. 9.33 is applied to a 500 μF capacitor. Sketch the voltage $v(t)$ across the capacitor knowing that $i(t) = C\frac{dv(t)}{dt}$ or $v(t) = \frac{1}{C}\int i(t)\,dt$. Assume that the capacitor is completely discharged at $t = 0$ (i.e., $v(0)$ is 0 V).

Figure P9.33   Sawtooth current applied to a capacitor in problem P9-33.

9-34. The sawtooth voltage $v(t)$ shown in Fig. 9.34 is applied across a 100 mH inductor. Sketch the current $i(t)$ passing through the inductor knowing that $v(t) = L\frac{di(t)}{dt}$ or $i(t) = \frac{1}{L}\int v(t)\,dt$. Assume that the current flowing through the inductor at t is zero (i.e., $i(0) = 0$ A).

Figure P9.34   Sawtooth voltage applied across an inductor.

9-35. A current $i(t)$ is applied to a capacitor of $C = 100\ \mu$F, as shown in the Figure 9.35.

Figure P9.35   A current applied to a 100 μF capacitor.

(a) Knowing that $i(t) = C\frac{dv(t)}{dt}$, sketch the voltage across the capacitor $v(t)$.

Assume $v(0) = 0$ V. Also note that the time t is measured in *ms* and the current $i(t)$ is measured in *mA*.

(b) Given results of part (a), sketch the power $p(t) = v(t)i(t)$.

9-36. A biomedical engineer is evaluating an energy-absorbing aviation seat on a vertical deceleration tower, as shown in Fig. P9.36. The acceleration profile of the drop cage is given by

$$a(t) = 400 \sin(50 \pi t) \text{ m/s}^2.$$

(a) Knowing that $v(t) = v(0) + \int_0^t a(t)dt$, find and plot the velocity $v(t)$ of the drop cage. Assume the drop cage starts from rest at $t = 0$ s.

(b) What is the impact velocity v_{impact} of the drop cage if it takes 20 ms to hit the ground?

(c) The total impulse I is equal to the change in momentum, i.e., $I = \Delta p = p_f - p_i = mv_{impact} - mv_0$, where m is the mass of the system, v_{impact} is the final velocity, v_o is the initial velocity, p is the momentum, and I is the impulse. Find the total impulse after 20 ms. Assume that the total mass of the drop cage, seat, and crash test dummy is 1200 kg.

9-37. A biomedical engineer measures the velocity profiles of a belted and unbelted occupant during a 45 mph (≈ 20 m/s) frontal collison, as shown in Fig. P9.37.

(a) Knowing that $x(t) = x(0) + \int_0^t v(t) dt$, find and plot the displacement $x(t)$ of the belted occupant for time 0 to 40 ms. Assume that the initial displacement at $t = 0$ is 0 m (i.e., $x(0) = 0$ m).

(b) Find and plot the displacement $x(t)$ of the unbelted occupant for time 0 to 40 ms. Assume $x(0) = 0$ m.

(c) Based on the results of parts (a) and (b), how much farther did the unbelted occupant travel compared to the belted occupant?

Belted Occupant:

Unbelted Occupant:

Figure P9.36 Energy-absorbing aviation seat.

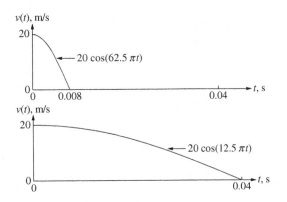

Figure P9.37 Velocities of the belted and unbelted occupants during frontal collison.

9-38. A civil engineer designs a building overhang to withstand a triangular snow loading per unit length $p(x) = p_o \left(1 - \frac{x}{L}\right)$, as shown in Fig. P9.38.

(a) Compute the resulting force $V = \int_0^L p(x)dx$.

(b) Compute the corresponding moment $M = \int_0^L x p(x)dx$.

(c) Locate the position of the centroid
$$\bar{x} = \frac{\int_0^L x p(x)\, dx}{\int_0^L p(x)\, dx} = \frac{M}{V}.$$

Figure P9.38   Triangular snow loading on a building overhang.

9-39. A building overhang is subjected to a parabolic snow loading $p(x)$, as shown in Fig. P9.39. The overhang is constructed from two metal face plate separated by a nonmetallic core of thickness h. This sandwich beam construction deforms primarily due to shear deformation with deflection $y(x)$ satisfying
$$y(x) = \frac{1}{h\,G} \int_0^x \left[\int_0^x p(x)\, dx - V\right] dx,$$

where $V = \hat{p}\frac{L}{3}$, $p(x) = \hat{p}\left(1 - \frac{x}{L}\right)^2$ and G is the shear modulus.

(a) Show that the deflection of the sandwich beam is given by
$$y(x) = -\frac{\hat{p}\,x}{h\,G}\left\{\frac{L}{3} - x\left[\frac{1}{2} - \frac{x}{3L}\left(1 - \frac{x}{4L}\right)\right]\right\}.$$

(b) Find the location and value of maximum deflection $y(x)$.

(a) Cantilever overhang.

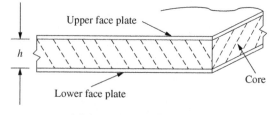

(b) Sandwich construction.

Figure P9.39   Parabolic snow loading of sandwich-constructed building overhang.

Differential Equations in Engineering

The objective of this chapter is to familiarize engineering students with the solution of differential equations (DEQ) as needed for first- and second-year engineering courses such as physics, circuits, and dynamics. A differential equation relates an output variable and its derivatives to an input variable or forcing function. There are several different types of differential equations. This chapter discusses first- and second-order linear differential equations with constant coefficients. These are the most common type of differential equations found in undergraduate engineering classes.

10.1 INTRODUCTION: THE LEAKING BUCKET

Consider a bucket of cross-sectional area A being filled with water at a volume flow rate Q_{in}, as shown in Fig. 10.1. If $h(t)$ is the height and $V = A\,h(t)$ is the volume of water in the bucket, the rate of change of the volume is given by

$$\frac{dV}{dt} = A\,\frac{dh(t)}{dt}. \tag{10.1}$$

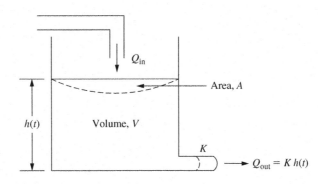

Figure 10.1   A leaking bucket with a small hole.

Suppose the bucket has a small hole on the side through which water is leaking at a rate

$$Q_{out} = K h(t) \tag{10.2}$$

where K is a constant. In reality, Q_{out} is not a linear function of $h(t)$, but it is assumed here for simplicity. The constant K is an engineering design parameter that depends on the size and shape of the hole, as well as the properties of the fluid. By conservation of volume, the volume of water in the bucket is given by

$$\frac{dV}{dt} = Q_{in} - Q_{out}. \tag{10.3}$$

Substituting equations (10.1) and (10.2) into equation (10.3) gives

$$A \frac{dh(t)}{dt} = Q_{in} - K h(t)$$

or

$$A \frac{dh(t)}{dt} + K h(t) = Q_{in}. \tag{10.4}$$

Equation (10.4) is a first-order linear differential equation with constant coefficients. The objective is to solve the differential equation; in other words, determine the height $h(t)$ of the water when an input Q_{in} and the initial condition $h(0)$ are given. Before presenting the solution of this equation, a general discussion of differential equations and the solution of linear differential equations with constant coefficients is given.

10.2 DIFFERENTIAL EQUATIONS

An nth order linear differential equation relating an output variable $y(t)$ and its derivatives to some input function $f(t)$ can be written as

$$A_n \frac{d^n y(t)}{dt^n} + A_{n-1} \frac{d^{n-1} y(t)}{dt^{n-1}} + \ldots + A_1 \frac{dy(t)}{dt} + A_0 y(t) = f(t) \tag{10.5}$$

where the coefficients $A_n, A_{n-1}, \ldots, A_0$ can be constants, functions of y, or function of t. The input function $f(t)$ (also called the forcing function) represents everything on the right-hand side of the differential equation. The solution of the differential equation is the output variable, $y(t)$.

For a second-order system involving position $y(t)$, velocity $\frac{dy(t)}{dt}$, and acceleration $\frac{d^2 y(t)}{dt^2}$, equation (10.5) takes the form

$$A_2 \frac{d^2 y(t)}{dt^2} + A_1 \frac{dy(t)}{dt} + A_0 y(t) = f(t). \tag{10.6}$$

Note that engineers often use a dot notation when referring to derivatives with respect to time, for example, $\dot{y}(t) = \frac{dy(t)}{dt}$, $\ddot{y}(t) = \frac{d^2 y(t)}{dt^2}$, and so on. In this case, equation

(10.6) can be written as

$$A_2\,\ddot{y}(t) + A_1\,\dot{y}(t) + A_0\,y(t) = f(t). \tag{10.7}$$

In many engineering applications, the coefficients $A_n, A_{n-1}, \ldots, A_0$ are constants (not functions of y or t). In this case, the differential equation given by equation (10.5) is known as a linear differential equation with constant coefficients. For example, in the case of a spring-mass system subjected to an applied force $f(t)$, equation (10.8) is a second-order differential equation given by

$$m\,\ddot{y}(t) + k\,y(t) = f(t) \tag{10.8}$$

where m is the mass and k is the spring constant. If the coefficients $A_n, A_{n-1}, \ldots, A_0$ are functions of y or t, exact solutions can be difficult to obtain. In many cases, exact solutions do not exist, and the solution $y(t)$ must be obtained numerically (e.g., using the differential equation solvers in MATLAB). However, in the case of constant coefficients, the solution $y(t)$ can be obtained by following the step-by-step procedure outlined below.

10.3   SOLUTION OF LINEAR DEQ WITH CONSTANT COEFFICIENTS

In general, the total solution for the output variable $y(t)$ is the sum of two solutions: the **transient** solution and the **steady-state** solution.

1. **Transient Solution, $y_{tran}(t)$ (also called the Homogeneous or Complementary Solution)**: The transient solution is obtained using the following steps:

    a. Set the forcing function $f(t) = 0$. This makes the right-hand side of equation (10.5) zero, i.e.,

    $$A_n\,\frac{d^n y(t)}{dt^n} + A_{n-1}\,\frac{d^{n-1}y(t)}{dt^{n-1}} + \cdots + A_1\,\frac{dy(t)}{dt} + A_0\,y(t) = 0. \tag{10.9}$$

    b. Assume a transient solution of the form $y(t) = ce^{st}$, and substitute it into (10.9). Note that $\frac{dy(t)}{dt} = cse^{st}$, $\frac{d^2y(t)}{dt^2} = cs^2\,e^{st}$, and so on, so that each term will contain ce^{st}. Since the right-hand side of equation (10.9) is zero, canceling the ce^{st} will result in a polynomial in s:

    $$A_n\,s^n + A_{n-1}\,s^{n-1} + \cdots + A_1\,s + A_0 = 0. \tag{10.10}$$

    c. Solve for the roots of the above equation, which is known as the characteristic equation. The roots are the n values of s that make the characteristic equation equal to zero. Call these values $s_1, s_2, \ldots, s_n$.

    d. For the case of n distinct roots, the transient solution of the differential equation has the general form

    $$y_{tran}(t) = c_1\,e^{s_1 t} + c_2\,e^{s_2 t} + \ldots + c_n\,e^{s_n t}$$

    where the constants $c_1, c_2, \ldots, c_n$ are determined later from the initial conditions of the system.

    e. For the special case of **repeated** roots (i.e., two of the roots are the same), the solution can be made general by multiplying one of the roots by t. For example, for a second-order system with $s_1 = s_2 = s$, the transient solution is

$$y_{tran}(t) = c_1 e^{st} + c_2 t e^{st}. \tag{10.11}$$

2. **Steady-State Solution, $y_{ss}(t)$ (also called the Particular Solution):**
   The steady-state solution can be found using the **Method of Undetermined Coefficients:**

   a. Assume (guess) the form of the steady-state solution, y_{ss}. This will usually have the same general form as the forcing function and its derivatives but will contain unknown constants (i.e., undetermined coefficients). Example guesses are shown in Table 10.1, where K, A, B, and C are constants.

   TABLE 10.1   Assumed solutions $y_{ss}(t)$ for common input finctions $f(t)$.

If input $f(t)$ is	Assume $y_{ss}(t)$
K	A
Kt	$At + B$
Kt^2	$At^2 + Bt + C$
$K\sin\omega t$ or $K\cos\omega t$	$A\sin\omega t + B\cos\omega t$

   b. Substitute the assumed steady-state solution $y_{ss}(t)$ and its derivatives into the original differential equation.
   c. Solve for the unknown (undetermined) coefficients (A, B, C, etc.). This can usually be done by equating the coefficients of like terms on the left- and right-hand sides of equations.

3. Find the total solution, $y(t)$: The total solution is just the sum of the transient and steady-state solutions,

$$y(t) = y_{tran}(t) + y_{ss}.$$

4. Apply the initial conditions on $y(t)$ and its derivatives. A differential equation of order n must have exactly n initial conditions, which will result in an $n \times n$ system of equations for n constants $c_1, c_2, \ldots, c_n$.

10.4   FIRST-ORDER DIFFERENTIAL EQUATIONS

This section illustrates the application of the method described in Section 10.3 to a variety of first-order differential equations in engineering.

Example 10-1

The Leaking Bucket Problem Consider again the leaking bucket of Section 10.1, which satisfies the following first-order differential equation:

$$A\frac{dh(t)}{dt} + Kh(t) = Q_{in}. \tag{10.12}$$

Find the total solution of $h(t)$ if the input $Q_{in} = B$ is a constant. Assume that the initial height of the water is zero (i.e., $h(0) = 0$).

Solution (a) **Transient (Complementary or Homogeneous) Solution:** Since the transient solution is the zero-input solution, the input on the right-hand side (RHS) of equation (10.12) is set to zero. Thus, the homogeneous differential equation of the leaking bucket is given by

$$A \frac{dh(t)}{dt} + Kh(t) = 0. \tag{10.13}$$

Assume that the transient solution of the height $h_{tran}(t)$ is of the form given by equation (10.14):

$$h_{tran}(t) = ce^{st}. \tag{10.14}$$

The constant s is determined by substituting $h_{tran}(t)$ and its derivative into the homogeneous differential equation (10.13). The derivative of $h_{tran}(t)$ is given by

$$\frac{dh_{tran}(t)}{dt} = \frac{d}{dt}(ce^{st})$$
$$= cse^{st}. \tag{10.15}$$

Substituting equations (10.14) and (10.15) in equation (10.13) yields

$$A(cse^{st}) + K(ce^{st}) = 0.$$

Factoring out ce^{st} gives

$$ce^{st}(As + K) = 0$$

Since $ce^{st} \neq 0$, it follows that

$$As + K = 0. \tag{10.16}$$

Equation (10.16) is the **characteristic equation** for the leaking bucket. Solving equation (10.16) for s gives

$$s = -\frac{K}{A}. \tag{10.17}$$

Substituting the above value of s into equation (10.14), the transient solution for the leaking bucket is given by

$$h_{tran}(t) = ce^{-\frac{K}{A}t} \tag{10.18}$$

The constant c depends on the initial height of the water and cannot be determined until the initial condition is applied to the total solution in part (d).

(b) **Steady-State (Particular) Solution:** The steady-state solution of a differential equation is the solution to a particular input. Since the given input is $Q_{in} = B$, the differential equation (10.12) can be written as

$$A \frac{dh(t)}{dt} + Kh(t) = B. \tag{10.19}$$

According to the method of undetermined coefficients (Table 10.1), the steady-state solution will have the same general form as the input and its derivatives.

Since the input in this example is constant, the steady-state solution is assumed constant to be

$$h_{ss}(t) = E, \tag{10.20}$$

where E is a constant. The value of E can be determined by substituting $h_{ss}(t)$ and its derivative into equation (10.19). The derivative of $h_{ss}(t)$ is

$$\frac{dh_{ss}(t)}{dt} = \frac{d}{dt}(E)$$
$$= 0. \tag{10.21}$$

Substituting equations (10.20) and (10.21) into equation (10.19) gives

$$A(0) + KE = B.$$

Solving for E gives

$$E = \frac{B}{K}.$$

Therefore, the steady-state solution of the leaking bucket subjected to a constant input $Q_{in} = B$ is given by

$$h_{ss}(t) = \frac{B}{K}. \tag{10.22}$$

(c) **Total Solution:** The total solution for $h(t)$ is obtained by adding the transient and the steady-state solutions as

$$h(t) = c\,e^{-\frac{K}{A}t} + \frac{B}{K}. \tag{10.23}$$

(d) **Initial Conditions:** The constant c can now be obtained by substituting the initial condition $h(0) = 0$ into equation (10.23) as

$$h(0) = c\,e^{-\frac{K}{A}(0)} + \frac{B}{K} = 0$$

or

$$c\,(1) + \frac{B}{K} = 0,$$

which gives

$$c = -\frac{B}{K}. \tag{10.24}$$

Substituting the above value of c into equation (10.23) yields

$$h(t) = -\frac{B}{K}e^{-\frac{K}{A}t} + \frac{B}{K}$$

or

$$h(t) = \frac{B}{K}\left(1 - e^{-\frac{K}{A}t}\right). \tag{10.25}$$

Note that as $t \to \infty$, $h(t) \to \frac{B}{K}$; in other words, the total solution reaches the steady-state solution. Thus, at steady state the height $h(t)$ reaches a constant value of $\frac{B}{K}$. Physically speaking, the bucket continues to fill until the pressure is great enough that $Q_{out} = Q_{in}$ (i.e., $\frac{dh(t)}{dt} = 0$). That value depends only on $\frac{B}{K} = \frac{Q_{in}}{K}$. At time $t = \frac{A}{K}$ sec, the bucket fills to a height of

$$h\left(\frac{A}{K}\right) = \frac{B}{K}\left(1 - e^{-\frac{K}{A}\left(\frac{A}{K}\right)}\right)$$

$$= \frac{B}{K}(1 - e^{-1})$$

$$= \frac{B}{K}(1 - 0.368)$$

or

$$h\left(\frac{A}{K}\right) = 0.632\,\frac{B}{K}.$$

At time $t = 5A/K$ s, the bucket fills to a height of

$$h\left(5\frac{A}{K}\right) = \frac{B}{K}\left(1 - e^{-\frac{K}{A}\left(\frac{5A}{K}\right)}\right)$$

$$= \frac{B}{K}(1 - e^{-5})$$

$$= \frac{B}{K}(1 - 0.0067)$$

or

$$h\left(5\frac{A}{K}\right) = 0.9933\,\frac{B}{K}.$$

Thus, it takes $t = A/K$ s for the height to reach 63.2% of the steady-state value and $t = 5A/K$ s to reach 99.33% of the steady-state value. The time $t = A/K$ s is known as the **time constant** of the response and is usually denoted by the Greek letter τ. The response of a first-order system (for example, the leaking bucket) can generally be written as

$$y(t) = \text{steady-state solution}\left(1 - e^{-\frac{t}{\tau}}\right). \tag{10.26}$$

The plot of the height $h(t)$ for input $Q_{in} = B$ is shown in Fig. 10.2. It can be seen from this figure that after $t = 5\tau$ s, the water level has, for all practical purposes, reached its steady-state value.

Figure 10.2 Solution for $h(t)$ for $Q_{in} = B$ and $h(0) = 0$.

Example 10-2

Leaking Bucket with No Input Suppose now that $Q_{in} = 0$, and that the initial height of the water is h_0 (Fig. 10.3). The height $h(t)$ of the water is governed by the first-order differential equation

$$A \frac{dh(t)}{dt} + Kh(t) = 0. \tag{10.27}$$

Determine the total solution for $h(t)$. Also, find the time it takes for the water to completely leak out of the bucket.

Figure 10.3 Leaking bucket with no input for Example 10-2.

Solution   (a) **Transient Solution:** The transient solution is identical to that of the previous example and is given by

$$h_{tran}(t) = c_1 \, e^{-\frac{K}{A}t}.$$

(b) **Steady-State Solution:** Since the RHS of the differential equation (10.27) is zero (i.e., the input is zero), the steady-state solution is also zero:

$$h_{ss}(t) = 0.$$

(c) **Total Solution:** The total solution for the height $h(t)$ is given by

$$h(t) = h_{tran}(t) + h_{ss}(t)$$

$$= c_1 \, e^{-\frac{K}{A}t} + 0$$

or

$$h(t) = c_1 \, e^{-\frac{K}{A}t}. \tag{10.28}$$

(d) **Initial Conditions:** The constant c_1 is determined by substituting the initial height $h(0) = h_0$ into equation (10.28) as

$$h(0) = c_1 \, e^{-\frac{K}{A}(0)} = h_0$$

or

$$c_1 \, (1) = h_0,$$

which gives

$$c_1 = h_0.$$

Thus, the total solution for $h(t)$ is

$$h(t) = h_0 \, e^{-\frac{K}{A}t}. \tag{10.29}$$

The height $h(t)$ given in equation (10.29) is a decaying exponential function with time constant $\tau = A/K$. At time $t = A/K$ s, the bucket empties to a height of

$$h\left(\frac{A}{K}\right) = h_0 \, e^{-\frac{K}{A}\left(\frac{A}{K}\right)}$$

$$= h_0 \, e^{-1}$$

or

$$h\left(\frac{A}{K}\right) = 0.368 \, h_0.$$

At time $t = \dfrac{5A}{K}$ s, the bucket empties to a height of

$$h\left(\frac{5A}{K}\right) = h_0 \, e^{-\frac{K}{A}\left(\frac{5A}{K}\right)}$$

$$= h_0 \, e^{-5}$$

$$= 0.0067 \, h_0$$

or

$$h\left(\frac{5A}{K}\right) \approx 0.$$

The plot of the height is shown in Fig. 10.4. It can be seen from this figure that the height starts from the initial value h_0 and decays to 36.8% of the initial value in one time constant $\tau = A/K$, and is approximately zero after five time constants.

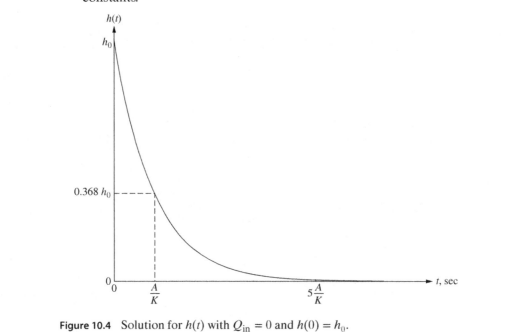

Figure 10.4   Solution for $h(t)$ with $Q_{in} = 0$ and $h(0) = h_0$.

Example 10-3

Voltage Applied to an RC Circuit Find the voltage $v(t)$ across the capacitor if a constant voltage source $v_s(t) = v_s$ is applied to the RC circuit shown in Fig. 10.5. Assume that the capacitor is initially completely discharged (i.e., $v(0) = 0$).

Figure 10.5   RC circuit with constant input for Example 10-3.

The governing equation for $v(t)$ follows from Kirchhoff's voltage law (KVL), which gives

$$v_R(t) + v(t) = v_s(t). \tag{10.30}$$

From Ohm's law, the voltage across the resistor is given by $v_R(t) = Ri(t)$. Since the resistor and capacitor are connected in series, the same current flows through the resistor and capacitor, $i(t) = C\frac{dv(t)}{dt}$. Therefore, $v_R(t) = RC\frac{dv(t)}{dt}$, and equation (10.30) can be written as

$$RC\frac{dv(t)}{dt} + v(t) = v_s(t). \tag{10.31}$$

Equation (10.31) is a first-order differential equation with constant coefficients. This equation can also be written as

$$RC\dot{v}(t) + v(t) = v_s(t), \tag{10.32}$$

where $\dot{v}(t) = \frac{dv(t)}{dt}$. The goal is to solve the voltage $v(t)$ if $v_s(t) = v_s$ is constant and $v(0) = 0$.

Solution (a) **Transient Solution:** The transient solution is obtained by setting the right-hand side of the differential equation equal to zero as

$$RC\dot{v}(t) + v(t) = 0, \tag{10.33}$$

and assuming a solution of the form

$$v_{tran}(t) = ce^{st}. \tag{10.34}$$

The constant s is determined by substituting $v_{tran}(t)$ and its derivative into equation (10.33). The derivative of $v_{tran}(t)$ is given by

$$\frac{dv_{tran}(t)}{dt} = \frac{d}{dt}(ce^{st}) = cse^{st}. \tag{10.35}$$

Substituting equations (10.34) and (10.35) in equation (10.33) yields

$$RC(cse^{st}) + (ce^{st}) = 0.$$

Factoring out ce^{st} gives

$$e^{st}(RCs + 1) = 0.$$

It follows that

$$RCs + 1 = 0, \tag{10.36}$$

which gives

$$s = -\frac{1}{RC}. \tag{10.37}$$

Substituting the above value of s into equation (10.34) gives

$$v_{tran}(t) = ce^{-\frac{1}{RC}t} \tag{10.38}$$

The constant c depends on the initial voltage across the capacitor, which is applied to the total solution in part (d).

(b) **Steady-State Solution:** For $v_s(t) = v_s$,

$$RC\,\dot{v}(t) + v(t) = v_s. \tag{10.39}$$

Since the input to the RC circuit is constant, the steady-state solution of the output voltage $v(t)$ is assumed to be

$$v_{ss}(t) = E, \tag{10.40}$$

where E is a constant. The value of E can be determined by substituting $v_{ss}(t)$ and its derivative in equation (10.39), which gives

$$RC(0) + E = v_s.$$

Solving for E yields

$$E = v_s.$$

Thus, the steady-state solution for the output voltage is

$$v_{ss}(t) = v_s. \tag{10.41}$$

(c) **Total Solution:** The total solution for $v(t)$ is obtained by adding the transient and the steady-state solutions given by equations (10.38) and (10.41) as

$$v(t) = ce^{-\frac{1}{RC}t} + v_s. \tag{10.42}$$

(d) **Initial Conditions:** The constant c can now be obtained by applying the initial condition as

$$v(0) = ce^{-\frac{1}{RC}(0)} + v_s = 0$$

or

$$c\,(1) + v_s = 0.$$

Solving for c gives

$$c = -v_s. \tag{10.43}$$

Substituting the above value of c into equation (10.42) yields

$$v(t) = -v_s\,e^{-\frac{1}{RC}t} + v_s$$

or

$$v(t) = v_s\left(1 - e^{-\frac{1}{RC}t}\right). \tag{10.44}$$

Note that as $t \to \infty$, $v(t) \to v_s$ (i.e., the total solution reaches the steady-state solution). At steady state, the capacitor is fully charged to a voltage equal to the input voltage. While the capacitor is charging, the voltage across the capacitor at time $t = RC$ s is given by

$$v(RC) = v_s \left(1 - e^{-\frac{1}{RC}(RC)}\right)$$
$$= v_s \left(1 - e^{-1}\right)$$
$$= v_s \left(1 - 0.368\right)$$

or

$$v = 0.632\, v_s.$$

Also, at time $t = 5\,RC$ s, the voltage across the capacitor is given by

$$v(5\,RC) = v_s \left(1 - e^{-\frac{1}{RC}(5\,RC)}\right)$$
$$= v_s \left(1 - e^{-5}\right)$$
$$= v_s \left(1 - 0.0067\right)$$
$$= 0.9933\, v_s$$

or

$$v \approx v_s.$$

Thus, it takes $t = RC$ s for the voltage to reach 63.2% of the input voltage and at $t = 5\,RC$ s, the voltage reaches 99.33% of the input value. The time $t = \tau = RC$ s is the **time constant** of the RC circuit, which is a measure of the time required for the capacitor to fully charge. Typically, to reduce the charge time of the capacitor, the resistance value of the resistor is reduced. The plot of the voltage $v(t)$ is shown in Fig. 10.6. It can be seen from this figure that it takes the response approximately 5τ to reach the steady state, which is identical to the result obtained for the leaking bucket with constant Q_{in}.

Figure 10.6   The voltage across the capacitor to a constant voltage in an RC circuit of Example 10-3.

Example 10-4

For the circuit shown in Fig. 10.7, the differential equation relating the output $v(t)$ and input $v_s(t)$ is given by

$$0.5 \dot{v}(t) + v(t) = v_s(t). \tag{10.45}$$

Find the output voltage $v(t)$ across the capacitor if the input voltage $v_s(t) = 10$ V. Assume that the initial voltage across the capacitor is zero (i.e., $v(0) = 0$).

Figure 10.7   RC circuit with input voltage $v_s(t)$.

Solution

(a) **Transient Solution:**   The transient solution is obtained by setting the right-hand side of the differential equation to zero as

$$0.5 \dot{v}(t) + v(t) = 0, \tag{10.46}$$

and assuming a solution of the form

$$v_{tran}(t) = ce^{st}. \tag{10.47}$$

Substituting the transient solution and its derivative into equation (10.46) and solving for s gives

$$0.5(cse^{st}) + (ce^{st}) = 0$$

$$ce^{st}(0.5\,s + 1) = 0$$

$$0.5\,s + 1 = 0$$

$$s = -2.$$

Thus, the transient solution of the output voltage is given by

$$v_{tran}(t) = ce^{-2t} \tag{10.48}$$

where c will be obtained from the initial condition.

(b) **Steady-State Solution:**   Since the input applied to the RC circuit is 10 V, equation (10.45) can be written as

$$0.5 \dot{v}(t) + v(t) = 10. \tag{10.49}$$

Since the input is constant, the steady-state solution of the output voltage $v(t)$ is assumed to be

$$v_{ss}(t) = E, \tag{10.50}$$

where E is a constant. The value of E can be determined by substituting $v_{ss}(t)$ and its derivative into equation (10.49) as

$$0.5\,(0) + E = 10$$

which gives

$$E = 10 \text{ V}.$$

Thus, the steady-state solution of the output voltage is given by

$$v_{ss}(t) = 10 \text{ V}. \tag{10.51}$$

(c) **Total Solution:** The total solution for $v(t)$ is obtained by adding the transient and steady-state solutions given by equations (10.48) and (10.51) as

$$v(t) = ce^{-2t} + 10. \tag{10.52}$$

(d) **Initial Conditions:** The constant c can now be obtained by applying the initial condition $(v(0) = 0)$ as

$$v(0) = ce^{-2\,(0)} + 10 = 0$$

or

$$c\,(1) + 10 = 0.$$

Solving for c gives

$$c = -10 \text{ V}. \tag{10.53}$$

Substituting the value of c from equation (10.53) into equation (10.52) yields

$$v(t) = -10\,e^{-2t} + 10$$

or

$$v(t) = 10\,(1 - e^{-2t}) \text{ V}. \tag{10.54}$$

Since the time constant is $\tau = 1/2 = 0.5$ s, it takes the capacitor 0.5 s to reach 63.2% of the input voltage and approximately $5(0.5) = 2.5$ s to fully charge to approximately 10 V. The plot of the output voltage $v(t)$ is shown in Fig. 10.8.

Figure 10.8   The voltage across the capacitor in Example 10-4.

Example 10-5

The differential equation for the capacitive circuit shown in Fig. 10.9 is given by

$$0.5\,\dot{v}(t) + v(t) = 0. \tag{10.55}$$

Find the output voltage $v(t)$ across the capacitor C as it discharges from an initial voltage of $v(0) = 10$ V.

Figure 10.9   Discharging of a capacitor in an RC circuit.

Solution

(a) **Transient Solution:** Since the left-hand side of the governing equation is the same as that for the previous example, the transient solution of the output voltage is given by equation (10.48) as

$$v_{tran}(t) = ce^{-2t}.$$

(b) **Steady-State Solution:** Since there is no input applied to the circuit, the steady-state value of the output voltage is zero:

$$v_{ss}(t) = 0.$$

(c) **Total Solution:** The total solution for $v(t)$ is obtained by the adding the transient and the steady-state solutions, which gives

$$v(t) = ce^{-2t}.$$

(d) **Initial Conditions:** The constant c can now be obtained by applying the initial condition $(v(0) = 10 \text{ V})$ as

$$v(0) = ce^{-2\,(0)} = 10,$$

which gives

$$c = 10 \text{ V}.$$

Thus, the output voltage is given by

$$v(t) = 10\,e^{-2t} \text{ V}.$$

While the capacitor is discharging, the voltage across the capacitor at time $t = 0.5$ s (one time constant) is given by

$$v(0.5) = 10\,e^{-2\,(0.5)}$$
$$= 10\,e^{-1}$$

or

$$v = 3.68 \text{ V}.$$

Also, at time $t = 2.5$ s (five time constants), the voltage across the capacitor is given by

$$v(2.5) = 10\,e^{-2\,(2.5)}$$
$$= 10\,e^{-5}$$
$$= 0.067$$

or

$$v \approx 0.$$

The plot of the output voltage, $v(t)$ is shown in Fig. 10.10. Mathematically speaking, the response of a capacitor discharging in an RC circuit is identical to the response of a leaking bucket with initial fluid height h_0!

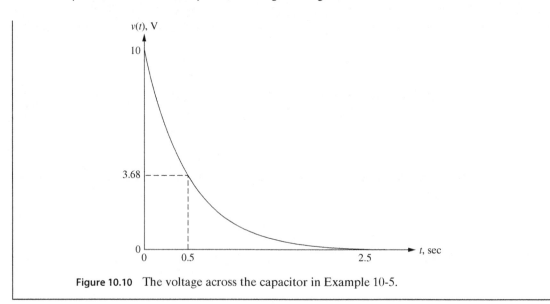

Figure 10.10 The voltage across the capacitor in Example 10-5.

<table>
</table>

| Example
10-6 | Consider a voltage $v_s(t)$ applied to an RL circuit, as shown in Fig. 10.11. Applying KVL yields |

Figure 10.11 Voltage applied to an RL circuit.

$$v_R(t) + v_L(t) = v_s(t),\qquad(10.56)$$

where $v_R(t) = R\,i(t)$ is the voltage across the resistor and $v_L(t) = L\,\dfrac{di(t)}{dt}$ is the voltage across the inductor. Thus, equation (10.56) can be written in terms of the current $i(t)$ as

$$L\,\frac{di(t)}{dt} + R\,i(t) = v_s(t).\qquad(10.57)$$

If the applied voltage source is $v_s(t) = v_s = $ constant, find the total solution for the current $i(t)$. Assume the initial current is zero ($i(0) = 0$).

Solution (a) **Transient Solution:** The transient solution is obtained by setting the right-hand side of the differential equation to zero as

$$L\frac{di(t)}{dt} + R\,i(t) = 0, \tag{10.58}$$

and assuming a transient solution of the form

$$i_{tran}(t) = ce^{st}. \tag{10.59}$$

The constant s is determined by substituting $i_{tran}(t)$ and its derivative into equation (10.58), which gives

$$L\,(cse^{st}) + R\,(ce^{st}) = 0.$$

Factoring out e^{st} gives

$$ce^{st}(Ls + R) = 0.$$

which implies

$$Ls + R = 0.$$

Solving for s gives

$$s = -\frac{R}{L}. \tag{10.60}$$

Substituting the above value of s into equation (10.59), the transient solution of the output voltage is given by

$$i_{tran}(t) = ce^{-\frac{R}{L}t}. \tag{10.61}$$

The constant c depends on the initial current flowing through the circuit and is found in part (d).

(b) **Steady-State Solution:** Since the input $v_s(t) = v_s$, equation (10.57) can be written as

$$L\frac{di(t)}{dt} + Ri(t) = v_s. \tag{10.62}$$

Because the voltage applied to the RL circuit is constant, the steady-state solution of the current $i(t)$ is assumed to be

$$i_{ss}(t) = E, \tag{10.63}$$

where E is a constant. The value of E can be determined by substituting $i_{ss}(t)$ and its derivative into equation (10.62), which gives

$$L(0) + R\,(E) = v_s.$$

Solving for E gives

$$E = \frac{v_s}{R}.$$

Thus, the steady-state solution of the current is given by

$$i_{ss}(t) = \frac{v_s}{R}. \tag{10.64}$$

(c) **Total Solution:** The total solution for the current $i(t)$ is obtained by adding the transient and the steady-state solutions given by equations (10.61) and (10.64) as

$$i(t) = ce^{-\frac{R}{L}t} + \frac{v_s}{R}. \tag{10.65}$$

(d) **Initial Conditions:** The constant c can now be obtained by applying the initial condition $(i(0) = 0)$ to equation (10.65) as

$$i(0) = ce^{-\frac{R}{L}(0)} + \frac{v_s}{R} = 0.$$

or

$$c\,(1) + \frac{v_s}{R} = 0$$

Solving for c gives

$$c = -\frac{v_s}{R}. \tag{10.66}$$

Substituting the above value of c into equation (10.65) gives

$$i(t) = -\frac{v_s}{R} e^{-\frac{R}{L}t} + \frac{v_s}{R}$$

or

$$i(t) = \frac{v_s}{R}(1 - e^{-\frac{R}{L}t}) \text{ A}. \tag{10.67}$$

As $t \to \infty$, $i(t) \to v_s/R$ (i.e., the steady-state solution). It takes the current $t = \tau = L/R$ s to reach 63.2% of its steady-state value v_s/R. The plot of the current $i(t)$ is shown in Fig. 10.12. It can be seen that the current $i(t)$ takes approximately 5τ to reach the steady-state value, as obtained for both the charging of a capacitor and the filling of a leaking bucket.

Figure 10.12   The current flowing through an RL circuit.

Example 10-7

A constant voltage $v_s(t) = 10$ V is applied to the RL circuit shown in Fig. 10.13. The circuit is described by the following differential equation:

$$0.1 \frac{di(t)}{dt} + 100\, i(t) = 10. \tag{10.68}$$

Find the current $i(t)$ if the initial current is 50 mA (i.e., $i(0) = 50 \times 10^{-3}$).

Figure 10.13 RL circuit for Example 10-7.

Solution

(a) **Transient Solution:** The transient solution is obtained by setting the right-hand side of the differential equation to zero as

$$0.1 \frac{di(t)}{dt} + 100\, i(t) = 0, \tag{10.69}$$

and assuming a solution of the form

$$i_{tran}(t) = ce^{st}. \tag{10.70}$$

The constant s is determined by substituting $i_{tran}(t)$ and its derivative into equation (10.69) and solving for s as

$$0.1\,(cse^{st}) + 100\,(ce^{st}) = 0$$

$$ce^{st}(0.1s + 100) = 0$$

$$0.1s + 100 = 0$$

$$s = -1000. \tag{10.71}$$

Substituting the value of s into equation (10.70), the transient solution of the current is given by

$$i_{tran}(t) = ce^{-1000t}. \tag{10.72}$$

The constant c depends on the initial current flowing through the circuit and will be obtained by applying the initial condition to the total solution in part (d).

(b) **Steady-State Solution:** Because the voltage applied to the RL circuit is constant (the right-hand side of equation (10.68) is constant), the steady-state solution is assumed to be

$$i_{ss}(t) = E, \tag{10.73}$$

where E is a constant. The value of E can be determined by substituting $i_{ss}(t)$ and its derivative into equation (10.68), which gives

$$0.1(0) + 100\,(E) = 10.$$

Solving for E gives

$$E = 0.1 \text{ A}.$$

Thus, the steady-state solution for the current is given by

$$i_{ss}(t) = 0.1 \text{ A}. \tag{10.74}$$

(c) **Total Solution:** The total solution is obtained by adding the transient and the steady-state solutions given by equations (10.72) and (10.74) as

$$i(t) = ce^{-1000t} + 0.1 \text{ A}. \tag{10.75}$$

(d) **Initial Conditions:** The constant c can now be obtained by applying the initial condition $i(0) = 50$ mA into equation (10.75) as

$$i(0) = ce^{-1000\,(0)} + 0.1 = 0.05$$

or

$$c\,(1) + 0.1 = 0.05.$$

Solving for c gives

$$c = -0.05. \tag{10.76}$$

Substituting the value of c into equation (10.75) gives

$$i(t) = -0.05\,e^{-1000t} + 0.1$$

or

$$i(t) = 0.1\left(1 - 0.5\,e^{-1000t}\right) \text{ A}. \tag{10.77}$$

Note that as $t \rightarrow \infty, i(t) \rightarrow 0.1 = 100\,\text{mA}$ (i.e., the current reaches its steady-state solution). It takes the current $t = \tau = 1/1000 = 1$ ms to reach $0.1(1 - 0.5 \times 0.368) = 0.0816$ A or $81.6\,\text{mA}$. The value of the current at $t = \tau$ can also be found from the expression: *Initial value* $+ 0.632 \times$ (*Steady-state value* − *Initial value*) or $50 + 0.632 \times (100 - 50) = 81.6$ mA. The plot of the current $i(t)$ is shown in Fig. 10.14. It can be seen from this figure that the current $i(t)$ takes approximately $5\tau = 5$ ms to reach the final value.

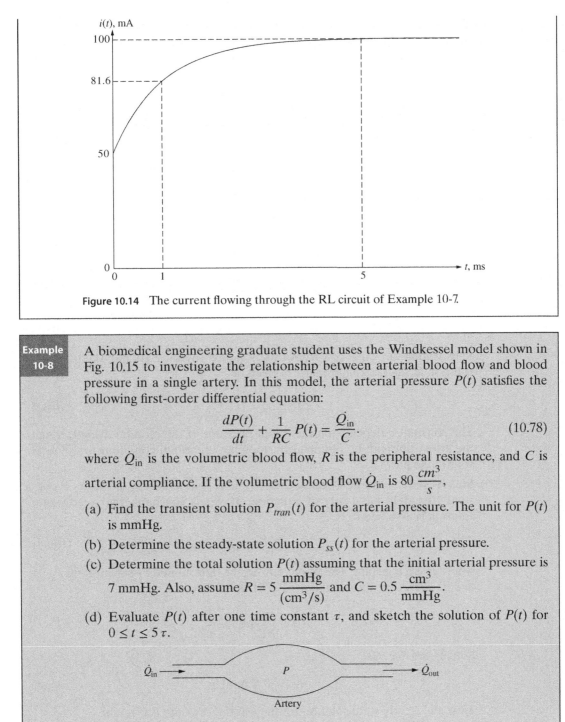

Figure 10.14 The current flowing through the RL circuit of Example 10-7.

Example 10-8

A biomedical engineering graduate student uses the Windkessel model shown in Fig. 10.15 to investigate the relationship between arterial blood flow and blood pressure in a single artery. In this model, the arterial pressure $P(t)$ satisfies the following first-order differential equation:

$$\frac{dP(t)}{dt} + \frac{1}{RC} P(t) = \frac{\dot{Q}_{\text{in}}}{C}. \tag{10.78}$$

where $\dot{Q}_{\text{in}}$ is the volumetric blood flow, R is the peripheral resistance, and C is arterial compliance. If the volumetric blood flow $\dot{Q}_{\text{in}}$ is $80 \ \frac{cm^3}{s}$,

(a) Find the transient solution $P_{tran}(t)$ for the arterial pressure. The unit for $P(t)$ is mmHg.

(b) Determine the steady-state solution $P_{ss}(t)$ for the arterial pressure.

(c) Determine the total solution $P(t)$ assuming that the initial arterial pressure is 7 mmHg. Also, assume $R = 5 \ \frac{\text{mmHg}}{(cm^3/s)}$ and $C = 0.5 \ \frac{cm^3}{\text{mmHg}}$.

(d) Evaluate $P(t)$ after one time constant τ, and sketch the solution of $P(t)$ for $0 \le t \le 5\tau$.

Figure 10.15 Windkessel model.

Solution (a) **Transient Solution:** The transient solution is obtained by setting the right-hand side of the differential equation (10.78) to zero as

$$\frac{dP(t)}{dt} + \frac{1}{RC} P(t) = 0 \tag{10.79}$$

and assuming a solution of the form

$$P_{tran}(t) = ce^{st}. \tag{10.80}$$

The constant s is determined by substituting $P_{tran}(t)$ and its derivative into equation (10.79) and solving for s as

$$(cse^{st}) + \frac{1}{RC}(ce^{st}) = 0$$

$$ce^{st}(s + \frac{1}{RC}) = 0$$

$$s + \frac{1}{RC} = 0$$

$$s = -\frac{1}{RC}. \tag{10.81}$$

Substituting the value of s from equation into equation (10.80), the transient solution of the arterial pressure is given by

$$P_{tran}(t) = ce^{-\frac{1}{RC}t}. \tag{10.82}$$

The constant c depends on the initial pressure of the blood flowing through the artery and will be obtained by applying the initial condition to the total solution in part (d).

(b) **Steady-State Solution:** Because the volumetric blood flow in the artery is constant (the right-hand side of equation (10.78) is constant), the steady-state solution is assumed to be

$$P_{ss}(t) = K \tag{10.83}$$

where K is a constant. The value of K can be determined by substituting $P_{ss}(t)$ and its derivative into equation (10.78), which gives

$$(0) + \frac{1}{RC}(K) = \frac{\dot{Q}}{C}.$$

Solving for K gives

$$K = 80\,R \text{ mmHg}.$$

Thus, the steady-state solution for the arterial pressure is given by

$$P_{ss}(t) = 80\,R \text{ mmHg}. \tag{10.84}$$

(c) **Total Solution:** The total solution is obtained by adding the transient and the steady-state solutions given by equations (10.82) and (10.84) as

$$P(t) = ce^{-\frac{1}{RC}t} + 80\,R \text{ mmHg}. \tag{10.85}$$

(d) **Initial Conditions:** The constant c can now be obtained by applying the initial condition $P(0) = 7$ mmHg into equation (10.85) as

$$P(0) = ce^{-\frac{1}{RC}(0)} + 80\,R = 7.$$

Substituting $R = 5$ and $C = 0.5$ gives

$$c(1) + 400 = 7.$$

Solving for c gives

$$c = -393. \tag{10.86}$$

Substituting the value of c into equation (10.85) gives

$$P(t) = -393\,e^{-0.4t} + 400$$

or

$$P(t) = 400\left(1 - 0.9825\,e^{-0.4t}\right) \text{ mmHg}. \tag{10.87}$$

Note that as $t \to \infty, P(t) \to 400$ mmHg (i.e., the pressure reaches its steady-state solution). It takes the pressure $t = \tau = 1/0.4 = 2.5$ s to reach $400(1 - 0.9825(0.368)) = 255.4$ mmHg. The value of the pressure at $t = \tau$ can also be found from the expression: *Initial value* $+ 0.632 \times$ (*Steady-state value* $-$ *Initial value*) or $7 + 0.632 \times (400 - 7) = 255.4$ mmHg. The plot of the current $P(t)$ is shown in Fig. 10.16. It can be seen from this figure that it takes the pressure $P(t)$ approximately $5\tau = 12.5$ s to reach its final value.

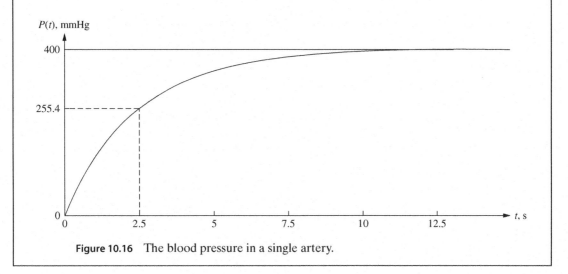

Figure 10.16 The blood pressure in a single artery.

Example
10-9
The differential equation for the RC circuit of Fig. 10.5 is given by

$$RC \frac{dv(t)}{dt} + v(t) = v_s(t). \tag{10.88}$$

Find the output voltage $v(t)$ if $v_s(t) = V \sin \omega t$ and the initial voltage is zero (i.e., $v(0) = 0$).

Solution

(a) **Transient Solution:** The transient solution $v_{tran}(t)$ for the differential equation (10.88) is the same as that found in Example 10-3 and is given by

$$v_{tran}(t) = ce^{-\frac{1}{RC}t}. \tag{10.89}$$

The constant c will be determined by applying the initial condition to the total solution in part (d).

(b) **Steady-State Solution:** Since $v_s(t) = V \sin \omega t$, equation (10.88) can be written as

$$RC \dot{v}(t) + v(t) = V \sin \omega t. \tag{10.90}$$

According to Table 10.1, the steady-state solution of the output voltage $v(t)$ has the form

$$v_{ss}(t) = A \sin \omega t + B \cos \omega t \tag{10.91}$$

where A and B are constants to be determined. The values of A and B can be found by substituting $v_{ss}(t)$ and its derivative into equation (10.90). The derivative of $v_{ss}(t)$ is obtained by differentiating equation (10.91), which gives

$$\dot{v}_{ss}(t) = A\omega \cos \omega t - B \omega \sin \omega t. \tag{10.92}$$

Substituting equations (10.91) and (10.92) into equation (10.90) gives

$$RC(A\omega \cos \omega t - B \omega \sin \omega t) + A \sin \omega t + B \cos \omega t = V \sin \omega t. \tag{10.93}$$

Grouping like terms in equation (10.93) yields

$$(-RCB\omega + A) \sin \omega t + (RCA\omega + B) \cos \omega t = V \sin \omega t. \tag{10.94}$$

Comparing the coefficients of $\sin \omega t$ on both sides of equation (10.94) gives

$$-RCB\omega + A = V. \tag{10.95}$$

Similarly, comparing the coefficients of $\cos \omega t$ on both sides of equation (10.94) gives

$$RCA\omega + B = 0. \tag{10.96}$$

Equations (10.95) and (10.96) represent a 2×2 system of equations for the two unknowns A and B. These equations can be solved using one of the methods discussed in Chapter 7 and are given by

$$A = \frac{V}{1 + (RC\omega)^2} \tag{10.97}$$

$$B = \frac{-RC\omega V}{1 + (RC\omega)^2}. \tag{10.98}$$

Substituting A and B from equations (10.97) and (10.98) into equation (10.92) yields

$$v_{ss}(t) = \left(\frac{V}{1 + (RC\omega)^2}\right) \sin \omega t + \left(\frac{-RC\omega V}{1 + (RC\omega)^2}\right) \cos \omega t$$

or

$$v_{ss}(t) = \frac{V}{1 + (RC\omega)^2} (\sin \omega t - RC\omega \cos \omega t). \tag{10.99}$$

As discussed in Chapter 6, summing sinusoids of the same frequency gives

$$\sin \omega t - RC\omega \cos \omega t = \sqrt{1 + (RC\omega)^2} \sin(\omega t + \phi), \tag{10.100}$$

where $\phi = \text{atan2}(-RC\omega, 1) = -\tan^{-1}(RC\omega)$. Substituting equation (10.100) into equation (10.99) gives the steady-state solution as

$$v_{ss}(t) = \left(\frac{V}{1 + (RC\omega)^2}\right) \left(\sqrt{1 + (RC\omega)^2} \sin(\omega t + \phi)\right)$$

or

$$v_{ss}(t) = \left(\frac{V}{\sqrt{1 + (RC\omega)^2}}\right) \sin(\omega t + \phi). \tag{10.101}$$

(c) **Total Solution:** The total solution is obtained by adding the transient and the steady-state solutions given by equations (10.89) and (10.101) as

$$v(t) = ce^{-\frac{1}{RC}t} + \left(\frac{V}{\sqrt{1 + (RC\omega)^2}}\right) \sin(\omega t + \phi). \tag{10.102}$$

(d) **Initial Conditions:** The constant c can now be obtained by applying the initial condition $v(0) = 0$ to equation (10.102) as

$$v(0) = c\,(1) + \left(\frac{V}{\sqrt{1 + (RC\omega)^2}}\right) \sin \phi = 0$$

or

$$c = -\left(\frac{V}{\sqrt{1 + (RC\omega)^2}}\right) \sin \phi. \tag{10.103}$$

Since $\phi = -\tan^{-1}(RC\omega)$, the value of $\sin \phi$ can be found from the fourth-quadrant triangle shown in Fig. 10.17 as

$$\sin \phi = \frac{-RC\omega}{\sqrt{1+(RC\omega)^2}} \qquad (10.104)$$

Figure 10.17 Fourth-quadrant triangle to find $\sin \phi$.

Substituting $\sin \phi$ from equation (10.104) into equation (10.103) yields

$$c = -\left(\frac{V}{\sqrt{1+(RC\omega)^2}}\right)\left(\frac{-RC\omega}{\sqrt{1+(RC\omega)^2}}\right)$$

or

$$c = \frac{RC\omega V}{1+(RC\omega)^2}. \qquad (10.105)$$

Substituting the value of c from equation (10.105) into equation (10.102) gives

$$v(t) = \frac{RC\omega V}{1+(RC\omega)^2} e^{-\frac{1}{RC}t} + \frac{V}{\sqrt{1+(RC\omega)^2}} \sin(\omega t + \phi). \qquad (10.106)$$

Note that as $t \to \infty$, the total solution reaches the steady-state solution. Thus, the amplitude of the output voltage as $t \to \infty$ is given by

$$|v(t)| = \frac{V}{\sqrt{1+(RC\omega)^2}}. \qquad (10.107)$$

The amplitude of the input voltage $v_s(t) = V \sin(\omega t)$ is given by

$$|v_s(t)| = V. \qquad (10.108)$$

Dividing the amplitude of the output at steady-state (equation (10.107)) by the amplitude of the input (equation (10.108)) gives

$$\frac{|v(t)|}{|v_s(t)|} = \frac{\dfrac{V}{\sqrt{1+(RC\omega)^2}}}{V}$$

or

$$\frac{|v(t)|}{|v_s(t)|} = \frac{1}{\sqrt{1+(RC\omega)^2}}. \qquad (10.109)$$

Note that as $\omega \rightarrow 0$,

$$\frac{|v(t)|}{|v_s(t)|} \rightarrow 1.$$

This means that for low-frequency input, the amplitude of the output is about the same as the input. However, as $\omega \rightarrow \infty$,

$$\frac{|v(t)|}{|v_s(t)|} \rightarrow 0.$$

This means that for high-frequency input, the amplitude of the output is close to zero.

The RC circuit shown in Fig. 10.5 is known as a **low-pass filter**, because it passes the low-frequency inputs but filters out the high-frequency inputs. This will be further illustrated in Example 10-10.

Example 10-10

Consider the low-pass filter of the previous example with $RC = 0.5$ and $V = 10$ V.

(a) Find the total solution $v(t)$.

(b) Find the ratio $\dfrac{|v(t)|}{|v_s(t)|}$ as $\omega \rightarrow \infty$. Also plot the steady-state output for both $\omega = 0.1$ rad/s and 10 rad/s.

Solution

(a) The total solution for the output voltage is obtained by substituting $RC = 0.5$ and $V = 10$ into equation (10.106) as

$$v(t) = \frac{5\omega}{1 + (0.5\omega)^2} e^{-2t} + \frac{10}{\sqrt{1 + (0.5\omega)^2}} \sin\left(\omega t - \tan^{-1} 0.5\,\omega\right). \quad (10.110)$$

(b) For $\omega = 0.1$ rad/s, the ratio $\dfrac{|v(t)|}{|v_s(t)|}$ can be found from equation (10.109) as

$$\frac{|v(t)|}{|v_s(t)|} = \frac{1}{\sqrt{1 + (0.5 * 0.1)^2}} = 0.9988.$$

For $\omega = 10$ rad/s, the ratio is given by

$$\frac{|v(t)|}{|v_s(t)|} = \frac{1}{\sqrt{1 + (0.5 * 10)^2}} = 0.1961.$$

As $\omega \rightarrow \infty$, the ratio $\dfrac{|v(t)|}{|v_s(t)|} \rightarrow 0.$

The plots of the output for $\omega = 0.1$ and 10 rad/s are shown in Figs. 10.18 and 10.19, respectively. It can be seen that as ω increases from 0.1 to 10 rad/s, the amplitude of the steady-state output decreases from 10 * (0.9988) = 9.988 V to 10 * (0.1961) =

1.961 V. It can be seen from equation (10.109) that if $\omega \rightarrow \infty$, the amplitude of the output will approach zero.

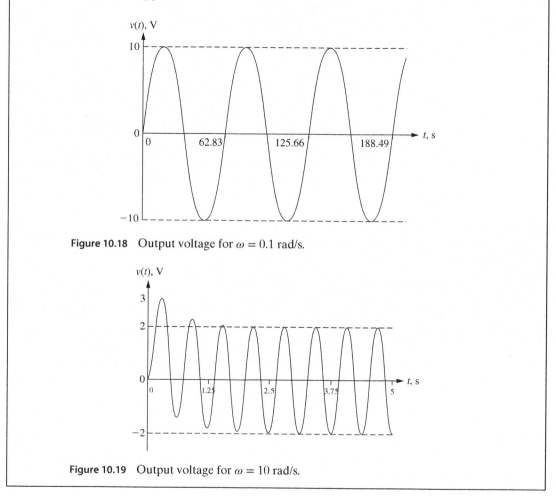

Figure 10.18 Output voltage for $\omega = 0.1$ rad/s.

Figure 10.19 Output voltage for $\omega = 10$ rad/s.

10.5 SECOND-ORDER DIFFERENTIAL EQUATIONS

10.5.1 Free Vibration of a Spring-Mass System

Consider a spring-mass system in the vertical plane, as shown in Fig. 10.20, where k is the spring constant, m is the mass, and $y(t)$ is the position measured from equilibrium.

At the equilibrium position, the external forces on the block are shown in the free-body diagram (FBD) of Fig. 10.21, where δ is the equilibrium elongation of the spring, mg is the force due to gravity, and $k\delta$ is the restoring force in the spring.

From equilibrium of forces in the y-direction,

$$k\delta = mg,$$

Figure 10.20   Mass-spring system.

Figure 10.21   Free-body diagram of the mass-spring system with no motion.

which gives

$$\delta = \frac{mg}{k}. \tag{10.111}$$

This equilibrium elongation δ is also called the static deflection. Now, if the mass is displaced from its equilibrium position by the amount $y(t)$, the FBD of the system is as shown in Fig. 10.22. Since the system is no longer in equilibrium, **Newton's second**

Figure 10.22   Free-body diagram of the mass-spring system displaced from equilibrium.

law ($\sum F = m\,a$) can be used to write the equation of motion as

$$\sum F_y = m\,a = m\,\ddot{y}(t).$$

Summing the forces in Fig. 10.22 gives

$$mg - k(\delta + y(t)) = m\,\ddot{y}(t)$$

or

$$mg - k\,\delta - k\,y(t) = m\,\ddot{y}(t). \tag{10.112}$$

Substituting δ from equation (10.111) gives

$$mg - k\left(\frac{mg}{k}\right) - k\,y(t) = m\,\ddot{y}(t)$$

or

$$-k\,y(t) = m\,\ddot{y}(t),$$

which gives

$$m\,\ddot{y}(t) + k\,y(t) = 0. \tag{10.113}$$

Equation (10.113) is a second-order differential equation for the displacement $y(t)$ of the spring-mass system shown in Fig. 10.20.

Example 10-11   Find the solution of equation (10.113) if the mass is subjected to an initial displacement of $y(0) = A$ and let go. Note that the initial velocity is zero ($\dot{y}(0) = 0$).

Solution   (a) **Transient Solution:** Since the right-hand side of the equation is zero, assume a transient solution of the form

$$y_{tran}(t) = ce^{st}.$$

The first and second derivatives of the transient solution are given by

$$\dot{y}_{tran}(t) = cse^{st}$$
$$\ddot{y}_{tran}(t) = cs^2e^{st}.$$

Substituting the transient solution and its derivatives into equation (10.113) yields

$$m(cs^2\,e^{st}) + k(ce^{st}) = 0.$$

Factoring out e^{st} gives

$$ce^{st}(ms^2 + k) = 0$$

which implies that

$$ms^2 + k = 0. \tag{10.114}$$

Solving for s yields

$$s^2 = -\frac{k}{m}$$

which gives

$$s = \pm\sqrt{-\frac{k}{m}}$$

or

$$s = 0 \pm j \sqrt{\frac{k}{m}}$$

where $j = \sqrt{-1}$. The two roots of the characteristic equation (10.114) are thus $s_1 = +j \sqrt{\frac{k}{m}}$ and $s_2 = -j \sqrt{\frac{k}{m}}$. Therefore, the transient solution is given by

$$y_{tran}(t) = c_1 e^{s_1 t} + c_2 e^{s_2 t}$$

or

$$y_{tran}(t) = c_1 e^{j \sqrt{\frac{k}{m}} t} + c_2 e^{-j \sqrt{\frac{k}{m}} t} \tag{10.115}$$

where c_1 and c_2 are constants. Using Euler's formula $e^{j\theta} = \cos \theta + j \sin \theta$, equation (10.115) can be written as

$$y_{tran}(t) = c_1 \left(\cos \sqrt{\frac{k}{m}} t + j \sin \sqrt{\frac{k}{m}} t \right) +$$

$$c_2 \left[\cos \left(-\sqrt{\frac{k}{m}} t \right) + j \sin \left(-\sqrt{\frac{k}{m}} t \right) \right]. \tag{10.116}$$

Since $\cos(-\theta) = \cos(\theta)$ and $\sin(-\theta) = -\sin(\theta)$, equation (10.116) can be written as

$$y_{tran}(t) = c_1 \left(\cos \sqrt{\frac{k}{m}} t + j \sin \sqrt{\frac{k}{m}} t \right) +$$

$$c_2 \left(\cos \sqrt{\frac{k}{m}} t - j \sin \sqrt{\frac{k}{m}} t \right)$$

or

$$y_{tran}(t) = (c_1 + c_2) \cos \sqrt{\frac{k}{m}} t + j(c_1 - c_2) \sin \sqrt{\frac{k}{m}} t.$$

This can be further simplified as

$$y_{tran}(t) = c_3 \cos \sqrt{\frac{k}{m}} t + c_4 \sin \sqrt{\frac{k}{m}} t \tag{10.117}$$

where $c_3 = c_1 + c_2$ and $c_4 = j(c_1 - c_2)$ are real constants. Note that the constants c_1 and c_2 must be complex conjugates for $y_{tran}(t)$ to be real. Therefore, the transient solution of a mass-spring system can be written in terms of sines and cosines with natural frequency $\omega_n = \sqrt{\frac{k}{m}}$ rad/s.

(b) **Steady-State Solution:** Since the RHS of equation (10.113) is already zero (no forcing function), the steady-state solution is zero, for example,

$$y_{ss}(t) = 0. \tag{10.118}$$

(c) **Total Solution:** The total solution for the displacement $y(t)$ can be found by adding the transient and steady-state solutions from equations (10.117) and (10.118), which gives

$$y(t) = c_3 \cos \sqrt{\frac{k}{m}} \, t + c_4 \sin \sqrt{\frac{k}{m}} \, t. \qquad (10.119)$$

(d) **Initial Conditions:** The constants c_3 and c_4 are determined from using the initial conditions $y(0) = A$ and $\dot{y}(0) = 0$. Substituting $y(0) = A$ in equation (10.119) gives

$$y(0) = c_3 \cos(0) + c_4 \sin(0) = A \qquad (10.120)$$

or

$$c_3 \, (1) + c_4 \, (0) = A,$$

which gives

$$c_3 = A.$$

Thus, the displacement of the mass is given by

$$y(t) = A \cos \sqrt{\frac{k}{m}} \, t + c_4 \sin \sqrt{\frac{k}{m}} \, t. \qquad (10.121)$$

The velocity of the mass can be found by differentiating $y(t)$ in equation (10.121) as

$$\dot{y}(t) = -A \sqrt{\frac{k}{m}} \sin \sqrt{\frac{k}{m}} \, t + c_4 \sqrt{\frac{k}{m}} \cos \sqrt{\frac{k}{m}} \, t. \qquad (10.122)$$

The constant c_4 can now be found by substituting $\dot{y}(0) = 0$ in equation (10.122) as

$$\dot{y}(0) = -A \sqrt{\frac{k}{m}} \sin(0) + c_4 \sqrt{\frac{k}{m}} \cos(0) = 0$$

or

$$-A \, (0) + c_4 \left(\sqrt{\frac{k}{m}} \right) = 0,$$

which gives

$$c_4 = 0.$$

Thus, the total solution for the displacement is given by

$$y(t) = A \cos \sqrt{\frac{k}{m}} \, t$$

or

$$y(t) = A \cos \omega_n t.$$

The plot of the displacement $y(t)$ is shown in Fig. 10.23. It can be seen that the amplitude of the displacement is simply the initial displacement A, and the block oscillates at a frequency of $\omega_n = \sqrt{\frac{k}{m}}$. Note that the natural frequency is proportional to the square root of the spring constant and is inversely proportional to the square root of the mass (i.e., the natural frequency increases with stiffness and decreases with mass). This is a general result for free vibration of mechanical systems.

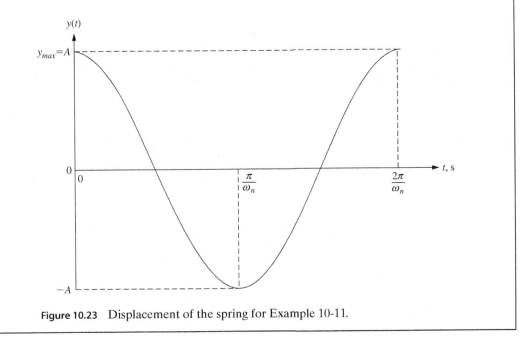

Figure 10.23 Displacement of the spring for Example 10-11.

10.5.2 Forced Vibration of a Spring-Mass System

Suppose the spring-mass system is subjected to an applied force $f(t)$, as shown in Fig. 10.24.

Figure 10.24 Spring-mass system subjected to applied force.

In this case, the derivation of the governing equation includes an additional force $f(t)$ on the right-hand side. Thus, the equation of motion of the system can be written as

$$m\,\ddot{y}(t) + k\,y(t) = f(t). \qquad (10.123)$$

Equation (10.123) is a second-order differential equation for the displacement $y(t)$ of a mass-spring system subjected to a force $f(t)$.

Example 10-12

Find the solution to equation (10.123) if $f(t) = F\cos\omega t$ and $y(0) = \dot{y}(0) = 0$. Also, investigate the response as $\omega \rightarrow \sqrt{\dfrac{k}{m}}$.

Solution

(a) **Transient Solution:** The transient solution is obtained by setting $f(t) = 0$, which gives

$$m\,\ddot{y}(t) + k\,y(t) = 0.$$

This is the same as equation (10.113) for free vibration. Hence, the transient solution is given by equation (10.117) as

$$y_{tran}(t) = c_3 \cos\sqrt{\frac{k}{m}}\,t + c_4 \sin\sqrt{\frac{k}{m}}\,t, \qquad (10.124)$$

where c_3 and c_4 are real constants to be determined.

(b) **Steady-State Solution:** Since the forcing function is $f(t) = F\cos\omega t$, the steady-state solution is of the form

$$y_{ss}(t) = A\sin\omega t + B\cos\omega t. \qquad (10.125)$$

The first and second derivatives of the steady-state solution are thus

$$\dot{y}_{ss}(t) = A\,\omega\cos\omega t - B\,\omega\sin\omega t$$
$$\ddot{y}_{ss}(t) = -A\,\omega^2 \sin\omega t - B\,\omega^2 \cos\omega t. \qquad (10.126)$$

Substituting $\ddot{y}_{ss}(t)$, $y_{ss}(t)$, and $f(t) = F\cos(\omega t)$ into equation (10.123) gives

$$m(-A\,\omega^2 \sin\omega t - B\,\omega^2 \cos\omega t) + k(A\sin\omega t + B\cos\omega t) = F\cos\omega t.$$

Grouping like terms yields

$$A(k - m\omega^2)\sin\omega t + B(k - m\omega^2)\cos\omega t = F\cos(\omega t). \qquad (10.127)$$

Equating the coefficients of $\sin\omega t$ on both sides of equation (10.127) yields

$$A(k - m\omega^2) = 0,$$

which gives

$$A = 0 \quad \left(\text{provided } \omega \neq \sqrt{\frac{k}{m}}\right).$$

Similarly, equating the coefficients of $\cos \omega t$ on both sides of equation (10.127) yields

$$B(k - m\omega^2) = F,$$

which gives

$$B = \frac{F}{k - m\omega^2} \qquad \left(\text{provided } \omega \neq \sqrt{\frac{k}{m}}\right).$$

Therefore, the steady-state solution is given by

$$y_{ss}(t) = \left(\frac{F}{k - m\omega^2}\right) \cos \omega t. \tag{10.128}$$

(c) **Total Solution:** The total solution for $y(t)$ is obtained by adding the transient and steady-state solutions from equations (10.124) and (10.128) as

$$y(t) = c_3 \cos \sqrt{\frac{k}{m}}\, t + c_4 \sin \sqrt{\frac{k}{m}}\, t + \left(\frac{F}{k - m\omega^2}\right) \cos \omega t. \tag{10.129}$$

(d) **Initial Conditions:** The constants c_3 and c_4 are determined from the initial conditions $y(0) = 0$ and $\dot{y}(0) = 0$. The velocity of the mass can be obtained by differentiating equation (10.129) as

$$\dot{y}(t) = -c_3 \sqrt{\frac{k}{m}} \sin \sqrt{\frac{k}{m}}\, t + c_4 \sqrt{\frac{k}{m}} \cos \sqrt{\frac{k}{m}}\, t -$$
$$\omega \left(\frac{F}{k - m\omega^2}\right) \sin \omega t. \tag{10.130}$$

Substituting $y(0) = 0$ in equation (10.129) gives

$$y(0) = c_3 \cos(0) + c_4 \sin(0) + \left(\frac{F}{k - m\omega^2}\right) \cos(0) = 0$$

or

$$c_3 (1) + c_4 (0) + \left(\frac{F}{k - m\omega^2}\right) (1) = 0,$$

which gives

$$c_3 = -\frac{F}{k - m\omega^2}.$$

Similarly, substituting $\dot{y}(0) = 0$ in equation (10.130) yields

$$\dot{y}(0) = -c_3 (0) + c_4 \sqrt{\frac{k}{m}} \cos(0) - \omega \left(\frac{F}{k - m\omega^2}\right) \sin(0) = 0$$

or

$$c_3 (0) + c_4 \sqrt{\frac{k}{m}} (1) - \omega \left(\frac{F}{k - m\omega^2}\right) (0) = 0$$

which gives

$$c_4 = 0.$$

Thus, the displacement of the mass is given by

$$y(t) = -\left(\frac{F}{k - m\omega^2}\right) \cos \sqrt{\frac{k}{m}}\, t + \left(\frac{F}{k - m\omega^2}\right) \cos \omega t \qquad (10.131)$$

or

$$y(t) = \left(\frac{F}{k - m\omega^2}\right)\left(\cos \omega t - \cos \sqrt{\frac{k}{m}}\, t\right). \qquad (10.132)$$

Note that the results obtained above assumed that $\omega \neq \sqrt{\frac{k}{m}}$. But nevertheless we can investigate the behavior as ω gets very close to $\sqrt{\frac{k}{m}}$.

What is the response of $y(t)$ as $\omega \to \sqrt{\frac{k}{m}}$?   As $\omega \to \sqrt{\frac{k}{m}}$,

$$y(t) \to \left(\frac{F}{0}\right)\left(\cos \sqrt{\frac{k}{m}}\, t - \cos \sqrt{\frac{k}{m}}\, t\right)$$

$$= \frac{0}{0}.$$

This is an "indeterminate" form and can be evaluated by methods of calculus not yet available to all students. However, the result can be investigated by picking values of ω close to $\sqrt{\frac{k}{m}}$ and plotting the results. For example, let $k = m = F = 1$, and choose the values of $\omega = 0.9\sqrt{\frac{k}{m}}$, $\omega = 0.99\sqrt{\frac{k}{m}}$, and $\omega = 0.9999\sqrt{\frac{k}{m}}$. The plots of equation (10.132) for these values are as shown in Figs. 10.25, 10.26, and 10.27,

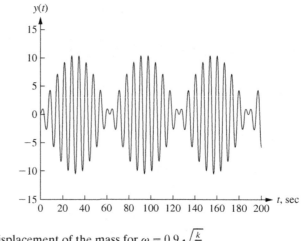

Figure 10.25   Displacement of the mass for $\omega = 0.9\sqrt{\frac{k}{m}}$.

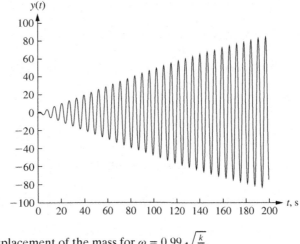

Figure 10.26   Displacement of the mass for $\omega = 0.99 \sqrt{\frac{k}{m}}$.

respectively. The plot for $\omega = 0.9 \sqrt{\frac{k}{m}}$ in Fig. 10.25 shows the "beating" phenomenon typical of problems where the forcing frequency ω is in the neighborhood of the natural frequency $\sqrt{\frac{k}{m}}$. As ω is increased to $0.99 \sqrt{\frac{k}{m}}$ and $0.9999 \sqrt{\frac{k}{m}}$, Figs. 10.26 and 10.27 show $y(t)$ increasing without bound. This is called **resonance** and is generally undesirable in mechanical systems.

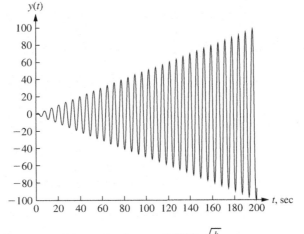

Figure 10.27   Displacement of the spring for $\omega = 0.9999 \sqrt{\frac{k}{m}}$.

Example 10-13

A biomedical engineer is designing a resistive training device to strengthen the latissimus dorsi muscle. The task can be represented as a spring-mass system, as shown in Fig. 10.28. The displacement $y(t)$ of the exercise bar satisfies the second-order differential equation

$$m\,\ddot{y}(t) + k\,y(t) = f(t) \tag{10.133}$$

subject to the initial condition $y(0) = E$ and $\dot{y}(0) = 0$.

(a) Determine the transient solution $y_{tran}(t)$.

(b) Determine the steady-state solution $y_{ss}(t)$ for the applied force shown in Fig. 10.29.

(c) Determine the total solution, subject to the initial conditions.

Figure 10.28 Spring-mass model of resistive training device.

Figure 10.29 Applied force for resistive training device.

Solution

(a) **Transient Solution:** The transient solution is obtained by setting the RHS of equation (10.133) equal to zero as

$$m\,\frac{d^2y(t)}{dt^2} + k\,y(t) = 0 \tag{10.134}$$

and assuming a solution of the form

$$y_{tran}(t) = e^{st}.$$

Substituting the transient solution and its second derivative into equation (10.134) yields

$$m(s^2 e^{st}) + k(e^{st}) = 0.$$

Factoring out e^{st} gives

$$e^{st}(m s^2 + k) = 0,$$

which implies

$$m s^2 + k = 0. \qquad (10.135)$$

Solving for s yields

$$s^2 = -\frac{k}{m}$$

or

$$s = \pm j \sqrt{\frac{k}{m}}.$$

The two roots of equation (10.135) are $s_1 = +j\sqrt{\frac{k}{m}}$ and $s_2 = -j\sqrt{\frac{k}{m}}$. Thus, the transient solution is given by

$$y_{tran}(t) = c_3 \cos\sqrt{\frac{k}{m}}\, t + c_4 \sin\sqrt{\frac{k}{m}}\, t, \qquad (10.136)$$

where c_3 and c_4 are real constants and $\sqrt{\frac{k}{m}}$ is the natural frequency ω_n.

(b) **Steady-State Solution:** Since the forcing function is $f(t) = H t/2$, the steady-state solution is of the form

$$y_{ss}(t) = A t + B. \qquad (10.137)$$

Substituting the $\ddot{y}_{ss}(t)$ and $y_{ss}(t)$ into equation (10.133) gives

$$m \times 0 + k(A t + B) = \frac{H}{2}t. \qquad (10.138)$$

Equating the coefficients of t on both sides of equation (10.138) yields

$$kA = \frac{H}{2}$$

which gives

$$A = \frac{H}{2k}.$$

Similarly, equating the constant coefficients on both sides of equation (10.138) yields

$$B = 0.$$

Thus, the steady-state solution is given by

$$y_{ss}(t) = \left(\frac{H}{2k}\right) t. \tag{10.139}$$

(c) **Total Solution:** The total solution of the displacement $y(t)$ can be found by adding the transient and steady-state solutions from equations (10.136) and (10.139), which gives

$$y(t) = c_3 \cos \sqrt{\frac{k}{m}}\, t + c_4 \sin \sqrt{\frac{k}{m}}\, t + \left(\frac{H}{2k}\right) t. \tag{10.140}$$

The constants c_3 and c_4 are determined using the initial conditions $y(0) = E$ and $\dot{y}(0) = 0$. Substituting $y(0) = E$ into equation (10.140) yields

$$y(0) = c_3 \cos(0) + c_4 \sin(0) + \left(\frac{H}{2k}\right)(0) = E$$

or

$$c_3(1) + c_4(0) + 0 = E,$$

which gives

$$c_3 = E.$$

The derivative of $y(t)$ is obtained by differentiating equation (10.140) as

$$\dot{y}(t) = -c_3 \sqrt{\frac{k}{m}} \sin \sqrt{\frac{k}{m}}\, t + c_4 \sqrt{\frac{k}{m}} \cos \sqrt{\frac{k}{m}}\, t + \frac{H}{2k} \tag{10.141}$$

Substituting $\dot{y}(0) = 0$ in equation (10.141) yields

$$0 = -c_3(0) + c_4 \sqrt{\frac{k}{m}} + \left(\frac{H}{2k}\right)$$

which gives

$$c_4 = -\frac{H}{2k\sqrt{k/m}}.$$

Thus, the displacement of the exercise bar is given by

$$y(t) = E \cos \sqrt{\frac{k}{m}}\, t - \left(\frac{H}{2k\sqrt{k/m}}\right) \sin \sqrt{\frac{k}{m}}\, t + \frac{H}{2k}\, t.$$

10.5.3   Second-Order LC Circuit

A source voltage $v_s(t)$ is applied to an LC circuit, as shown in Fig. 10.30. Applying KVL to the circuit gives

$$v_L(t) + v(t) = v_s \tag{10.142}$$

Figure 10.30   Voltage applied to an LC circuit.

where $v_L(t) = L \frac{di(t)}{dt}$ is the voltage across the inductor. Since the current flowing through the circuit is given by $i(t) = C \frac{dv(t)}{dt}$, $v_L(t)$ can be written as $v_L(t) = LC \frac{d^2v(t)}{dt^2}$. Substituting $v_L(t)$ into equation (10.142) yields

$$LC \frac{d^2v(t)}{dt^2} + v(t) = v_s(t). \tag{10.143}$$

Equation (10.143) is a second-order differential equation for an LC circuit subjected to forcing function $v_s(t)$.

Example 10-14	Suppose the LC circuit of Fig. 10.30 is subjected to a voltage source $v_s(t) = V \cos \omega t$. Solve the resulting differential equation $$LC \ddot{v}(t) + v(t) = V \cos \omega t$$ subject to the initial condition $v(0) = \dot{v}(0) = 0$. Note that since $i(t) = C \dfrac{dv}{dt}$, the condition $\dot{v}(0) = 0$ means the initial current is zero.

Solution   (a) **Transient Solution:** The transient solution is the solution obtained by setting the RHS of equation (10.143) equal to zero

$$LC \frac{d^2v(t)}{dt^2} + v(t) = 0, \tag{10.144}$$

and assuming a solution of the form

$$y_{tran}(t) = e^{st}.$$

Substituting the transient solution and its second derivative into the equation (10.144) yields

$$LC(s^2 e^{st}) + (e^{st}) = 0.$$

Factoring out e^{st} gives

$$e^{st}(LCs^2 + 1) = 0,$$

which implies

$$LCs^2 + 1 = 0. \tag{10.145}$$

Solving for s yields

$$s^2 = -\frac{1}{LC}$$

or

$$s = \pm j\sqrt{\frac{1}{LC}}.$$

The two roots of equation (10.145) are $s_1 = +j\sqrt{\frac{1}{LC}}$ and $s_2 = -j\sqrt{\frac{1}{LC}}$. Thus, the transient solution is given by

$$v_{tran}(t) = c_3 \cos \sqrt{\frac{1}{LC}}\, t + c_4 \sin \sqrt{\frac{1}{LC}}\, t, \qquad (10.146)$$

where c_3 and c_4 are real constants and $\sqrt{\frac{1}{LC}}$ is the natural frequency ω_n in rad/s.

(b) **Steady-State Solution:** Since the forcing function is $v_s(t) = V \cos \omega t$, the steady-state solution is of the form

$$v_{ss}(t) = A \sin \omega t + B \cos \omega t. \qquad (10.147)$$

Substituting the $\ddot{v}_{ss}(t)$ and $v_{ss}(t)$ into equation (10.143) gives

$$LC(-A\omega^2 \sin \omega t - B\omega^2 \cos \omega t) + (A \sin \omega t + B \cos \omega t) = V \cos\omega t.$$

Grouping like terms yields

$$A(1 - LC\omega^2)\sin\omega t + B(1 - LC\omega^2)\cos\omega t = V \cos\omega t. \qquad (10.148)$$

Equating the coefficients of $\sin\omega t$ on both sides of equation (10.148) yields

$$A(1 - LC\omega^2) = 0,$$

which gives

$$A = 0 \quad \left(\text{provided } \omega \neq \sqrt{\frac{1}{LC}}\right).$$

Similarly, equating the coefficients of $\cos\omega t$ on both sides of equation (10.148) yields

$$B(1 - LC\omega^2) = V$$

or

$$B = \frac{V}{1 - LC\omega^2} \quad \left(\text{provided } \omega \neq \sqrt{\frac{1}{LC}}\right).$$

Thus, the steady-state solution is given by

$$v_{ss}(t) = \left(\frac{V}{1 - LC\omega^2}\right)\cos \omega t. \qquad (10.149)$$

(c) **Total Solution:** The total solution for the voltage $v(t)$ can be found by adding the transient and steady-state solutions from equations (10.146) and (10.149), which gives

$$v(t) = c_3 \cos \sqrt{\frac{1}{LC}} \, t + c_4 \sin \sqrt{\frac{1}{LC}} \, t + \left(\frac{V}{1 - LC\omega^2} \right) \cos \omega t. \qquad (10.150)$$

(d) **Initial Conditions:** The constants c_3 and c_4 are determined using the initial conditions $v(0) = 0$ and $\dot{v}(0) = 0$. Substituting $v(0) = 0$ into equation (10.150) yields

$$v(0) = c_3 \cos(0) + c_4 \sin(0) + \left(\frac{V}{1 - LC\omega^2} \right) \cos(0) = 0$$

or

$$c_3 \, (1) + c_4 \, (0) + \left(\frac{V}{1 - LC\omega^2} \right) (1) = 0,$$

which gives

$$c_3 = -\frac{V}{1 - LC\omega^2}.$$

The derivative of $v(t)$ is obtained by differentiating equation (10.150) as

$$\dot{v}(t) = -c_3 \sqrt{\frac{1}{LC}} \sin \sqrt{\frac{1}{LC}} \, t + c_4 \sqrt{\frac{1}{LC}} \cos \sqrt{\frac{1}{LC}} \, t$$

$$-\omega \left(\frac{V}{1 - LC\omega^2} \right) \sin \omega t. \qquad (10.151)$$

Substituting $\dot{v}(0) = 0$ in equation (10.151) yields

$$\dot{v}(0) = -c_3 \, (0) + c_4 \sqrt{\frac{1}{LC}} \cos(0) - \omega \left(\frac{V}{1 - LC\omega^2} \right) \sin(0) = 0$$

or

$$c_3 \, (0) + c_4 \sqrt{\frac{1}{LC}} \, (1) - \omega \left(\frac{V}{1 - LC\omega^2} \right) (0) = 0,$$

which gives

$$c_4 = 0.$$

Thus, the voltage across the capacitor is given by

$$v(t) = - \left(\frac{V}{1 - LC\omega^2} \right) \cos \sqrt{\frac{1}{LC}} \, t + \left(\frac{V}{1 - LC\omega^2} \right) \cos \omega t$$

or

$$v(t) = \left(\frac{V}{1 - LC\omega^2} \right) \left(\cos \omega t - \cos \sqrt{\frac{1}{LC}} \, t \right) V. \qquad (10.152)$$

Note: A comparison of Examples 10-12 (spring-mass) and 10-14 (LC circuit) reveals that the solutions are identical, with the following corresponding quantities:

Spring-Mass	LC Circuit
$y(t)$	$v(t)$
m	LC
k	1
F	V

Although the two physical systems are entirely different, the math is exactly the same. Such is the case for a wide range of problems across all disciplines of engineering. Make no mistake ... if you want to study engineering, then a little bit of math can go an awfully long way.

PROBLEMS

10-1. A faucet supplies fluid to a container of cross-sectional area A at a volume flow rate Q_{in}, as shown in Fig. P10.1. At the same time, the fluid leaks out the bottom at a rate $Q_{out} = k\,h(t)$, where k is a constant. If the container is initially empty, the fluid height $h(t)$ satisfies the following first-order differential equation and initial condition:

$$A\frac{dh(t)}{dt} + k\,h(t) = Q_{in}, \qquad h(0) = 0$$

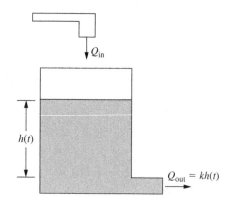

Figure P10.1 Leaking tank for problem P10-1.

(a) Determine the transient solution, $h_{tran}(t)$.

(b) Suppose the faucet is turned on and off in a sinusoidal fashion, so that $Q_{in} = \dfrac{Q}{2}(1 - \sin\omega t)$. Determine the steady-state solution, $h_{ss}(t)$.

(c) Determine the total solution $h(t)$, subject to the initial condition.

10-2. The initial temperature of the hot coffee cup shown in Fig. P10.2 is $T(0) = 175°F$. The cup is placed in a room temperature of $T_\infty = 70°F$. The temperature $T(t)$ of the coffee at time t can be approximated by Newton's law of cooling as

$$\frac{dT}{dt} + k\,T(t) = h\,T_\infty$$

where h is the convective heat transfer coefficient in $\dfrac{Btu}{ft^2 h°F}$.

(a) Find the transient solution, $T_{tran}(t)$. What is the time constant of the response?

(b) Find the steady-state solution $T_{ss}(t)$.

(c) Determine the total solution $T(t)$.

Figure P10.2   A hot coffee cup placed in a room temperature of 70°F.

(d) Sketch the total solution $T(t)$. How long does it take for the temperature to reach 99% of the room temperature?

(e) Would increasing the value of h increase or decrease the time when the temperature of coffee reaches the room temp?

(f) Would a lower or higher value of h be best for a cup of coffee?

10-3. A constant voltage $v_s(t) = 10$ V is applied to the RC circuit shown in Fig. P10.3. Assume that the switch has been in position 1 for a long time. At $t = 0$, the switch is moved instantaneously to position 2. For $t \geq 0$, the voltage $v(t)$ across the capacitor satisfies the following differential equation and initial condition:

$$RC\frac{dv(t)}{dt} + v(t) = 0, \qquad v(0) = 10\text{V}.$$

(a) Find the transient solution, $v_{tran}(t)$. What is the time constant of the response?

(b) Find the steady-state solution $v_{ss}(t)$.

(c) Determine the total solution $v(t)$.

(d) Sketch the total solution $v(t)$. How long does it take for the response to reach 99% of its steady-state value?

(e) Mark each of the following statements as true (T) or false (F):

_____ Increasing the value of resistance R will increase the time the voltage $v(t)$ reaches 99% of its steady-state value.

_____ Increasing the value of capacitance C will decrease the time the voltage $v(t)$ reaches 99% of its steady-state value.

_____ Doubling the value of resistance R will double the time constant of the response.

_____ Doubling the value of capacitance C will double the time constant of the response.

Figure P10.3   RC circuit for problem P10-3.

10-4. Repeat parts (a)–(d) of problem P10-3 if $R = 10$ kΩ and $C = 50$ μF.

10-5. Repeat parts (a)–(d) of problem P10-3 if $R = 20$ kΩ and $C = 100$ μF.

10-6. A constant voltage $v_s(t) = 10$ V is applied to the RC circuit shown in Fig. P10.6. The voltage $v(t)$ across the capacitor satisfies the first-order differential equation

$$RC\frac{dv(t)}{dt} + v(t) = v_s(t).$$

(a) Find the transient solution, $v_{tran}(t)$. What is the time constant of the response?

(b) Find the steady-state solution $v_{ss}(t)$.

(c) If the initial voltage across the capacitor is $v(0) = 5$ V, determine the total solution $v(t)$.

(d) Sketch the total solution $v(t)$. How long does it take for the response to reach 99% of its steady-state value?

Figure P10.6 RC circuit for problem P10-6.

Repeat problem P10-6 if $R = 200$ kΩ and $C = 100$ μF.

3. Repeat problem P10-6 if $R = 100$ kΩ, $C = 50$ μF, $v_s(t) = 20$ V, and $v(0) = 10$ V.

10-9. A sinusoidal voltage $v_s(t) = 10\sin(0.01\,t)$ V is applied to the RC circuit shown in Fig. P10.6. The voltage $v(t)$ across the capacitor satisfies the first-order differential equation

$$RC\frac{dv(t)}{dt} + v(t) = 10\sin(0.01\,t).$$

(a) Find the transient solution $v_{tran}(t)$. What is the time constant of the response?

(b) Find the steady-state solution $v_{ss}(t)$ and plot one cycle of the response. **Note:** One of the two terms in the steady-state solution is small enough to be neglected.

(c) If the initial voltage across the capacitor is $v(0) = 0$, determine the total solution $v(t)$.

10-10. Repeat problem P10-9 if $R = 10$ kΩ and $C = 10$ μF.

10-11. Repeat problem P10-9 if $R = 20$ kΩ and $C = 20$ μF.

10-12. The circuit shown in Fig. P10.12 consists of a resistor and capacitor in parallel that are subjected to a *constant* current source I. At time $t = 0$, the initial voltage across the capacitor is zero. For time $t \geq 0$, the voltage across the capacitor satisfies the following first-order differential equation and initial condition:

$$C\frac{dv(t)}{dt} + \frac{v(t)}{R} = I.$$

(a) Determine the transient solution, $v_{tran}(t)$.

(b) Determine the steady-state solution, $v_{ss}(t)$.

(c) Determine the total solution for $v(t)$ if $v(0) = 0$ V.

(d) Calculate the voltage at times $t = RC, 2RC, 4RC$, and as $t \to \infty$. Use your results to sketch $v(t)$.

Figure P10.12 RC circuit for problem P10-12.

10-13. Repeat problem P10-12 if $R = 1$ kΩ, $C = 10$ μF, $I = 10$ mA, and $v(0) = 0$ V.

10-14. Repeat problem P10-12 if $R = 2$ kΩ, $C = 100$ μF, $I = 5$ mA, and $v(0) = 10$ V.

10-15. A constant current $i_s(t) = 100$ mA is applied to the RL circuit shown in Fig. P10.15. Assume that the switch has been closed for a long time. At $t = 0$, the switch is opened instantaneously. For $t \geq 0$, the current $i(t)$ flowing through the resistor satisfies the following differential equation and initial condition:

$$\frac{L}{R}\frac{di(t)}{dt} + i(t) = 0, \qquad i(0) = 50\text{mA}.$$

(a) Find the transient solution, $i_{tran}(t)$. What is the time constant of the response?

(b) Find the steady-state solution $i_{ss}(t)$.

(c) Determine the total solution $i(t)$.

(d) Sketch the total solution $i(t)$. How long does it take for the response to reach 99% of its steady-state value?

(e) Mark each of the following statements as true (T) or false (F):

_____ Increasing the value of resistance R will increase the time the voltage $v(t)$ reaches 99% of its steady-state value.

_____ Increasing the value of inductance L will decrease the time the voltage $v(t)$ reaches 99% of its steady-state value.

_____ Doubling the value of resistance R will double the time constant of the response.

_____ Doubling the value of inductance L will double the time constant of the response.

10-16. Repeat problem P10-15 if $R = 100\ \Omega$ and $L = 100$ mH.

10-17. Repeat problem P10-15 if $R = 100\ \Omega$ and $L = 10$ mH.

10-18. At time $t = 0$, an input voltage v_{in} is applied to the RL circuit shown in Fig. P10.18. The output voltage $v(t)$ satisfies the following first-order differential equation:

$$\frac{dv(t)}{dt} + \frac{R}{L}v(t) = \frac{R}{L}v_{in}(t).$$

If the input voltage $v_{in}(t) = 10$ volts,

(a) Determine the transient solution, $v_{tran}(t)$.

(b) Determine the steady-state solution, $v_{ss}(t)$.

(c) Determine the total solution for $v(t)$, assuming the initial voltage is zero.

(d) Calculate the output voltage $v(t)$ at times $t = \dfrac{L}{R}, \dfrac{2L}{R}, \dfrac{4L}{R}$ s, and as $t \to \infty$. Use your results to sketch $v(t)$.

Figure P10.18   *RL* circuit for problem P10-18.

10-19. Repeat problem P10-18 if $R = 50\ \Omega$ and $L = 500$ mH.

10-20. Repeat problem P10-18 if $R = 10\ \Omega$, $L = 200$ mH, and $v_{in}(t) = 20$ V.

10-21. The switch in the circuit shown in Fig. P10.21 has been in position 1 for a long time. At $t = 0$, the switch is moved instantaneously to position 2. For $t \ge 0$, the current $i(t)$ flowing through the

Figure P10.15   RL circuit for problem P10-15.

inductor satisfies the following differential equation and initial condition:

$$0.2\frac{di(t)}{dt} + 10\,i(t) = 5, \quad i(0) = -1\text{A}.$$

(a) Find the transient solution, $i_{tran}(t)$. What is the time constant of the response?

(b) Find the steady-state solution $i_{ss}(t)$.

(c) Determine the total solution $i(t)$.

(d) Sketch the total solution $i(t)$. How long does it take for the response to reach 99% of its steady-state value?

Figure P10.21   RL circuit for problem P10-21.

10-22. A constant voltage source $v_{in}(t) = 10$ volts is applied to the Op–Amp circuit shown in Fig. P10.22. The output voltage $v_o(t)$ satisfies the following first-order differential equation and initial conditions:

$$0.01\frac{dv_o(t)}{dt} + v_o(t) = -v_{in}(t),$$

$$v_o(0) = 0 \text{ V}.$$

(a) Find the transient solution, $v_{o,tran}(t)$.

(b) Find the steady-state solution $v_{o,ss}$ if $v_{in} = 10$ V.

(c) If the initial output voltage is $v_o(0) = 0$ V, determine the total response.

(d) What is the time constant τ of the response? Plot the response $v_o(t)$ and give the values of v_o at $t = \tau$, $2\,\tau$, and $5\,\tau$.

Figure P10.22   Op–Amp circuit for problem P10-22.

10-23. A sinusoidal voltage source $v_{in}(t) = 10\sin(10\,t)$ volts is applied to the Op–Amp circuit shown in Fig. P10.23. The output voltage $v_o(t)$ satisfies the following first-order differential equation and initial conditions:

$$0.2\frac{dv_o(t)}{dt} + v_o(t) = -2v_{in}(t),$$

$$v_o(0) = 0 \text{ V}.$$

Figure P10.23   Op–Amp circuit for problem P10-23.

(a) Find the transient solution, $v_{o,tran}(t)$. What is the time constant τ of the response?

(b) Find the steady-state solution $v_{o,ss}$ if $v_{in} = 10\sin(10t)$ V.

(c) If the initial output voltage is $v_o(0) = 0$ V, determine the total response.

10-24. The relationship between arterial blood flow and blood pressure in a single artery satisfies the following first-order differential equation:

$$\frac{dP(t)}{dt} + \frac{1}{RC}P(t) = \frac{\dot{Q}_{in}}{C}$$

where $\dot{Q}_{in}$ is the volumetric blood flow, R is the peripheral resistance, and C is arterial compliance (all constant).

(a) Find the transient solution $P_{tran}(t)$ for the arterial pressure. The unit for $P(t)$ is mmHg. What is the time constant of the arterial pressure?

(b) Determine the steady-state solution $P_{ss}(t)$ for the arterial pressure.

(c) Determine the total solution $P(t)$ assuming that the initial arterial pressure is 0.

(d) Evaluate $P(t)$ at times $t = RC$, $t = 2RC$, $t = 4RC$, and $t \to \infty$ and use your results to sketch $P(t)$.

(e) Mark each of the following statements as true (T) or false (F):

_____ Increasing the value of resistance R will increase the time it takes for the arterial pressure $P(t)$ to reach 99% of its steady-state value.

_____ Increasing the value of capacitance C will decrease the time it takes for the arterial pressure $P(t)$ to reach 99% of its steady-state value.

_____ Doubling the value of resistance R will double the time constant of the response.

_____ Doubling the value of capacitance C will double the time constant of the response.

10-25. Repeat problem P10-24 if the volumetric blood flow $\dot{Q}_{in}$ is $60 \frac{cm^3}{s}$ and the initial arterial pressure is 6 mmHg. Also, assume $R = 4 \frac{mmHg}{(cm^3/s)}$ and $C = 0.4 \frac{cm^3}{mmHg}$.

10-26. The displacement $y(t)$ of a spring-mass system shown in Fig. P10.26 is given by

$$0.25\,\ddot{y}(t) + 10\,y(t) = 0.$$

(a) Find the transient solution, $y_{tran}(t)$.

(b) Find the steady-state solution of the displacement y_{ss}.

(c) Determine the total displacement $y(t)$ if the initial displacement $y(0) = 0.2$ m and the initial velocity $\dot{y}(0) = 0$ m/s.

(d) Sketch the total displacement $y(t)$.

Figure P10.26 Mass-spring system for problem P10-26.

10-27. Repeat problem P10-26 if the displacement $y(t)$ of a spring-mass ($m = 500$ gm and $k = 5$ N/m) system shown is given by $0.5\,\ddot{y}(t) + 5\,y(t) = 0$ and $y(0) = 10$ cm.

10-28. The displacement $y(t)$ of a spring-mass system shown in Fig. P10.28 is given by

$$\ddot{y}(t) + 25\,y(t) = f(t).$$

(a) Find the transient solution, $y_{tran}(t)$.
(b) Find the steady-state solution of the displacement y_{ss} if $f(t) = 10$ N.
(c) Determine the total displacement $y(t)$ if the initial displacement $y(0) = 0$ m and the initial velocity $\dot{y}(0) = 0$ m/s.
(d) Sketch the total displacement $y(t)$.

Figure P10.28 Mass-spring systems for problem P10-28.

10-29. Repeat problem P10-28 if the displacement $y(t)$ of a spring-mass ($m = 2$ kg and $k = 32$ N/m) system is given by

$$2\ddot{y}(t) + 32\,y(t) = f(t)$$

and $f(t) = 6$ N, $y(0) = 20$ cm, and $\dot{y}(0) = 0.1$ m/s.

10-30. A block of mass m is dropped from a height h above a spring k, as shown in Fig. P10.30. Beginning at the time of impact ($t = 0$), the position $x(t)$ of the block satisfies the following second-order differential equation and initial conditions:

$$m\ddot{x}(t) + k\,x(t) = m\,g \quad x(0) = 0,$$

$\dot{x}(0) = \sqrt{2gh}$.
(a) Determine the transient solution $x_{tran}(t)$, and determine the frequency of oscillation.

(b) Determine the steady-state solution, $x_{ss}(t)$.
(c) Determine the total solution $x(t)$, subject to the initial conditions.
(d) Mark each of the following statements as true (T) or false (F):

_____ Increasing the stiffness k will increase the frequency of $x(t)$.

_____ Increasing the height h will increase the frequency of $x(t)$.

_____ Increasing the mass m will decrease the frequency of $x(t)$.

_____ Doubling the height h will double the maximum value of $x(t)$.

_____ Doubling the mass m will double the maximum value of $x(t)$.

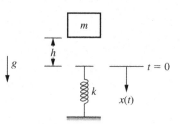

Figure P10.30 Mass dropped on a spring.

10-31. Repeat parts (a)–(c) of problem P10-30 if $m = 500$ g, $k = 20$ N/m, and $g = 9.8$ m/s^2.

10-32. Repeat parts (a)–(c) of problem P10-30 if $m = 1$ kg, $k = 30$ N/m, and $g = 9.8$ m/s^2.

10-33. The displacement $y(t)$ of the spring-mass system shown in Fig. P10.33 is measured from the equilibrium configuration of the spring, where the static deflection is $\delta = \dfrac{mg}{k}$.

Equilibrium Configuration: Undeformed Configuration:

$\delta = mg/k$

$y(t)$

$F\cos(\omega t)$

Figure P10.33 Mass-spring system for problem P10-33.

If $m = 4$ kg, $k = 1.5$ N/m, $\omega = 0.5$ rad/s, and a sinusoidal force of $0.25 \cos(0.5\,t)$ is applied at $t = 0$, the displacement $y(t)$ satisfies the second-order differential equation

$$4\,\ddot{y}(t) + 1.5\,y(t) = 0.25 \cos(0.5\,t).$$

(a) Find the transient response $y_{tran}(t)$ of the system. What is the natural frequency of the response?

(b) Show that the steady-state solution is given by

$$y_{ss}(t) = 0.5 \cos(0.5\,t).$$

Show all steps.

(c) Assume the total solution is given by

$$y(t) = c_3 \cos(\sqrt{0.375}\,t)$$
$$+ c_4 \sin(\sqrt{0.375}\,t) + 0.5 \cos(0.5\,t),$$

Given $y(0) = 0.5$ and $\dot{y}(0) = 0$, show that $c_3 = 0$ and $c_4 = 0$. Show all steps.

(d) Given the total solution $y(t) = 0.5 \cos(0.5\,t)$, determine the minimum value of the deflection $y(t)$ and the first time it reaches that value.

10-34. Under static loading by a weight of mass m, a rod of length L and axial rigidity AE deforms by an amount $\delta = \dfrac{mgL}{AE}$, where g is the acceleration due to gravity. However, if the mass is applied suddenly (dynamic loading), vibration of the mass will ensue. If the mass m is initially at rest, the deflection $x(t)$ satisfies the following second-order differential equation and initial conditions:

$$m\,\ddot{x}(t) + \frac{AE}{L}\,x(t) = mg,$$

$$x(0) = 0, \; \dot{x}(0) = 0.$$

(a) Determine the transient solution, $x_{tran}(t)$.

(b) Determine the steady-state solution, $x_{ss}(t)$.

(c) Determine the total solution for $x(t)$, subject to the given initial conditions.

(d) Calculate the maximum value of the deflection $x(t)$. How does your result compare to the static deflection δ?

Static Loading: Dynamic Loading:

$\delta = \dfrac{mgL}{AE}$

$x(t)$

Figure P10.34 Rod under axial loading by a weight of mass m.

10-35. At time $t = 0$, a cart of mass m moving at an initial velocity v_0 impacts a spring of stiffness k, as illustrated in Fig. P10.35.

Figure P10.35 Moving cart for problem P10-35.

The resulting deformation $x(t)$ of the spring satisfies the following second-order differential equation and initial conditions:

$$m\ddot{x}(t) + kx(t) = 0, \quad x(0) = 0, \quad \dot{x}(0) = v_0.$$

(a) Determine the total solution for $x(t)$, subject to the initial conditions.

(b) Plot one-half cycle of the deformation $x(t)$, and clearly label both its maximum value and the time it takes to get there.

(c) Mark each of the following statements as true (T) or false (F):

_____ Increasing the spring stiffness k will increase the frequency of $x(t)$.

_____ Increasing the initial velocity v_0 will increase the frequency of $x(t)$.

_____ Increasing the mass m will decrease the frequency of $x(t)$.

_____ Doubling the velocity v_0 will double the amplitude of $x(t)$.

_____ Doubling the mass m will double the amplitude of $x(t)$.

10-36. An LC circuit is subjected to a constant voltage source V_s that is suddenly applied at time $t = 0$.

The current $i(t)$ satisfies the following second-order differential equation and initial conditions:

$$LC\frac{d^2i(t)}{dt^2} + i(t) = 0, \quad i(0) = 0, \quad \frac{di}{dt}(0) = \frac{V_s}{L}.$$

Figure P10.36 LC circuit for problem P10-36.

(a) Determine the total solution for $i(t)$, subject to the given initial conditions.

(b) Plot one-half cycle of the current $i(t)$, and clearly label both its maximum value and the time it takes to get there.

(c) Mark each of the following statements as true (T) or false (F):

_____ Increasing the capacitance C will increase the frequency of $i(t)$.

_____ Increasing the inductance L will increase the frequency of $i(t)$.

_____ Increasing the capacitance C will increase the amplitude of $i(t)$.

_____ Increasing the voltage V_s will increase the amplitude of $i(t)$.

_____ Increasing the inductance L will increase the amplitude of $i(t)$.

10-37. Repeat parts (a) and (b) of problem P10-36 if $L = 200$ mH, $C = 500$ μF, and $V_s = 2$ V.

10-38. An LC circuit is subjected to input voltage v_{in} that is suddenly applied at time $t = 0$.

The voltage $v(t)$ satisfies the following second-order differential equation and initial conditions:

$$LC \frac{d^2v(t)}{dt^2} + v(t) = v_{in}, \quad v(0) = 0, \quad \dot{v}(0) = 0.$$

Figure P10.38 LC circuit for problem P10-38.

(a) Determine the transient solution, $v_{tran}(t)$. What is the frequency of the response?

(b) If $v_{in} = 10.0$ V, determine the steady-state solution, $v_{ss}(t)$.

(c) Determine the total solution for $v(t)$, subject to the given initial conditions.

(d) Calculate the maximum value of $v(t)$. Does your result depend on the values of L and C?

10-39. Repeat parts (a)–(c) of problem P10-38 if $L = 100$ mH, $C = 40$ μF, and $v_{in} = 10$ V.

10-40. Repeat parts (a)–(c) of problem P10-38 if $L = 40$ mH, $C = 400$ μF, and $v_{in} = 10\sin(100\,t)$ V.

10-41. Repeat parts (a)–(c) of problem P10-38 if $L = 40$ mH, $C = 400$ μF, and $v_{in} = 10\sin(245\,t)$ V.

10-42. A biomedical engineer is designing a resistive training device to strengthen the latissimus dorsi muscle. The task can be represented as a spring-mass system, as shown in Fig. P10.28. The displacement $y(t)$ of the exercise bar satisfies the second-order differential equation

$$m\ddot{y}(t) + k\,y(t) = f(t),$$

subject to the initial condition $y(0) = 0.01$ m and $\dot{y}(0) = 0$ m/s.

(a) Determine the transient solution $y_{tran}(t)$.

(b) Determine the steady-state solution $y_{ss}(t)$ for the applied force shown in Fig. P10.42.

(c) Determine the total solution, subject to the initial conditions.

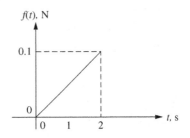

Figure P10.42 Applied force for resistive training device for problem P10-42.

10-43. A rod of mass m and length l is pinned at the bottom and supported by a spring of stiffness k at the top. When displaced by a force $f(t)$, its position is described by the angle $\theta(t)$, as shown in Fig. P10.43. For relatively small oscillations, the angle $\theta(t)$ satisfies the second-order differential equation

$$\frac{1}{3}\,m\,l\,\ddot{\theta}(t) + k\,l\,\theta(t) = f(t).$$

Figure P10.43 Rod supported by a spring.

(a) Find the transient solution $\theta_{tran}(t)$, and determine the frequency of oscillation.

(b) Suppose $f(t) = F\cos\omega t$. Determine the steady-state solution, $\theta_{ss}(t)$.

(c) Suppose now that $f(t) = 0$, and that the initial conditions are $\theta(0) = \frac{\pi}{18}$ rad and $\dot{\theta}(0) = 0$ rad/s. Determine the total solution, $\theta(t)$.

(d) Given your solution to part (c), mark each of the following statements as true or false:

_____ Increasing the length l would decrease the frequency of the oscillation.

_____ Increasing the mass m would increase the frequency of the oscillation.

_____ Increasing the stiffness k would increase the frequency of the oscillation.

_____ Increasing the mass m would increase the amplitude of the oscillation.

_____ Increasing the stiffness k would decrease the amplitude of the oscillation.

Answers to Selected Problems

Chapter 1

P1-1 $\qquad k = 10$ N/m

P1-3 (a) $F = 3.48 x + 48$ N

P1-5 (a) $F_b(P) = 0.5 P + 160$ lb

P1-7 (a) $v(t) = -9.8 t + 4$ m/s

P1-9 (a) $v(t) = -32.2 t + 161.0$ ft/s

P1-11 (a) $v(t) = -75 t + 150$ ft/s

P1-13 (a) $v(t) = 10 t$ m/s

 (b) $v(t) = 10$ m/s

 (c) $v(t) = -10 t + 40$ m/s

P1-15 (a) $T(x) = 26.67 x + 30°$F

 (b) $T(0.75) = 50°$F

P1-17 (b) $I = 2$ A

P1-19 (a) $V_s = 3 I + 3$ V

P1-21 $\qquad V_s = 50 I + 0.7$ V

P1-23 $\qquad v_o = 1.5 v_{in} - 4$ V

P1-25 $\qquad e_a = 2.667 i_a + 0.667$ V

P1-27 $\qquad v_o = -1000\ i_D + 12$ V

P1-29 (a) $F = 18$ N

 (b) $V = 0.6$ V

P1-31 (a) $A(F) = 4.55 \times 10^{-6} F$ V

 (c) $A(200) = 9.091 \times 10^{-4}$ V

P1-33 (a) $T(V) = 0.04\ V - 25°$C

P1-35 (a) $T(V) = 18.1\ V + 4.6°$C

 (c) $\Delta T = 65.17°$C

P1-37 (a) $R = 100\ \varepsilon + 50$ Ω

P1-39 (a) $a(c) = 0.0667\ c + 0.0833$

 (b) $c = 6.04\ \frac{\mu g}{ml}$

 (c) $a(0.00419) = 0.0836$

P1-41 (a) $T(°F) = 1.8\ T(°C) + 32°$F

 (c) $68°F \leq T \leq 176°$F

P1-43 (a) $p(x) = -12\ x + 30$ psi, $l = 2.5$ in.

 (b) $p(1.0) = 18$ psi

Chapter 2

P2-1 (a) $I^2 + 2I - 15 = 0$

    (b) $I = 3\,\text{A}$ or $I = -5\,\text{A}$

P2-3 (a) $I^2 + 4I - 21 = 0$

    (b) $I = 3\,\text{A}$ or $I = -7\,\text{A}$

P2-5 (a) $t = 8\,\text{s}$

    (b) $t = 9\,\text{s}$

P2-7 (a) $R_1^2 - 1067\,R_1 - 266667 = 0$

    (b) $R_1 = 1275\,\Omega$, $R_2 = 1775\,\Omega$

P2-9 (a) $C_1^2 - 140\,C_1 - 12000 = 0$

    (b) $C_1 = 200\,\mu\text{F}$, $C_2 = 300\,\mu\text{F}$

P2-11 (a) $C_1^2 - 100\,C_1 - 10000 = 0$

    (b) $C_1 = 161.8\,\mu\text{F}$, $C_2 = 261.8\,\mu\text{F}$

P2-13 (a) $L_2^2 + 100\,L_2 - 60000 = 0$

    (b) $L_1 = 600\,\text{mH}$, $L_2 = 200\,\text{mH}$

P2-15 (a) $t = 1\,\text{s}$ and $t = 3\,\text{s}$

    (b) $t = 1.5\,\text{s}$ and $t = 2.5\,\text{s}$

    (c) Time when rocket hit ground $= 4\,\text{s}$

    (d) Maximum height $= 64\,\text{ft}$

P2-17 (a) $t = 0\,\text{s}$ and $t = 3.5\,\text{s}$

    (b) Time when ball hit ground $= 4\,\text{s}$

    (c) Maximum height $= 81\,\text{ft}$

P2-19 (a) $3\,k_2^2 + 0.5\,k_2 - 3.5 = 0$

    (b) $k_1 = 3\,\text{lb/in.}$, $k_2 = 1\,\text{lb/in.}$

P2-21 (a) $\omega^2 + \omega - 1000 = 0$

      $\omega = 31.13\,\text{rad/s}$

P2-23 (a) $\omega^2 - 30\omega - 400 = 0$

    (b) $\omega = 40\,\text{rad/s}$

P2-25 (a) $R^2 - 10R - 200 = 0$

    (b) $R = 20\,\Omega$

P2-27 (a) $R^2 + 40R - 1200 = 0$

    (b) $R = 20\,\Omega$

P2-29 $s = -5,000$ or $s = -20,000$

P2-31 $L = 3\,\text{m}$ and $W = 12\,\text{m}$ or
          $L = 12\,\text{m}$ and $W = 3\,\text{m}$

P2-33 (a) $x^2 - 70x + 600 = 0$

    (b) $x_A = 10\,\text{m}$ and $x_B = 60\,\text{m}$

    (c) Length of tunnel $= 50\,\text{m}$

P2-35 (a) $0.9\,P^2 - 1.9\,P + 0.3125 = 0$

    (b) $P = 0.178$

    (c) $0.9\,P^2 - 1.9\,P + 0.9178 = 0$
        $P = 0.748$

P2-37 (a) $x^2 - 5x + 2 = 0$

    (b) $x = 0.438$

    (c) CO $= 0.562$, $H_2O = 1.562$,
        $CO_2 = 0.438$, $H_2 = 0.438$

P2-39 (a) $b^2 + 10b - 600 = 0$

    (b) $b = 20\,\text{ft}$

    (c) height $= 30\,\text{ft}$, width $= 20\,\text{ft}$

Chapter 3

P3-1 $h = 66.14\,\text{m}$ and $\theta = 41.4°$

P3-3 $l = 163.2\,\text{in.}$ and $\theta = 17.1°$

P3-5   A $= 48{,}453\,\text{m}^2$

P3-7 $\theta = 77.36°$

P3-9 $h = 415\,\text{m}$ and $\theta = 53.13°$

P3-11 $x = -1.061\,\text{m}$ and $y = 1.061\,\text{m}$

P3-13 $x = -1.061\,\text{m}$ and $y = -1.061\,\text{m}$

P3-15 (a) $(x, y) = (-4.33, 2.5)$ cm

(b) $(x, y) = (-2.5, -4.33)$ cm

(c) $(x, y) = (2.5, 4.33)$ cm

(d) $(x, y) = (3.536, 3.536)$ cm

P3-17 (a) $l = 4.472$ in. and $\theta = 26.57°$

(b) $l = 4.472$ in. and $\theta = 116.57°$

(c) $l = 9.014$ in. and $\theta = 233.75°$

(b) $l = 8.485$ in. and $\theta = -45°$

P3-19 $\theta_1 = 16.31°$ and $\theta_2 = 49.8°$

P3-21 $\theta_1 = -57.3°$ and $\theta_2 = 73.6°$

P3-23 $\theta_1 = -145°$ and $\theta_2 = -63.26°$

P3-25 $\theta_1 = 127.8°$ and $\theta_2 = 21.4°$

P3-27 $\theta = -60°$ and $\theta_2 = 90°$

P3-29 $V = 8.77$ mph and $\theta = 136.8°$

P3-31 $Z = 104.4\ \Omega$ and $\theta = 16.7°$

P3-33 $X_c = 500.0\ \Omega$ and $\theta = 26.56°$

P3-35 $V = 20.62$ V and $\theta = 14.04°$

P3-37 (a) $\theta = 11.54°$ and grade $= 20.41\%$

(b) $\Delta_x = 489.9$ m

P3-39 (a) $\theta = -30°$ and grade $= -57.74\%$

(b) $\Delta_x = 173.2$ m

P3-41 $h = 16.28$ m

P3-43 (a) Resultant Force $= 35.5$ N

(b) $\theta = 43.05°$

P3-45 (a) $P(x, y) = (18.97, -15.44)$ in.

(b) $\theta_1 = -60°$ and $\theta_2 = 30°$

Chapter 4

P4-1 $\vec{P} = 6\vec{i} + 10.39\vec{j}$ in., or
$\vec{P} = 12 \angle 60°$ in.

P4-3 $\vec{P} = -0.50\vec{i} + 0.866\vec{j}$ m, or
$\vec{P} = 1 \angle 120°$ m

P4-5 $\vec{P} = 8.66\vec{i} + 5.0\vec{j}$ in., or
$\vec{P} = 10 \angle 30°$ in.

P4-7 $\vec{P} = -7.5\vec{i} - 13.0\vec{j}$ cm, or
$\vec{P} = 15 \angle -120°$ cm

P4-9 $\vec{P} = 2\vec{i} + 3\vec{j}$ cm, or
$\vec{P} = 3.61 \angle 56.3°$ cm

P4-11 $\vec{P} = -1.5\vec{i} + 2\vec{j}$ cm, or
$\vec{P} = 2.5 \angle 126.87°$ cm

P4-13 $F_x = 14.14$ lb and $F_y = -14.14$ lb

P4-15 $\vec{V} = 10\vec{i} + 5\vec{j}$ V, or
$\vec{V} = 11.18 \angle 26.57°$ V

P4-17 $\vec{V} = 20\vec{i} - 5\vec{j}$ V, or
$\vec{V} = 20.62 \angle -14.04°$ V

P4-19 $\vec{V} = 23.66\vec{i} - 5\vec{j}$ V

P4-21 $\vec{V} = 246.77\vec{i} + 192.13\vec{j}$ mph, or
$\vec{V} = 312.75 \angle 37.9°$ mph

P4-23 (a) $\vec{V} = -8\vec{i} + 7.5\vec{j}$ mph
(b) magnitude $= 10.97$ mph
direction $\theta = 136.8°$

P4-25 (a) $\vec{V} = -4.485\vec{i} - 8.485\vec{j}$ mph
(b) magnitude $= 9.6$ mph
direction $\theta = -117.9°$

P4-27 (a) $\vec{P} = -2.28\vec{i} + 10.55\vec{j}$ in.
(b) $\vec{P} = 10.8 \angle 102.2°$ in.

P4-29 (a) $\vec{P} = -9.6\vec{i} - 14\vec{j}$ cm
(b) $\vec{P} = 16.98 \angle -124.4°$ cm

P4-31 $\vec{T}_1 = 424.75\vec{i} + 735.67\vec{j}$ N
$\vec{T}_2 = -424.75\vec{i} + 245.25\vec{j}$ N

P4-33 (a) $\theta = 30°$

(b) $\vec{F} = -0.866\,F\vec{i} + 0.5\,F\vec{j}$ lb
$\vec{N} = 0.5\,N\vec{i} + 0.866\,N\vec{j}$ lb
$\vec{W} = 0\vec{i} - 2000\vec{j}$ lb

(c) F = 1000 lb
N = 1732.1 lb

P4-35 (a) $\theta = 30°$

(b) $\vec{F} = 0.866\,F\vec{i} + 0.5\,F\vec{j}$ N
$\vec{N} = -0.5\,N\vec{i} + 0.866\,N\vec{j}$ N

(c) F = 250 N, N = 433 N

P4-37 (a) $\vec{F}_1 = 0.7705\,F_1\vec{i} + 0.6374\,F_1\vec{j}$ lb
$\vec{F}_2 = -0.4633\,F_2\vec{i} + 0.8862\,F_2\vec{j}$ lb
$\vec{P} = 0\vec{i} - 346\vec{j}$ lb

(b) $F_1 = 163.8$ lb, $F_2 = 272.5$ lb

P4-39 (a) $\vec{F}_1 = -F_1\vec{i} + 0\vec{j}$ lb
$\vec{F}_2 = 0.7466\,F_2\vec{i} + 0.6652\,F_2\vec{j}$ lb
$\vec{W} = 0\vec{i} - 125\vec{j}$ lb

(b) $F_1 = 140.3$ lb, $F_2 = 187.9$ lb

P4-41 (a) $\vec{F}_m = -43.3\vec{i} + 25\vec{j}$ lb
$\vec{W}_a = 0\vec{i} - 7\vec{j}$ lb
$\vec{W}_p = 0\vec{i} - 3\vec{j}$ lb

(b) $R_x = 43.3$ lb, $R_y = -15$ lb
$R = 45.82$ lb, Direction, $\theta = -19.11°$

P4-43 (a) $\vec{P} = 0.319\vec{i} + 2.156\vec{j}$ ft

(b) $P = 2.18$ ft, Direction, $\theta = 81.58°$

Chapter 5

P5-1 (a) $V_R = 1 + j0$ V and $V_L = 0 + j1$ V

(b) $V = 1 + j1$ V, $V = 1.414\ \angle45°$V

P5-3 (a) $V_R = 8.05 - j4.03$ V
$V_L = 2.02 + j4.02$ V

(b) $V = 10.07 + j0$ V, $V = 10.06\ \angle0°$V

(c) $Re(V) = 10.06$ V, $Im(V) = 0$ V

P5-5 (a) $V_R = 9 + j3$ V, $V_C = 1 - j3$ V

(b) $V = 10 + j0$ V, $V = 10\ \angle0°$V

(c) $Re(V) = 10$ V, $Im(V) = 0$ V

P5-7 (a) $I_R = 100 - j0$ mA and
$I_L = 0 - j200$ mA

(b) $I = 100 - j200$ mA
$I = 223.6\ \angle-63.43°$mA

(c) $Re(I) = 100$ mA, $Im(I) = -200$ mA

P5-9 (a) $I_R = 75 + j43.3\ \mu$A and
$I_L = 25 - j43.3\ \mu$A

(b) $I = 100 + j0\ \mu$A, $I = 100\ \angle0°\,\mu$A

(c) $Re(I) = 100\ \mu$A, $Im(I) = 0\ \mu$A

P5-11 (a) $I_R = 0.5 + j0$ mA and
$I_C = 0 + j0.2$ mA

(b) $I = 0.5 + j0.2$ mA
$I = 0.539\ \angle21.8°$mA

(c) $Re(I) = 0.5$ mA, $Im(I) = 0.2$ mA

P5-13 $v_o(t) = Re\left(15\,e^{j(120\pi t - 60°)}\right)$ V

P5-15 (a) $Z = 100 + j82.4\ \Omega$

(b) $Z = 129.6\ \angle39.5°\Omega$

(c) $Z^* = 100 - j82.4\ \Omega$
$Z\,Z^* = 16{,}789.9$

P5-17 (a) $I = 26.83\ \angle56.57°$ A, or
$I = 14.75 + j22.39$ A

(b) $V = 31.3\ \angle18.43°$V
$V = 29.69 + j9.9$ V

P5-19 (a) $Z_2 + Z_3 = 0 - j100.45\ \Omega$
$Z_2 + Z_3 = 100.45\ \angle-90°\Omega$

(b) $Z_1 + Z_2 + Z_3 = 100 - j100.45\ \Omega$
$Z_1 + Z_2 + Z_3 = 141.74\ \angle-45.13°\Omega$

(c) $H = 0.502 - j0.50$ or
$H = 0.709\ \angle-44.87°$

(d) $H^* = 0.502 + j0.50$
$H\,H^* = 0.503$

P5-21 (a) $Z_R = 100 + j0 \, \Omega$ or
$\qquad Z_R = 100.0 \, \angle 0°\Omega$

$\qquad Z_L = 0 + j60 \, \Omega$ or
$\qquad Z_L = 60.0 \, \angle 90°\Omega$

$\qquad Z_C = 0 - j50 \, \Omega$ or
$\qquad Z_C = 50.0 \, \angle -90°\Omega$

(b) $Z = 100 + j10 \, \Omega$
$\qquad Z = 100.5 \, \angle 5.71°\Omega$

(c) $Z^* = 100 - j10 \, \Omega$, $Z\,Z^* = 10,100$

(d) $V_C = -5.45 - j54.46 \, V$
$\qquad V_C = 54.73 \, \angle -95.71°V$

P5-23 (a) $Z_R = 9 + j0 \, \Omega$ or
$\qquad Z_R = 9 \, \angle 0°\Omega$

$\qquad Z_L = 0 + j3 \, \Omega$ or
$\qquad Z_L = 3 \, \angle 90°\Omega$

$\qquad Z_C = 0 - j4 \, \Omega$ or
$\qquad Z_C = 4 \, \angle -90°\Omega$

(b) $Z_{LC} = 0 + j12 \, \Omega$ or
$\qquad Z_{LC} = 12 \, \angle 90°\Omega$

(c) $Z = 9 + j12 \, \Omega$ or
$\qquad Z = 15 \, \angle 53.13°\Omega$

(d) $Z^* = 9 - j12 \, \Omega$, $Z\,Z^* = 225$

P5-25 (a) $Z_R = 100 + j0 \, \Omega$ or
$\qquad Z_R = 100 \, \angle 0°\Omega$

$\qquad Z_C = 0 - j50 \, \Omega$ or
$\qquad Z_C = 50 \, \angle -90°\Omega$

(b) $Z = 100 - j50 \, \Omega$ or
$\qquad Z = 111.8 \, \angle -26.57°\Omega$

(c) $I = 0.800 + j0.400 \, A$ or
$\qquad I = 0.894 \, \angle 26.57°A$

(d) $V_R = 80.0 + j40.0 \, V$ or
$\qquad V_R = 89.4 \, \angle 26.57°V$

$\qquad V_C = 20.0 - j40.0 \, V$ or
$\qquad V_C = 44.7 \, \angle -63.43°V$

P5-25 (e) $V = 100.0 - j0 \, V$ or
$\qquad V = 100 \, \angle 0°V$

P5-27 (a) $Z_R = 100 + j0 \, \Omega$ or
$\qquad Z_R = 100 \, \angle 0°\Omega$

$\qquad Z_L = 0 + j188.5 \, \Omega$ or
$\qquad Z_L = 188.5 \, \angle 90°\Omega$

$\qquad Z_C = 0 - j106.1 \, \Omega$ or
$\qquad Z_C = 106.1 \, \angle -90°\Omega$

(b) $Z = 100 + j82.4 \, \Omega$ or
$\qquad Z = 129.6 \, \angle 39.5°\Omega$

(c) $I = 0.596 - j0.491 \, A$ or
$\qquad I = 0.772 \, \angle -39.5°A$

(d) $V_R = 59.57 - j49.11 \, V$ or
$\qquad V_R = 77.2 \, \angle -39.5°V$

$\qquad V_L = 92.56 + j112.3 \, V$ or
$\qquad V_L = 145.5 \, \angle 50.5°V$

$\qquad V_C = -52.1 - j63.2 \, V$ or
$\qquad V_C = 81.91 \, \angle -129.5°V$

(e) $V = 100.0 + j0 \, V$ or
$\qquad V = 100 \, \angle 0°V$

P5-29 (a) $Z_R = 1000 + j0 \, \Omega$ or
$\qquad Z_R = 1000 \, \angle 0°\Omega$

$\qquad Z_C = 0 - j1000 \, \Omega$ or
$\qquad Z_C = 1000 \, \angle -90°\Omega$

(b) $I_1 = 0.5 - j0.5 \, mA$ or
$\qquad I_1 = 0.7071 \, \angle -45°mA$

$\qquad I_2 = 0.5 + j0.5 \, mA$ or
$\qquad I_2 = 0.7071 \, \angle 45°mA$

(c) $I_1 + I_2 = 1 + j0 \, mA$ or
$\qquad I_1 + I_2 = 1 \, \angle 0°mA$

P5-31 (a) $Z_R = 1000 + j0 \, \Omega$ or
$\qquad Z_R = 1000 \, \angle 0°\Omega$

$\qquad Z_L = 0 + j1000 \, \Omega$ or
$\qquad Z_L = 1000 \, \angle 90°\Omega$

P5-31 (b) $I_1 = 0.5 + j0.5$ mA or
$I_1 = 0.7071 \angle 45°$mA

$I_2 = 0.5 - j0.5$ mA or
$I_2 = 0.7071 \angle -45°$mA

(c) $I_1 + I_2 = 1 + j0$ mA or
$I_1 + I_2 = 1 \angle 0°$mA

P5-33 (a) $Z_R = 1 + j0$ kΩ or
$Z_R = 1 \angle 0°$kΩ

$Z_C = 0 - j2$ kΩ or
$Z_C = 2 \angle -90°$kΩ

(b) $V_o = 2.5 + j0$ V or
$V_o = 2.5 \angle 0°$V

P5-35 (a) $Z_{R1} = 1.5 + j0 = 1.5 \angle 0°k\Omega$
$Z_{R2} = 1 + j0 = 1 \angle 0°k\Omega$
$Z_C = 0 - j0.5 = 0.5 \angle -90°k\Omega$

(b) $V_o = -0.3578 + j0.1789$ V or
$V_o = 0.4 \angle 153.43°$V

P5-37 (a) $Z_a = 20 + j0 = 20 \angle 0°\Omega$
$Z_b = 0 - j10 = 10 \angle -90°\Omega$
$Z_c = 20 + j50 = 53.85 \angle 68.2°\Omega$

(b) $Z_1 = -2.5 - j2.5 = 3.54 \angle -135°\Omega$
$Z_2 = 17.5 + j7.5 = 19.04 \angle 23.2°\Omega$
$Z_3 = 3.75 - j8.75 = 9.52 \angle -66.8°\Omega$

P5-39 (a) $Z_1 = 2.5 - j2.5 = 3.53 \angle -45°\Omega$
$Z_2 = 17.5 + j7.5 = 19.04 \angle 23.2°\Omega$
$Z_3 = 3.75 - j8.75 = 9.52 \angle -66.8°\Omega$

(b) $Z_a = 72.4 - j0 = 72.4 \angle 0°\Omega$
$Z_b = 5 - j12.5 = 13.46 \angle -68.2°\Omega$
$Z_c = 25 + j10 = 26.92 \angle 21.8°\Omega$

Chapter 6

P6-1 Amplitude $= l = 5$ in.
Frequency $f = 1$ Hz, $\omega = 2\pi$ rad/s
Period $= 1$ s
Phase angle $\phi = 0$ rad
Time shift $= 0$ s

P6-3 Amplitude $= l = 20$ cm
Frequency $f = 1$ Hz, $\omega = 2\pi$ rad/s
Period $= 1$ s
Phase angle $\phi = -\pi/4$ rad
Time shift $= 1/8$ s (to right)

P6-5 Amplitude $= l = 15$ cm
Frequency $f = 0.5$ Hz, $\omega = \pi$ rad/s
Period $= 2$ s
Phase angle $\phi = 3\pi/4$ rad
Time shift $= 3/4$ s (to left)

P6-7 Amplitude $= A = 2$ cm
Frequency $f = 2$ Hz, $\omega = 4\pi$ rad/s
Period $= 0.5$ s
Phase angle $\phi = 0$ rad
$y(t) = 2 \sin(4\pi t)$ cm

P6-9 Amplitude $= A = 10$ in.
Frequency $f = 1$ Hz, $\omega = 2\pi$ rad/s
Period $= 1$ s
Phase angle $\phi = 0.2\pi$ rad
$x(t) = 10 \sin(2\pi t + 0.2\pi)$ in.

P6-11 (a) Amplitude $= A = 10$ cm
Frequency $f = 2/\pi$ Hz, $\omega = 4$ rad/s
Period $= \pi/2$ s

(b) $t = \pi/4$ s

P6-13 (a) Amplitude $= A = 8$ cm
Frequency $f = 1$ Hz, $\omega = 2\pi$ rad/s
Period $= 1$ s
Time shift $= 1/8$ s (to left)

(b) $t = 1/8$ s

P6-15 (a) Amplitude $= A = 5$ cm
Frequency $f = 1/2$ Hz, $\omega = \pi$ rad/s
Period $= 2$ s

(b) $t = 1$ s

P6-17 (a) Amplitude $= A = 5$ rad
Frequency $f = 1$ Hz, $\omega = 2\pi$ rad/s
Period $= 1$ s

(b) $t = 0.25$ s

P6-19 (a) $v(t) = 41.38 \cos(120 \pi t)$ V

(b) Amplitude = 41.38 V
Frequency $f = 60$ Hz
Period = $T = 1/60$ s
Phase angle = $0°$
Time shift = 0 s

P6-21 (a) Amplitude = 0.1 V
Frequency $f = 10$ Hz
Period = $T = 1/10$ s
Phase angle = $\pi/2$ rad
Time shift = $1/40$ s (to left)

(c) $v(t) = 0.51 \cos(20 \pi t - 78.7°)$ V

P6-23 (a) $i(t) = 14.14 \sin(120 \pi t - 45°)$ A

(b) Amplitude = 14.14 A
Frequency $f = 60$ Hz
Period = $T = 1/60$ s
Phase angle = $-\pi/4$ rad
Time shift = $1/480$ s (to right)

P6-25 (a) $i(t) = 4 \sin(120 \pi t + 60°)$ A

(b) Amplitude = 4 A
Frequency $f = 60$ Hz
Period = $T = 1/60$ s
Phase angle = $\pi/3$ rad = $60°$
Time shift = $1/360$ s (to left)

P6-27 (a) Amplitude = 10 V
Frequency $f = 50$ Hz
Period = $T = 1/50$ s
Phase angle = $-\pi/4$ rad = $-45°$
Time shift = $1/400$ s (to right)

(c) $v_o(t) = 22.36 \cos(100 \pi t + 63.4°)$ V

P6-29 (a) Amplitude = $10\sqrt{2}$ V
Frequency $f = 250$ Hz
Period = $T = 1/250$ s
Phase angle = $-3\pi/4$ rad = $-135°$
Time shift = $3/2000$ s (to right)

(c) $v_o(t) = 11.18 \sin(500 \pi t - 153.4°)$ V

P6-31 (a) Amplitude = 8 in.
Frequency $f = 2$ Hz
Period = $T = 1/2$ s
Phase angle = $\pi/3$ rad = $60°$
Time shift = $1/12$ s (to left)

(c) $\delta(t) = 4.11 \cos(4 \pi t - 76.918°)$ in.

P6-33 (a) Amplitude = 8 in.
Frequency $f = 2$ Hz
Period = $T = 1/2$ s
Phase angle = $\pi/4$ rad = $45°$
Time shift = $1/8$ s (to left)

(c) $\delta(t) = 11.79 \cos(4 \pi t + 28.68°)$ in.

P6-35 (a) Amplitude = 164 V
Frequency $f = 60$ Hz
Period = $T = 1/60$ s
Phase angle = $7\pi/18$ rad = $70°$
Time shift = $7/2160$ s (to left)

(c) $v_T(t) = 220 \cos(120 \pi t - 45°)$ V

P6-37 (a) Amplitude = 120 V
Frequency $f = 60$ Hz
Period = $T = 1/60$ s
Phase angle = $2\pi/3$ rad = $120°$
Time shift = $1/180$ s (to left)

(c) $v_{bc}(t) = 207.8 \cos(120 \pi t + 0°)$ V

P6-39 (a) Amplitude = 200 N
Frequency $f = 5/2$ Hz
Period = $T = 2/5$ s
Phase angle = 0 rad = $0°$
Vertical shift = 1000 N

Chapter 7

P7-1 (a) $I_1 = 5.063$ A, $I_2 = -3.22$ A

(b) $\begin{bmatrix} 16 & -9 \\ -9 & 20 \end{bmatrix} \begin{bmatrix} I_1 \\ I_2 \end{bmatrix} = \begin{bmatrix} 110 \\ -110 \end{bmatrix}$

(c) $I_1 = 5.063$ A, $I_2 = -3.22$ A

(d) $I_1 = 5.063$ A, $I_2 = -3.22$ A

P7-3 (a) $I_1 = 0.0471$ A, $I_2 = -0.0429$ A

(b) $\begin{bmatrix} 1100 & 1000 \\ 1000 & 1100 \end{bmatrix} \begin{bmatrix} I_1 \\ I_2 \end{bmatrix} = \begin{bmatrix} 9 \\ 0 \end{bmatrix}$

(c) $I_1 = 0.0471$ A, $I_2 = -0.0429$ A

(d) $I_1 = 0.0471$ A, $I_2 = -0.0429$ A

P7-5 (a) $V_1 = 7.64$ V, $V_2 = 7.98$ V

(b) $\begin{bmatrix} 17 & -10 \\ -6 & 11 \end{bmatrix} \begin{bmatrix} V_1 \\ V_2 \end{bmatrix} = \begin{bmatrix} 50 \\ 42 \end{bmatrix}$

(c) $V_1 = 7.64$ V, $V_2 = 7.98$ V

(d) $V_1 = 7.64$ V, $V_2 = 7.98$ V

P7-7 (a) $V_1 = 0.976$ V, $V_2 = 1.707$ V

(b) $\begin{bmatrix} 7 & -4 \\ 2 & -7 \end{bmatrix} \begin{bmatrix} V_1 \\ V_2 \end{bmatrix} = \begin{bmatrix} 0 \\ -10 \end{bmatrix}$

(c) $V_1 = 0.976$ V, $V_2 = 1.707$ V

(d) $V_1 = 0.976$ V, $V_2 = 1.707$ V

P7-9 (a) $G_1 = 0.0175$ ℧, $G_2 = 0.0025$ ℧

(b) $\begin{bmatrix} 10 & 10 \\ 5 & 15 \end{bmatrix} \begin{bmatrix} G_1 \\ G_2 \end{bmatrix} = \begin{bmatrix} 0.2 \\ 0.125 \end{bmatrix}$

(c) $G_1 = 0.0175$ ℧, $G_2 = 0.0025$ ℧

(d) $G_1 = 0.0175$ ℧, $G_2 = 0.0025$ ℧

P7-11 (a) $\begin{bmatrix} 0.6 & 0.8 \\ 0.8 & -0.6 \end{bmatrix} \begin{bmatrix} T_1 \\ T_2 \end{bmatrix} = \begin{bmatrix} 100 \\ 0 \end{bmatrix}$

(b) $T_1 = 60$ lb, $T_2 = 80$ lb

P7-13 (a) $\begin{bmatrix} 0.707 & -1 \\ 0.707 & 0 \end{bmatrix} \begin{bmatrix} F_1 \\ F_2 \end{bmatrix} = \begin{bmatrix} 0 \\ 98.1 \end{bmatrix}$

(b) $F_1 = 138.7$ N, $F_2 = 98.1$ N

(c) $F_1 = 138.7$ N, $F_2 = 98.1$ N

P7-15 (a) $\begin{bmatrix} 0.866 & 0.707 \\ 0.5 & -0.707 \end{bmatrix} \begin{bmatrix} F_1 \\ F_2 \end{bmatrix} = \begin{bmatrix} 200 \\ 0 \end{bmatrix}$

(b) $F_1 = 146.4$ N, $F_2 = 103.5$ N

(c) $F_1 = 146.4$ N, $F_2 = 103.5$ N

P7-17 (a) $R_1 = 900$ lb, $R_2 = 2100$ lb

(b) $\begin{bmatrix} 1 & 1 \\ 7 & -3 \end{bmatrix} \begin{bmatrix} R_1 \\ R_2 \end{bmatrix} = \begin{bmatrix} 3000 \\ 0 \end{bmatrix}$

(c) $R_1 = 900$ lb, $R_2 = 2100$ lb

(d) $R_1 = 900$ lb, $R_2 = 2100$ lb

P7-19 (a) $\begin{bmatrix} 0.9285 & -0.3714 \\ 0.3714 & 0.9285 \end{bmatrix} \begin{bmatrix} F \\ N \end{bmatrix} = \begin{bmatrix} 0 \\ 10 \end{bmatrix}$

(b) $F = 3.714$ kN, $N = 9.285$ kN

(c) $F = 3.714$ kN, $N = 9.285$ kN

P7-21 (a) $\begin{bmatrix} 0.866 & -0.5 \\ 0.5 & 0.866 \end{bmatrix} \begin{bmatrix} F \\ N \end{bmatrix} = \begin{bmatrix} -30 \\ 110 \end{bmatrix}$

(b) $F = 24.0$ lb, $N = 101.6$ lb

(c) $F = 24.0$ lb, $N = 101.6$ lb

P7-23 (a) $N_F = 415.4$ N, $a = 4$ m/s^2

(b) $\begin{bmatrix} 0.4 & 100 \\ 1 & 0 \end{bmatrix} \begin{bmatrix} N_F \\ a \end{bmatrix} = \begin{bmatrix} 565.6 \\ 415.4 \end{bmatrix}$

(c) $N_F = 415.4$ N, $a = 4$ m/s^2

P7-25 (a) $R_1 + R_2 = 9810$
$2R_1 - 1.5R_2 = 7500$

(b) $R_1 = 6347.1$ N, $R_2 = 3462.9$ N

(c) $\begin{bmatrix} 1 & 1 \\ 2 & -1.5 \end{bmatrix} \begin{bmatrix} R_1 \\ R_2 \end{bmatrix} = \begin{bmatrix} 9810 \\ 7500 \end{bmatrix}$

P7-27 (a) $R_1 + R_2 = 11772$
$2R_1 - 2R_2 = 8100$

(b) $R_1 = 7911$ N, $R_2 = 3861$ N

(c) $\begin{bmatrix} 1 & 1 \\ 2 & -2 \end{bmatrix} \begin{bmatrix} R_1 \\ R_2 \end{bmatrix} = \begin{bmatrix} 11772 \\ 8100 \end{bmatrix}$

(d) $R_1 = 7911$ N, $R_2 = 3861$ N

(e) $R_1 = 7911$ N, $R_2 = 3861$ N

P7-29 (a) $0.707 F_m - 0.866 W_f = 49.5$
$0.707 F_m - 0.5 W_f = 80$

(b) $W_f = 83.29$ N, $F_m = 172.03$ N

(c) $\begin{bmatrix} 0.707 & -0.866 \\ 0.707 & -0.5 \end{bmatrix} \begin{bmatrix} F_m \\ W_f \end{bmatrix} = \begin{bmatrix} 49.5 \\ 80 \end{bmatrix}$

(d) $W_f = 83.29$ N, $F_m = 172.03$ N

(e) $W_f = 83.29$ N, $F_m = 172.03$ N

P7-31 (a) $V_1 + V_2 = 400$
$0.2 V_1 + 0.4 V_2 = 100$

(b) $V_1 = 300$ L, $V_2 = 100$ L

(c) $\begin{bmatrix} 1 & 1 \\ 0.2 & 0.4 \end{bmatrix} \begin{bmatrix} V_1 \\ V_2 \end{bmatrix} = \begin{bmatrix} 400 \\ 100 \end{bmatrix}$

(d) $V_1 = 300$ L, $V_2 = 100$ L

(e) $V_1 = 300$ L, $V_2 = 100$ L

P7-33 (a) $I_1 = \dfrac{100 \left(1 + \dfrac{50}{s^2}\right)}{10 + 10s + \dfrac{500}{s}}$ A,

$I_2 = \dfrac{100}{10 + 10s + \dfrac{500}{s}}$ A

(b) $\begin{bmatrix} (0.2s + 10) & -0.2s \\ -0.2s & \left(0.2s + \dfrac{10}{s}\right) \end{bmatrix} \begin{bmatrix} I_1 \\ I_2 \end{bmatrix} = \begin{bmatrix} \dfrac{100}{s} \\ 0 \end{bmatrix}$

P7-35 (a) $X_1 = \dfrac{1.25(s^2 + 65)}{s(s^2 + 25)}$,

$X_2 = \dfrac{50}{s(s^2 + 25)}$

(b) $\begin{bmatrix} 40 & -40 \\ -40 & s^2 + 65 \end{bmatrix} \begin{bmatrix} X_1 \\ X_2 \end{bmatrix} = \begin{bmatrix} 50/s \\ 0 \end{bmatrix}$

P7-37 (a) $X_A = 0.415$ lb, $X_B = 4.67$ lb

(b) $\begin{bmatrix} 0.15 & 3.2 \\ 1.6 & 2.0 \end{bmatrix} \begin{bmatrix} X_A \\ X_B \end{bmatrix} = \begin{bmatrix} 15 \\ 10 \end{bmatrix}$

(c) $X_A = 0.415$ lb, $X_B = 4.67$ lb

P7-39 (a) $u = -0.0508$ in., $v = 0.0239$ in.

(c) $\begin{bmatrix} 2.57 & 3.33 \\ 3.33 & 6.99 \end{bmatrix} \begin{bmatrix} u \\ v \end{bmatrix} = \begin{bmatrix} 0.05 \\ 0 \end{bmatrix}$

(d) $u = -0.0508$ in., $v = 0.0239$ in.

Chapter 8

P8-1 (a) $v(t) = 150 - 32.2 t$ ft/s

(b) $a(t) = -32.2$ ft/s^2

(c) $t_{max} = 4.66$, $y_{max} = 399$ ft

P8-3 (a) $v(t) = 250 - 9.81 t$ m/s

(b) $a(t) = -9.81$ m/s^2

(c) $t_{max} = 25.5$, $y_{max} = 3185.5$ m

P8-5 (a) $x(3) = 7.0$ m
$v(3) = 14.14$ m/s
$a(3) = -84.17$ m/s^2

(b) $x(3) = 689.8$ m
$v(3) = 1184.6$ m/s
$a(3) = 1610.4$ m/s^2

(c) $x(3) = 325.5$ km
$v(3) = 1302.1$ km/s
$a(3) = 5208.2$ km/s^2

P8-7 (a) $y(2) = 52$ m, $y(6) = 20$ m
$a(2) = -12$ m/s^2, $a(6) = 12$ m/s^2

(b) $y(0) = 20$ m, $y(9) = 101$ m

P8-9 (a) $v(t) = 0.125\, e^{-2t}\, (3t^2 - 2t^3)$ V

(b) $i(0) = 0$ A and $i(1.5) = 0.168$ A

P8-11 (a) $v(t) = -5.2\, e^{-200t}$ V
$p(t) = -67.6\, e^{-400t}$ W
$p_{max} = 67.6$ W at $t = 0$ s

(b) $v(t) = -4.8\,\pi\, \sin(120\,\pi\, t)$ V
$p(t) = -48 \sin(240\,\pi\, t)$ W
$p_{max} = 48\,\pi$ W at $t = \dfrac{3}{480}$ s

P8-13 (a) $i(t) = -4\,\pi \sin(200\,\pi\, t)$ A

(b) $p(t) = -1000\,\pi \sin(400\,\pi\, t)$ W
$p_{max} = 1000\,\pi$ W

P8-15 (a) $0 \le t \le 2$ s: $a(t) = -4$ m/s^2
$2 < t \le 6$ s: $a(t) = 4$ m/s^2
$6 < t \le 8$ s: $a(t) = -4$ m/s^2

(b) $0 \le t \le 2$ s: quadratic with decreasing slope/concave down
$x(t) = -2t^2$ m/s, $x(2) = -8$ m

$2 < t \le 6$ s: quadratic with increasing slope and zero slope at $t = 4$ s, therefore,
$x_{min} = -16$ m at $t = 4$ s
$x(6) = -16 + \dfrac{1}{2} \times 2 \times 8 = -8$ m

$6 < t \le 8$ s: quadratic with decreasing slope, zero slope at $t = 8$ s
$x(8) = -8 + \dfrac{1}{2} \times 2 \times 8 = 0$ m

P8-17 (a) $0 \le t \le 4$ s: $v(t) = 12 - 3t$ m/s
$4 < t \le 8$ s: $v(t) = -12 + 3t$ m/s

$t > 8$ s: $v(t) = 12$ m/s

(b) $0 \le t \le 4$ s: quadratic with decreasing slope/concave down
Slope $= 0$ at $t = 4$ s

$x(4) = 0 + \dfrac{1}{2} \times 4 \times 12 = 24$ m

$4 < t \le 8$ s: quadratic with increasing slope/concave up
$x(8) = 24 + \dfrac{1}{2} \times 4 \times 12 = 48$ m

P8-19 (a) $0 \le t \le 2$ s: linear with slope $= 5$ m/s^2, since $v(0) = 0$,
$v(t) = 5t$, and $v(2) = 10$ m/s

$2 < t \le 5$ s: linear with slope $= -5$ m/s^2, since $v(2) = 10$,
$v(t) = 10 - 5(t - 2) = 20 - 5t$
$v(5) = -5$ m/s

$5 < t \le 6$ s: linear with slope $= 5$ m/s^2, since $v(5) = -5$ m/s,
$v(t) = -5 + 5(t - 5) = -30 + 5t$
$v(6) = 0$ m/s

$t > 6$ s: $v(t) = 0$ m/s

(b) $0 \le t \le 2$ s: quadratic with increasing slope/concave up

$x(t) = 2.5\, t^2$, $x(2) = 10$ m. Also,

$x(2) = 0 + \dfrac{1}{2} \times 4 \times 10 = 10$ m

$2 < t \le 5$ s: quadratic with decreasing slope/concave down zero slope at $t = 4$ s, therefore,

$x(4) = 10 + (1/2) \times 2 \times 10 = 20$ m
$x(5) = 20 - (1/2) \times 1 \times 5 = 17.5$ m

$5 < t \le 6$ s: quadratic with increasing slope, zero slope at $t = 5$,
$x(6) = 17.5 - \dfrac{1}{2} \times 1 \times 5 = 15$ m

P8-21 Current:

$0 \le t \le 1$ s: linear with

slope $= \dfrac{1}{2} \times 6 = 3$ A/s

since $i(0) = 0$, $i(t) = 3t$ A
$i(1) = 3$ A

$1 < t \le 2$ s: linear with slope $= 0$ A/s, since $i(1) = 3$ A,
$i(t) = 3$ A, and $i(2) = 3$ A

$2 < t \le 3$ s: linear with slope $= 3$ A/s, since $i(2) = 3$,
$i(t) = 3 + 3(t - 2) = -3 + 3t$
$i(3) = 6$ A

P8-21 $3 < t \le 4$ s: linear with
slope $= 0$ A/s, since $i(3) = 6$ A,
$i(t) = 6$ A, and $i(4) = 6$ A

$4 < t \le 5$ s: linear with
slope $= 3$ A/s, since $i(4) = 6$,
$i(t) = 6 + 3(t - 4) = -6 + 3t$
$i(5) = 9$ A

Power:

$0 \le t \le 1$ s: $p(t) = 18t$ W
$1 < t \le 2$ s: $p(t) = 0$ W
$2 < t \le 3$ s: $p(t) = -18 + 18t$ W
$3 < t \le 4$ s: $p(t) = 0$ W
$4 < t \le 5$ s: $p(t) = -36 + 18t$ W

P8-23 Current:

$0 \le t \le 1$ ms: linear with
slope $= \dfrac{1}{2 \times 10^{-3}} \times 10$
$= 5000$ A/s
since $i(0) = 0$, $i(t) = 5000t$ A
$i(1 \times 10^{-3}) = 5$ A

$1 < t \le 2$ ms: linear with
slope $= \dfrac{1}{2 \times 10^{-3}}(-10)$
$= -5000$ A/s
since $i(1 \times 10^{-3}) = 5$ A,
$i(t) = 5 - 5000(t - 1 \times 10^{-3})$
$= 10 - 5000t$ A
$i(2 \times 10^{-3}) = 0$

$2 < t \le 3$ ms: linear with
slope $= \dfrac{1}{2 \times 10^{-3}} \times 10 = 5000$ A/s
since $i(2 \times 10^{-3}) = 0$,
$i(t) = 0 + 5000(t - 2 \times 10^{-3})$
$= -10 + 5000t$ A
$i(3 \times 10^{-3}) = 5$ A

Power:

$0 \le t \le 1$ ms: $p(t) = 50 \times 10^3 t$ W
$1 < t \le 3$ ms: $p(t) = -$ W
$-100 + 50 \times 10^3 t$

P8-25 Charge:

$0 \le t \le 1$ ms: $q(t) = t$ mC
$1 < t \le 2$ ms: $q(t) = 1$ mC
$2 < t \le 3$ ms: $q(t) = -3 + 3t$ mC
$3 < t \le 4$ ms: $q(t) = t$ mC
$4 < t \le 5$ ms: $q(t) = 12 - 2t$ mC
$t > 5$ ms: $q(t) = 2$ mC

Voltage:

$v(t) = 4000\, q(t)$ V

P8-27 (a) $x = \sqrt{\dfrac{l^2 - b^2}{3}}$

$y_{\max} = \dfrac{Pb}{6EIl} \sqrt{\dfrac{l^2 - b^2}{3}} \left(\dfrac{2b^2}{3} - \dfrac{2l^2}{3} \right)$

(b) $\theta(0) = \dfrac{Pb}{6EIl}(b^2 - l^2)$

$\theta(l) = \dfrac{Pb}{6EIl}(b^2 + 2l^2)$

P8-29 (a) $\theta(x) = \dfrac{M_o}{6EI\,L}\left(L^2 - \dfrac{3}{2}Lx - 3x^2\right)$

(b) $x = 0.345\,L$

$y_{\max} = \dfrac{0.0358\,M_o\,L^2}{EI}$

(c) $\theta(0) = \dfrac{M_o L}{6\,EI}, y(0) = 0$

$\theta\left(\dfrac{L}{2}\right) = -\dfrac{M_o L}{12\,EI}, y\left(\dfrac{L}{2}\right) = 0$

P8-31 (a) $x = L/\sqrt{5}$

$y_{\max} = -0.00239\dfrac{L^4}{EI}$

(b) $\theta(0) = -\dfrac{W_o L^3}{120\,EI}$

$\theta(L) = 0$

P8-33 (a) $\theta(x) = \dfrac{2\pi A}{L}\cos\left(\dfrac{2\pi x}{L}\right)$

(b) $\theta(0) = -\dfrac{2\pi A}{L}$

$\theta\left(\dfrac{L}{2}\right) = \dfrac{2\pi A}{L}$

$\theta(L) = -\dfrac{2\pi A}{L}$

P8-33 (c) $x = \dfrac{L}{4}$ and $x = \dfrac{3L}{4}$

$$y\left(\dfrac{L}{4}\right) = -A$$

$$y\left(\dfrac{3L}{4}\right) = A$$

P8-35 (a) $x = 0.578\,L$

$$y_{max} = -0.0054\dfrac{W\,L^4}{EI}$$

(b) $\theta(0) = 0$

$$\theta(L) = \dfrac{W\,L^3}{48\,EI}$$

P8-37 (a) $\theta = \dfrac{\pi}{4}$ rad

(b) $\theta = 0$ rad

P8-39 (a) $b = 0.15$

(b) $\hat{y} = -0.1\,x + 22$ m

(c) $(176,\ 4.4)$ m

(d) $y(x) = -0.00071\,x^2 + 0.15\,x$ m

Chapter 9

P9-1 (a) $A \approx 0.6481\,k\,l$

P9-3 (a) Distance covered $= A = 700$ ft

(b) Distance covered $= A = 750$ ft

(c) Distance covered $= A = 800$ ft

(d) $A = 800$ ft

P9-5 (a) $W = 172.95$ N-m

(b) $W = 201.7$ N-m

(c) $W = 3.82$ N-m

P9-7 (a) $y(x) = -0.75\,x + 9$ cm

(b) $A = 54$ cm^2

(c) $\bar{x} = 4$ cm, $\bar{y} = 3$ cm

(d) $\bar{y} = 3$ cm

P9-9 (a) $y(x) = -0.5\,x + 6$ ft

(b) $A = 27$ ft^2

(c) $\bar{x} = 2.667$ ft

(d) $\bar{y} = 2.333$ ft

P9-11 (a) $h = 4$ in.

$b = 2$ in.

(b) $A = 5.33$ in.^{2}

(c) $\bar{x} = 0.75$ in.

(d) $\bar{y} = 1.6$ in.

P9-13 (a) $h = 8$ in.

$b = 8$ in.

(b) $A = 48$ in.^{2}

(c) $\bar{x} = 3.56$ in.

(d) $\bar{y} = 5.61$ in.

P9-15 (a) $a = 2$ cm

$b = 6$ cm

(b) $A = 6.67$ cm^2

(c) $\bar{x} = 1.88$ cm

(d) $\bar{y} = 1.2$ cm

P9-17 (a) $R = \dfrac{2}{3}\,w_0\,l$

(b) $\bar{x} = \dfrac{3}{8}\,l$

P9-19 (a) $v(t) = 10\,t^4 + 10\,t^3 + 10\,t^2 + 10\,t$ m/s

$y(t) = 2\,t^5 + 2.5\,t^4 + \dfrac{10}{3}\,t^3 + 5\,t^2$ m

(b) $v(t) = \dfrac{1}{2}\,(1 - \cos(4\,t))$ m/s

$y(t) = \dfrac{1}{2}\,t - \dfrac{1}{8}\,\sin(4\,t)$ m

P9-21 Velocity:

$0 \le t \le 2$ s: $v(t) = 10\,t$ m/s

$2 < t \le 14$ s: $v(t) = 20$ m/s

$14 < t \le 16$ s: $v(t) = -10\,(t - 16)$

$t > 16$ s: $v(t) = 0$ m/s

P9-21   Position:

0 ≤ t ≤ 2 s: quadratic with
increasing slope/concave up
$x(t) = 5t^2$, $x(2) = 20$ m

2 < t ≤ 14 s: linear
$x(t) = 20(t - 1)$, $x(14) = 260$ m

14 < t ≤ 16 s: quadratic with
decreasing slope/concave down
$x(16) = 260 + \frac{1}{2}(2)(20) = 280$ m

P9-23   Velocity:

0 ≤ t ≤ 3 s: $v(t) = 10t$ m/s
3 < t ≤ 6 s: $v(t) = -10(t - 6)$
6 < t ≤ 12 s: $v(t) = (t - 6)$
t > 12 s: $v(t) = 30$ m/s

Position:

0 ≤ t ≤ 3 s: quadratic with
increasing slope/concave up
$x(t) = 5t^2$, $x(3) = 45$ m

3 < t ≤ 6 s: quadratic with
decreasing slope/concave down
$x(6) = 45 + \frac{1}{2}(3)(30) = 90$ m

6 < t ≤ 12 s: quadratic with
increasing slope/concave up
$x(16) = 90 + \frac{1}{2}(30)(6) = 180$ m

P9-25 $q(t) = \frac{1}{3}t^3 - t^2 + t$ C

P9-27 (a) $v_o(t) = 10(\cos(10t) - 1)$ V

(b) $w(t) = 7.5 + 2.5\cos(20t) - 10\cos(10t)$ mJ

P9-29 (a) $v_o(t) = \sin(100t) - 100t$ V

(b) $w(t) = 7.5 + 2.5\cos(20t) - 10\cos(10t)$ mJ

P9-31 (a) $v_o(t) = 10(1 - e^{-10t}) + 10$ V

(b) $w(t) = 0.01(1 - e^{-20t}) + 0.005$ J

P9-33 $v(0) = 0$ V, $v(1) = -10$ V and
$v(2) = 0$ V

0 ≤ t ≤ 2 s: quadratic with
increasing slope/concave up

Then simply repeat every 2 s

P9-35 (a) $v(0) = 0$ V, $v(1) = 10$ V,
$v(2) = 10$ V, $v(3) = 15$ V,
$v(4) = 5$ V and
$v(t) = 5$ V for t > 4 s

Since $i(t)$ is constant over each
interval, v(t) is linear between
each interval

P9-37 (a) 0 ≤ t ≤ 0.008 s:
$x_{belted}(t) = 0.102\sin(62.5\pi t)$ m

0.008 < t ≤ 0.04 s:
$x_{belted}(t) = 0.102$ m

(b) 0 ≤ t ≤ 0.04 s:
$x_{unbelted}(t) = 0.509\sin(12.5\pi t)$ m

(c) $x_{unbelted}(0.04) - x_{unbelted}(0.04) = 0.407$ m

P9-39 (a) $y(x) = -\frac{\hat{p}x}{hG}\left(\frac{L}{3} - \frac{x}{2} + \frac{x^2}{3L} - \frac{x^3}{12L^2}\right)$

P9-39 (b) $x = L$

$y_{max} = \frac{\hat{p}L^2}{12hG}$

Chapter 10

P10-1 (a) $h_{tran}(t) = Ce^{-\frac{k}{A}t}$

(b) $h_{ss}(t) = \frac{Q}{(A^2\omega^2 + k^2)} \times (k\sin\omega t - A\omega\cos\omega t)$

(c) $h(t) = \frac{Q}{A^2\omega^2 + k^2}(A\omega e^{-\frac{k}{A}t} + k\sin\omega t - A\omega\cos\omega t)$

P10-3 (a) $v_{tran}(t) = C e^{-\frac{1}{RC}t}$ V

(b) $v_{ss}(t) = 0$ V

(c) $v(t) = 10 e^{-\frac{1}{RC}t}$ V

(e) T, F, T, T

P10-5 (a) $v_{tran}(t) = C e^{-\frac{1}{2}t}$ V

(b) $v_{ss}(t) = 0$ V

(c) $v(t) = 10 e^{-\frac{1}{2}t}$ V

(e) T, F, T, T

P10-7 (a) $v_{tran}(t) = C e^{-\frac{1}{20}t}$ V

(b) $v_{ss}(t) = 10$ V

(c) $v(t) = 10 - 5 e^{-\frac{1}{20}t}$ V

P10-9 (a) $v_{tran}(t) = A e^{-\frac{1}{RC}t}$ V, $\tau = RC$ s

(b) $v_{ss}(t) = \frac{10}{(1+0.001\,R^2\,C^2)}\sin(0.01t)$ V

(c) $v(t) = \frac{10}{(1+0.001\,R^2\,C^2)}\sin(0.01t)$ V

P10-11 (a) $v_{tran}(t) = C e^{-2.5t}$ V, $\tau = 0.4$ s

(b) $v_{ss}(t) = 10\sin(0.01t)$ V

(c) $v(t) = 10\sin(0.01t)$ V

P10-13 (a) $v_{tran}(t) = C e^{-100t}$ V

(b) $v_{ss}(t) = 10$ V

(c) $v(t) = 10\left(1 - e^{-100t}\right)$ V

(d) $\tau = 0.01$ s

$v(0.01) = 6.321$ V

$v(0.02) = 8.647$ V

$v(0.04) = 9.817$ V

$v(\infty) = 10$ V

P10-15 (a) $i_{tran}(t) = C e^{-\frac{R}{L}t}$ A, $\tau = \frac{L}{R}$ s

(b) $i_{ss}(t) = 0$ A

(c) $i(t) = 50 e^{-\frac{R}{L}t}$ mA

(e) F, F, F, T

P10-17 (a) $i_{tran}(t) = C e^{-10,000t}$ A, $\tau = 0.1$ ms

(b) $i_{ss}(t) = 0$ A

(c) $i(t) = 50 e^{-10,000t}$ mA

(e) F, F, F, T

P10-19 (a) $v_{tran}(t) = C e^{-100t}$ V

(b) $v_{ss}(t) = 10$ V

(c) $v(t) = 10(1 - e^{-100t})$ V

(e) $\tau = 0.01$ s

$v(0.01) = 6.32$ V

$v(0.02) = 8.65$ V

$v(0.04) = 9.82$ V

$v(\infty) = 10$ V

P10-21 (a) $i_{tran}(t) = C e^{-50t}$ A, $\tau = 0.02$ s

(b) $i_{ss}(t) = 0.5$ A

(c) $i(t) = -1.5 e^{-50t} + 0.5$ A

P10-23 (a) $v_{o,tran}(t) = C e^{-5t}$ V, $\tau = 0.2$ s

(b) $v_{o,ss}(t) = 13.33\cos(10t)$
$+ 6.67\sin(10t)$ V

(c) $v_o(t) = -13.33 e^{-5t}$
$+13.33\cos(10t)+$
$6.67\sin(10t)$ V

P10-25 (a) $P_{tran}(t) = A e^{-\frac{1}{RC}t}$ mmHg

(b) $P_{ss}(t) = 60\,R$ mmHg

(c) $P(t) = 240 - 234 e^{-\frac{1}{8}t}$ mmHg

P10-27 (a) $y_{tran}(t) = C_3\cos(\sqrt{10}\,t)+$
$C_4\sin(\sqrt{10}\,t)$ m

(b) $y_{ss}(t) = 0$ m

(c) $y(t) = 0.2\cos(\sqrt{10}\,t)$ m

P10-29 (a) $y_{tran}(t) = C_3\cos(4t)+$
$C_4\sin(4t)$ m

(b) $y_{ss}(t) = 3/16$ m

(c) $y(t) = 0.028\cos(4t - 1.107)+$
$3/16$ m

P10-31 (a) $x_{tran}(t) = C_3\cos(\sqrt{40}\,t)+$
$C_4\sin(\sqrt{40}\,t)$ m

$\omega_n = \sqrt{40}$ rad/s or $f = 1.01$ Hz

(b) $x_{ss}(t) = 0.245$ m

(c) $x(t) = -0.245\cos(\sqrt{40}\,t)+$
$0.7\sqrt{h}\sin(\sqrt{40}\,t) + 0.245$ m

P10-33 (a) $y_{tran}(t) = C_3\cos(0.61\,t)+$
$C_4\sin(0.61\,t)$ m

$\omega_n = 0.61$ rad/s

(b) $y_{ss}(t) = 0.5\cos(0.5\,t)$ m

(c) $y_{min} = -0.5$ m at $t = 2\pi$ s

P10-35 (a) $x(t) = V_o\sqrt{\frac{m}{k}}\cos\left(\sqrt{\frac{k}{m}}\,t\right)$ m

(c) T, F, T, T, F

P10-37 $i_{tran}(t) = C_3\cos(100\,t)+$
$C_4\sin(100\,t)$ A

$i_{ss}(t) = 0$ A

$i(t) = 0.1\sin(100\,t)$ A

P10-39 (a) $v_{tran}(t) = C_3\cos(500\,t)+$
$C_4\sin(500\,t)$ V

$\omega_n = 500$ rad/s, $f = 250/\pi$ Hz

(b) $v_{ss}(t) = 10$ V

(c) $v(t) = 10\,(1 - \cos(500\,t))$ V

P10-41 (a) $v_{tran}(t) = C_3\cos(250\,t)+$
$C_4\sin(250\,t)$ V

$\omega_n = 250$ rad/s, $f = 125/\pi$ Hz

(b) $v_{ss}(t) = 5.1\sin(245\,t)$ V

(c) $v(t) = 5\sin(250\,t)+$
$5.1\sin(245\,t)$ V

P10-43 (a) $\theta_{tran}(t) = C_3\cos\left(\sqrt{\frac{3k}{m}}\,t\right)+$

$C_4\sin\left(\sqrt{\frac{3k}{m}}\,t\right)$ rad

$\omega_n = \sqrt{\frac{3k}{m}}$ rad/s

(b) $\theta_{ss}(t) = \dfrac{F}{kl - \frac{1}{3}ml\omega^2}\cos(\omega t)$ rad

(c) $\theta(t) = \dfrac{\pi}{18}\cos\left(\dfrac{3k}{m}\,t\right)$ rad

(d) F, F, T, F, F

INDEX

Useful Mathematical Relations

Algebra and Geometry

Arithmetic Operations

$$a(b + c) = ab + ac$$

$$\frac{a}{b} + \frac{c}{d} = \frac{ad + bc}{bd}$$

$$\frac{a + c}{b} = \frac{a}{b} + \frac{c}{b}$$

$$\frac{\frac{a}{b}}{\frac{c}{d}} = \frac{a}{b} \times \frac{d}{c}$$

$$= \frac{ad}{bc}$$

Exponents and Radicals

$$x^m x^n = x^{m+n}$$

$$\frac{x^m}{x^n} = x^{m-n}$$

$$(x^m)^n = = x^{mn}$$

$$x^{-n} = \frac{1}{x^n}$$

$$(xy)^n = x^n y^n$$

$$\left(\frac{x}{y}\right)^n = \frac{x^n}{y^n}$$

$$x^{1/n} = \sqrt[n]{x}$$

$$x^{m/n} = \sqrt[n]{x^m}$$

$$= \left(\sqrt[n]{x}\right)^m$$

$$\sqrt[n]{xy} = \sqrt[n]{x}\,\sqrt[n]{y}$$

$$\sqrt[n]{\frac{x}{y}} = \frac{\sqrt[n]{x}}{\sqrt[n]{y}}$$

Factoring Special Polynomials

$$x^2 - y^2 = (x + y)(x - y)$$

$$x^3 + y^3 = (x + y)(x^2 - xy + y^2)$$

$$x^3 - y^3 = (x - y)(x^2 + xy + y^2)$$

Quadratic Formula

If $ax^2 + bx + c = 0$. then

$$x = \frac{-b \pm \sqrt{b^2 - 4ac}}{2a}$$

Lines

Equation of line with slope m and y-intercept b:

$$y = mx + b$$

Slope of line through points $P_1(x_1, y_1)$ and $P_2(x_2, y_2)$:

$$m = \frac{y_2 - y_1}{x_2 - x_1}$$

Point-slope equation of line through point $P_1(x_1, y_1)$ with slope m:

$$y - y_1 = m(x - x_1)$$

Point-slope equation of line through point $P_2(x_2, y_2)$ with slope m:

$$y - y_2 = m(x - x_2)$$

Distance Formula

Distance between points $P_1(x_1, y_1)$ and $P_2(x_2, y_2)$:

$$d = \sqrt{(x_2 - x_1)^2 + (y_2 - y_1)^2}$$

Area of a Triangle:

$$A = \frac{1}{2} bh$$

$$= \frac{1}{2} ab \sin(\theta)$$

Equation, Area and Circumference of Circle:

Equation:

$$(x - k)^2 + (y - k)^2 = r^2$$

$$A = \pi r^2$$

$$C = 2\pi r$$

Trigonometry

Angle Measurement

π radians = 180°

$1° = \dfrac{\pi}{180}$ rad

1 rad $= \dfrac{180}{\pi}$ deg

$s = r\,\theta$

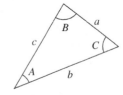

Right-Angle Trigonometry

$\sin\theta = \dfrac{\text{opp}}{\text{hyp}}$

$\cos\theta = \dfrac{\text{adj}}{\text{hyp}}$

$\tan\theta = \dfrac{\text{opp}}{\text{adj}}$

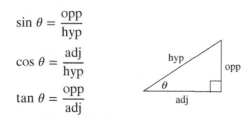

Trigonometric Functions

$\sin\theta = \dfrac{y}{r}$

$\cos\theta = \dfrac{x}{r}$

$\tan\theta = \dfrac{y}{x}$

$r = \sqrt{x^2 + y^2}$

$\theta = atan2(y, x)$

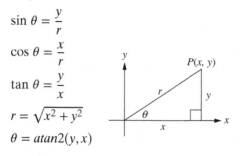

Fundamental Identities

$\csc\theta = \dfrac{1}{\sin\theta}$ $\qquad$ $\cot\theta = \dfrac{1}{\tan\theta}$

$\sec\theta = \dfrac{1}{\cos\theta}$ $\qquad$ $\sin^2\theta + \cos^2\theta = 1$

$\tan\theta = \dfrac{\sin\theta}{\cos\theta}$ $\qquad$ $1 + \tan^2\theta = \sec^2\theta$

$\qquad\qquad\qquad\qquad 1 + \cot^2\theta = \csc^2\theta$

$\sin(-\theta) = -\sin(\theta)$

$\sin\left(\theta - \frac{\pi}{2}\right) = -\cos(\theta)$

$\cos(-\theta) = \cos(\theta)$

$\cos\left(\theta - \frac{\pi}{2}\right) = \sin(\theta)$

$\tan(-\theta) = -\tan(\theta)$

$\tan\left(\theta - \frac{\pi}{2}\right) = -\cot(\theta)$

Double-Angle Formulas

$\sin(2\theta) = 2\sin(\theta)\,\cos(\theta)$

$\cos(2\theta) = \cos^2(\theta) - \sin^2(\theta)$

$\qquad = 2\cos^2(\theta) - 1 = 1 - 2\sin^2(\theta)$

Half-Angle Formulas

$\sin\left(\dfrac{\theta}{2}\right) = \pm\sqrt{\dfrac{(1 - \cos\theta)}{2}}$

$\cos\left(\dfrac{\theta}{2}\right) = \pm\sqrt{\dfrac{(1 + \cos\theta)}{2}}$

Addition and Subtraction Formulas

$\sin(\theta_1 \pm \theta_2) = \sin\theta_1\cos\theta_2 \pm \cos\theta_1\sin\theta_2$

$\cos(\theta_1 \pm \theta_2) = \cos\theta_1\cos\theta_2 \mp \sin\theta_1\sin\theta_2$

Law of Sines

$\dfrac{\sin A}{a} = \dfrac{\sin B}{b} = \dfrac{\sin C}{c}$

Law of Cosines

$a^2 = b^2 + c^2 - 2\,b\,c\cos A$

Differentiation Rules

$$\frac{d}{dt}(c) = 0$$

$$\frac{d}{dt}[cf(t)] = c\dot{f}(t)$$

$$\frac{d}{dt}(c_1 f(t) + c_2 g(t)) = c_1 \dot{f}(t) + c_2 \dot{g}(t)$$

$$\frac{d}{dt}(c_1 f(t) - c_2 g(t)) = c_1 \dot{f}(t) - c_2 \dot{g}(t)$$

$$\frac{d}{dt}(t^n) = n\, t^{n-1}$$

Product Rule

$$\frac{d}{dt}[f(t)\, g(t)] = f(t)\, \dot{g}(t) + \dot{f}(t)\, g(t)$$

Quotient Rule

$$\frac{d}{dt}\left[\frac{f(t)}{g(t)}\right] = \frac{g(t)\dot{f}(t) - f(t)\dot{g}(t)}{[g(t)]^2}$$

Chain Rule

$$\frac{d}{dt}f(g(t)) = \frac{df}{dg} \times \frac{dg}{dt}$$

Power Rule

$$\frac{d}{dt}(t^n) = n\, t^{n-1}$$

Integration Rules

$$\int y\, dx = yx - \int x\, dy$$

$$\int x^n\, dx = \frac{x^{n+1}}{n+1} + C, \quad n \neq -1$$

$$\int e^{ax}\, dx = \frac{e^{ax}}{a} + C$$

$$\int \sin(ax)\, dx = -\frac{\cos(ax)}{a} + C$$

$$\int \cos(ax)\, dx = \frac{\sin(ax)}{a} + C$$

$$\int cf(x)\, dx = c \int f(x)\, dx$$

$$\int [c_1 f(x) + c_2 g(x)]dx = c_1 \int f(x)dx + c_2 \int g(x)dx$$

Exponential Functions

$$\frac{d}{dt}(e^{at}) = a\, e^{at}$$

Trigonometric Functions

$$\frac{d}{dt}[\sin(at)] = a\cos(at)$$

$$\frac{d}{dt}[\cos(at)] = -a\sin(at)$$

Commonly Used Units in Engineering

Unit	English	SI	Conversion Factor
Length	in. or ft	m	1 in. = 0.0254m
Area	in.^{2} or ft^2	m^2	1550 in.^{2} = 1 m^2
Volume	in.^{3} or ft^3	m^3	61024 in.^{3} = 1 m^3
Velocity	in./s or ft/s	m/s	1 in./s = 0.0254 m/s
Acceleration	in./s^2 or ft/s^2	m/s^2	1 in./s^2 = 0.0254m/s^2
Force	lb	N	1 lb = 4.45 N
Pressure (Stress)	lb/in.^{2} (psi)	N/m^2 (Pa)	1 psi = 6890 Pa
Mass	lbm	kg	1 lbm = 0.454 kg
Energy	in.-lb or ft-lb	N-m (J)	1 in.-lb = 0.113 J
Power	in.-lb/s or ft-lb/s	W (J/s)	1 in.-lb/s = 0.113 W
Voltage	Volts (V)	Volts (V)	
Current	Amps (A)	Amps (A)	
Resistance	Ohms (Ω)	Ohms (Ω)	
in.ductance	Henrys (H)	Henrys (H)	
Capacitance	Farads (F)	Farads (F)	

Commonly Used Prefixes in Engineering

Nano (n) 10^{-9}

Micro (μ) 10^{-6}

Milli (m) 10^{-3}

Centi (c) 10^{-2}

Kilo (K) 10^{3}

Mega (M) 10^{6}

Giga (G) 10^{9}

CPSIA information can be obtained
at www.ICGtesting.com
Printed in the USA
BVOW08s0812200817

492455BV00004B/8/P